Studies in Computational Intelligence

Volume 634

Series editor

Janusz Kacprzyk, Polish Academy of Sciences, Warsaw, Poland
e-mail: kacprzyk@ibspan.waw.pl

About this Series

The series "Studies in Computational Intelligence" (SCI) publishes new developments and advances in the various areas of computational intelligence—quickly and with a high quality. The intent is to cover the theory, applications, and design methods of computational intelligence, as embedded in the fields of engineering, computer science, physics and life sciences, as well as the methodologies behind them. The series contains monographs, lecture notes and edited volumes in computational intelligence spanning the areas of neural networks, connectionist systems, genetic algorithms, evolutionary computation, artificial intelligence, cellular automata, self-organizing systems, soft computing, fuzzy systems, and hybrid intelligent systems. Of particular value to both the contributors and the readership are the short publication timeframe and the worldwide distribution, which enable both wide and rapid dissemination of research output.

More information about this series at http://www.springer.com/series/7092

Guy De Tré · Przemysław Grzegorzewski
Janusz Kacprzyk · Jan W. Owsiński
Wojciech Penczek · Sławomir Zadrożny
Editors

Challenging Problems and Solutions in Intelligent Systems

 Springer

Editors
Guy De Tré
Department of Telecommunications
  and Information Processing
Ghent University
Gent
Belgium

Przemysław Grzegorzewski
Faculty of Mathematics and Information
  Science
Warsaw University of Technology
Warsaw
Poland

Janusz Kacprzyk
Systems Research Institute
Polish Academy of Sciences
Warsaw
Poland

Jan W. Owsiński
Systems Research Institute
Polish Academy of Sciences
Warsaw
Poland

Wojciech Penczek
Institute of Computer Science
Polish Academy of Sciences
Warsaw
Poland

Sławomir Zadrożny
Systems Research Institute
Polish Academy of Sciences
Warsaw
Poland

ISSN 1860-949X   ISSN 1860-9503   (electronic)
Studies in Computational Intelligence
ISBN 978-3-319-30164-8   ISBN 978-3-319-30165-5   (eBook)
DOI 10.1007/978-3-319-30165-5

Library of Congress Control Number: 2016933117

Printed on acid-free paper

This Springer imprint is published by Springer Nature
The registered company is Springer International Publishing AG Switzerland

Preface

Intelligent computing is a new emerging computing paradigm inspired by a remarkable ability of animate systems, meant both as individuals and groups of individuals, in solving complex problems, notably related to decision making and support. This broad definition covers many specific disciplines, notably—in the context of this volume—including the following ones: artificial and computational intelligence, fuzzy logic, granular computing, intelligent database systems, information retrieval, information fusion, intelligent search (engines), data mining, cluster analysis, unsupervised learning, machine learning, intelligent data analysis, (group) decision support systems, decision theory, collective intelligence, case-based reasoning, intelligent agents and multi-agent systems, artificial neural networks, genetic algorithms, evolutionary computation, particle swarm optimization, artificial immune system, knowledge-based systems, approximate reasoning, knowledge engineering, expert systems, imprecision and uncertainty handling, human–computer interface, Internet computing, semantic web, electronic commerce, e-learning and Web-intelligence, cognitive systems, distributed systems, intelligent control, advanced computer modeling and simulation, bioinformatics, etc. Thus the paradigm of intelligent computing is meant to cover the areas of interest of traditional artificial intelligence as well as many other related topics, notably those inspired by the observation of biological systems and those which emerged with the development of advanced IT. Such a wide set of tools and techniques available does clearly promise that some challenging problems solved so far mainly by human beings may be formalized and solved by the "machine".

The growing interest and importance of this branch of research made a group of researchers, including the editors of this volume, to think of joining forces and working together on their topics of interest belonging to the above-mentioned broadly meant area. One goal was to define a common ground for sometimes seemingly disparate problems dealt with by particular teams involved and to address the identified problems using the paradigm of broadly perceived intelligent computing and/or intelligent systems. Another important goal was to get young researchers involved, to give them a chance to develop their careers in the modern and prospective field. As a result, in 2010, a consortium of three Polish academic institutions: the Systems Research Institute of the Polish Academy of Sciences, the Institute of Computer Science, also of the Polish Academy of Sciences, and the Faculty of Mathematics and Information Science, Warsaw University of Technology, started a large-scale project, entitled "International Ph.D. Projects in Intelligent Computing", which was supported by the Foundation for Polish Science, financed from the European Union within the Innovative Economy Operational Programme 2007–2013 and the European Regional Development Fund. The project had been carried out with intensive collaboration with 17 foreign partners, leading academic and research institutions from Australia, Belgium, Canada, Denmark, Finland, France, Germany, Japan, Luxembourg, Spain, the United Kingdom, and the USA. The core of the group, formed around senior researchers of high international stature in the field, had been the Ph.D. students selected in a highly competitive contest. Thanks to the support of the International Ph.D. Projects Programme, operated by the Foundation for Polish Science as mentioned above, these Ph.D. students had been supervised by well-known scientists and had an opportunity to work at the foreign partner institutions via a system of scientific internships.

The research program prepared for the Ph.D. students covered both theoretical and practical aspects of the intelligent computing paradigm. In particular, the theoretical foundations and the practical problems of implementation of the advanced IT/ICT tools and techniques were addressed. Five areas were in focus: foundations of intelligent computing, intelligent techniques in data mining, intelligent computing for supporting decision making, multi-agent-based technologies, and intelligent text and data retrieval.

Now, when the project is close to completion, we are happy to publish this volume, which constitutes another opportunity to present and promote the results obtained during the project. The participants have published numerous papers in scientific journals and presented their results during many international scientific conferences. Nevertheless, this volume provides a broad panorama of research carried out in the framework of the project, and a comprehensive collection of modern views and approaches that can be of use to a broader audience. The majority of the papers are authored by the Ph.D. students participating in the project and co-authored by their supervisors. In a few cases, we have also invited other valuable contributions from the supervisors and their younger collaborators: graduate students, Ph.D. students, and postdocs, showing thereby a broader view of the community, and the scale and depth of the research endeavor associated with the project.

Some of the papers collected in this volume are extended versions of the presentations given by the participants at special sessions on intelligent computing organized within many respected international conferences, notably the recent FQAS 2015, the Flexible Query Answering Systems 2015 conference that took place in Cracow during October 26–28, 2015, which constituted another opportunity to promote and summarize the results of the project to a large international audience, including a number of foreign members of the project consortium.

Special thanks are due to the Foundation for Polish Science for their continuous support for our efforts to provide the Ph.D. students and their supervisors with the best opportunity to carry out their research.

We express our particular gratitude to all the participants of the project's community, whose contribution, cooperation, ideas, as well as purely human qualities enriched all of us, and made the project for most of us a memorable stage in life, with quite essential consequences for many, especially the Ph.D. students. With this, we extend our best wishes to all those involved.

Peer reviewers deserve also deep appreciations because their insightful and constructive remarks and suggestions have considerably improved the contribution.

And last but not least, we wish to thank Dr. Tom Ditzinger, Dr. Leontina di Cecco, and Mr. Holger Schaepe for their dedication and help to implement and finish this large publication project on time maintaining the highest publication standards.

November 2015

Guy De Tré
Przemysław Grzegorzewski
Janusz Kacprzyk
Jan W. Owsiński
Wojciech Penczek
Sławomir Zadrożny

Contents

Part I
Foundations of
Intelligent Computing

SMT-Based Parameter Synthesis for Parametric Timed Automata

Michał Knapik and Wojciech Penczek

Abstract We present a simple method for representing finite executions of Parametric Timed Automata using Satisfiability Modulo Theories (SMT). The transition relation of an automaton is translated to a formula of SMT, which is used to represent all the prefixes of a given length of all the executions. This enables to underapproximate the set of parameter valuations for the undecidable problem of parametric reachability. We introduce a freely available, open-source tool PTA2SMT and show its application to the synthesis of parameter valuations under which a timed mutual exclusion protocol fails.

1 Introduction

As digital systems become ubiquitous in the modern world, so grows the need for ensuring the correctness of their behaviour. The sheer number of possible interactions between distributed and independent components often makes testing of such designs very difficult, if not impossible. In areas such as avionics or medical applications the errors are usually non-acceptable, thus every measure should be taken to prevent them from occuring in both the hardware and software. Increasingly, formal methods begin to play an important role in the development of critical systems. Among these, model checking is probably the most applied technique. In model checking we expect a simple binary (*yes/no*) answer to the question whether the model of a system is compliant with its specification. This is certainly a good approach to the verification of existing systems, where the details of the design are known. On the other hand, at the initial stage of system prototyping many of its particularities are usually unknown. In this case model checking is of limited use. Parameter synthesis attempts to alleviate

M. Knapik (✉) · W. Penczek
Institute of Computer Science, PAS, Warsaw, Poland
e-mail: mknapik@ipipan.waw.pl

W. Penczek
University of Natural Sciences and Humanities, II, Siedlce, Poland
e-mail: penczek@ipipan.waw.pl

© Springer International Publishing Switzerland 2016   3
G. De Tré et al. (eds.), *Challenging Problems and Solutions
in Intelligent Systems*, Studies in Computational Intelligence 634,
DOI 10.1007/978-3-319-30165-5_1

this problem by allowing free variables (i.e., *parameters*) that correspond to the unknown values. The goal is to synthesise at least a part of the set of all the valuations of parameters, under which the model becomes compliant with its specification.

In this paper we extend the approach of bounded model checking, where the computation tree of a verified model is unfolded up to a finite depth, to parameter synthesis for Parametric Timed Automata (PTA, for short). As we recall later, in general the task of parameter synthesis is undecidable for PTA [1], it however becomes decidable if the parameters can be divided into two disjoint groups that correspond to upper and lower bounds on clock values. The class of PTA that conforms to these restrictions is called Lower/Upper Bound Parametric Timed Automata (L/U PTA, for short). However, even in this case the synthesis of the full description of the set of solutions is problematic [8], in particular it cannot be represented as a finite union of polyhedra. We therefore focus on the underapproximation of the set of parameter valuations under which a given state is reachable for L/U PTA. To this end we decided to employ SMT-solvers that allow to conveniently represent linear, real-time constraints. To the best knowledge of the authors, the presented tool, PTA2SMT, is the first one that applies SMT-solvers to parameter synthesis for PTA.

Parametric Timed Automata were introduced by Alur et al. in the seminal paper [1]. As they show, the emptiness problem, i.e., the existence of a parameter valuation under which a given state is reachable, is undecidable for PTA. They also tackle the synthesis problem and illustrate how to derive constraints on timing parameters for Fischer's mutual exclusion protocol. In [20] Wang introduces a parametric extension of Timed Automata called Static Parametric Timed Automata (SPTA, for short). In this model, unlike in Parametric Timed Automata, the introduced parameters are not compared with clocks. The emptiness problem is decidable for SPTA. A general framework for the parametric analysis of timed systems, based on partition refinement, is introduced in [18] by Spelberg and Toetenel. In [7] Hune et al. introduce parametric difference bound matrices and show how to use them for underapproximating the parametric reachability problem. The theory has been implemented in an experimental version of the UPPAAL model checker [15]. In the paper the authors also define an important class of PTA, called Lower/Upper Bound PTA where a given parameter is allowed exclusively as an upper or a lower bound. As they show, unlike for PTA, the problem of emptiness is decidable for L/U PTA. Decision problems for L/U PTA are investigated in detail by Bozzelli and La Torre in [5]. In [2] André et al. introduce a conceptually new approach to parameter synthesis, called the *Inverse Method* (IM, for short). IM starts with a reference valuation of the parameters v_0. The goal is to synthesise all the other parameter valuations under which the given PTA admits the same set of time-abstract traces as under v_0. The theory is implemented in the stand-alone tool IMITATOR. A simple semi-algorithm for solving the parametric reachability problem for PTA, based on breadth-first search in the space of symbolic states, is proposed by Jovanovic et al. in [8]. Its specialised version for synthesis of bounded parameter valuations has been implemented in the tool for verification of (Parametric, Timed) Petri Nets, called Romeo.

This paper builds on our earlier work [10], where we presented a method for bounded model checking for PTA and is a revised and extended version of [11]. It is also conceptually related to our paper [12], where a method for the synthesis of time parameters for a branching time logic RTCTL is presented. We explored SAT-based methods for verification of parameter-free formulae of RTCTL and its quantified extension PRTCTL in [13, 14]. In [16], Penczek et al. show how to synthesise, for a selected class of Petri nets, the minimal (real) time in which a given state can be reached. This is done by means of witness generation and analysis interleaved with net modification.

The rest of the paper is organized as follows. In the next section we introduce basic definitions and results concerning Parametric Timed Automata. In Sect. 3 we recall the notion of Lower/Upper Bound Parametric Timed Automata and provide the framework for representing the parametric reachability problem for these models as the model synthesis problem for QF_LRA, a logic supported by most SMT-solvers. In Sect. 4 we present the PTA2SMT tool and evaluate its applicability on a classical timed Fischer's mutual exclusion benchmark. Section 5 contains conclusions and final remarks.

In what follows, by $\mathbb{N}$, $\mathbb{Z}$, $\mathbb{R}$, and $\mathbb{R}_{\geq 0}$ we denote natural, integer, real, and nonnegative real numbers, respectively.

2 Parametric Timed Automata

Timed Automata (TA, for short) constitute the most popular and applied class of formal models for representing real-time systems. Intuitively, TA are graphs extended with *clocks*—the special variables used to measure the passage of time. Parametric Timed Automata extend TA by allowing free variables in clock expressions.

2.1 Timed Automata

In what follows, by $\mathbb{N}$, $\mathbb{Z}$, $\mathbb{R}$, and $\mathbb{R}_{\geq 0}$ we denote natural, integer, real, and nonnegative reals, respectively.

We define the set of clocks as $\mathcal{K} = \{x_1, \ldots, x_n\}$, for some $n \in \mathbb{N}$. The clocks range over nonnegative real values, i.e., we define a *clock valuation* as a function $\omega \colon \mathcal{K} \to \mathbb{R}_{\geq 0}$. We denote the set of all the clock valuations by Ω. The initial clock valuation ω^0 satisfies $\omega^0(x_i) = 0$ for all $x_i \in \mathcal{K}$.

Let $\omega \in \Omega$. We introduce two operations that can be executed on the clocks: incrementation and reset. If $\delta \in \mathbb{R}_{\geq 0}$, then by $\omega + \delta$ we denote the clock valuation that satisfies:

$$(\omega + \delta)(x_i) = \omega(x_i) + \delta \quad \text{for all } 1 \leq i \leq n.$$

This operation is interpreted as a time 'tick', with all the clocks increasing at the same rate. A set of the expressions of the form $x_i := b_i$, where $b_i \in \mathbb{N}$ and x_i appears at most once for each $1 \leq i \leq n$, is called a *reset*. The set of all the resets is denoted by $\mathcal{R}$. If $r \in \mathcal{R}$, then by $\omega[r]$ we denote such a clock valuation that:

$$\omega[r](x_i) \stackrel{def}{=} \begin{cases} b_i & \text{if } (x_i := b_i) \in r, \\ \omega(x_i) & \text{otherwise.} \end{cases}$$

Intuitively, resetting amounts to setting the selected clocks to some fixed values, while leaving the remaining ones intact.

Let $x_i, x_j \in \mathcal{K}$ and $x_i \neq x_j$. The expressions of the form $x_i - x_j \prec b$ and $x_i \prec b$, where $\prec \in \{\leq, <\}$ and $b \in \mathbb{Z}$ are called the *simple guards*. The set of all the simple guards is denoted by $\mathcal{G}'$. The conjunctions of simple guards are called the *guards* and the set of all the guards is denoted by $\mathcal{G}$. We assume that the empty conjunction is always true. By $\mathcal{G}^u$ we mean the subset of $\mathcal{G}$ consisting of the conjunctions of simple guards of type $x_i \prec b$. Observe that $\mathcal{G}^u$ represent the upper bounds on clock values. If $\omega \in \Omega$ and $g = (x_i - x_j \prec b) \in \mathcal{G}'$, then we write $\omega \models g$ iff $\omega(x_i) - \omega(x_j) \prec b$. Let $g \in \mathcal{G}$ be such that $g = \bigwedge_{i=0}^{k} g_i$ for some $k \in \mathbb{N}$ and $g_i \in \mathcal{G}'$, where $0 \leq i \leq k$. We write $\omega \models g$ iff $\omega \models g_i$ for all $i \leq k$.

Definition 1 (*TA*) A *Timed Automaton* is a 6-tuple $\mathcal{AU} = (\mathcal{Q}, l_0, \mathcal{A}, \mathcal{K}, \rightarrow, \mathcal{I})$, where:

- $\mathcal{Q}$ is a finite set of locations,
- $l_0 \in \mathcal{Q}$ is the initial location,
- $\mathcal{A}$ is a finite set of actions,
- $\mathcal{K}$ is a finite sets of clocks,
- $\rightarrow \subseteq \mathcal{Q} \times \mathcal{A} \times \mathcal{G} \times \mathcal{R} \times \mathcal{Q}$ is a transition relation,
- $\mathcal{I} : \mathcal{Q} \rightarrow \mathcal{G}$ is an invariant function.

The transition $(l, a, g, r, l') \in \rightarrow$ is denoted by $l \stackrel{a,g,r}{\rightarrow} l'$.

In order to analyse the behaviours of Timed Automata we need to track both the current location and the clock values. The semantics of a TA $\mathcal{AU}$ is given in the form of the associated labeled transition system $TS(\mathcal{AU})$ defined as follows.

Definition 2 (*Concrete semantics of TA*) Let $\mathcal{AU} = (\mathcal{Q}, l_0, \mathcal{A}, \mathcal{K}, \rightarrow, \mathcal{I})$ be a TA. The labeled transition system $TS(\mathcal{AU})$ is defined as the tuple $TS(\mathcal{AU}) = (\mathcal{S}, s^0, \mathbb{R}_{\geq 0} \cup \mathcal{A}, \hookrightarrow)$, where:

- $\mathcal{S} = \{(l, \omega) \mid l \in \mathcal{Q}$ and ω is a clock valuation such that $\omega \models \mathcal{I}(l)\}$ is the set of concrete states,
- $s^0 = (l_0, \omega^0)$ is the initial state (we assume that $\omega^0 \models \mathcal{I}(l_0)$),
- $\hookrightarrow \subseteq \mathcal{S} \times (\mathbb{R}_{\geq 0} \cup \mathcal{A}) \times \mathcal{S}$ is the transition relation labeled by time delay values and actions, and defined as follows. If $(l, \omega), (l', \omega') \in \mathcal{S}$, then:

  - if $d \in \mathbb{R}_{\geq 0}$, then $(l, \omega) \stackrel{d}{\hookrightarrow} (l', \omega')$ iff $l = l'$, $\omega' = \omega + d$, and $\omega' \models \mathcal{I}(l)$,
  - if $d \in \mathcal{A}$, then $(l, \omega) \stackrel{d}{\hookrightarrow} (l', \omega')$ iff for some $g \in \mathcal{G}$ and $r \in \mathcal{R}$ we have $l \stackrel{a,g,r}{\rightarrow} l'$, $\omega \models g$, $\omega' = \omega[r]$, and $\omega' \models \mathcal{I}(l')$.

Intuitively, waiting does not change the current location, but it is allowed only if the location's invariant is satisfied. An action leaving the current location can be fired if the present clock values satisfy its guard and, after resetting, the invariant of the target location holds.

A *path* is a sequence $\pi = (s_0, s_1, \ldots)$ of states from S such that $s_i \overset{d}{\hookrightarrow} s_{i+1}$ for some $d \in \mathbb{R}_{\geq 0} \cup \mathcal{A}$, for each $0 \leq i < |\pi|$. The paths can be finite or infinite. The number of the states of the path π is denoted by $|\pi|$ and for each $0 \leq i < |\pi|$ we denote $\pi_i = s_i$. The set of all the paths in $TS(\mathcal{AU})$ is denoted by $\Pi(\mathcal{AU})$ and the set of all the paths starting from a state $s \in S$ is denoted by $\Pi(\mathcal{AU}, s)$. We omit the automaton symbol whenever it is evident from the context.

2.2 Parametric Timed Automata

Timed Automata use concrete values to specify bounds on the clocks given in the guards of the transitions and invariants. Parametric Timed Automata allow for the linear expressions in place of the numeric values.

In what follows we assume that we have fixed the set of parameters $\mathcal{X} = \{p_1, p_2, \ldots, p_m\}$, for some $m \in \mathbb{N}$. An expression of the form $\sum_{i=1}^{m} t_i \cdot p_i + t_0$, where $t_i \in \mathbb{Z}$ for all $0 \leq i \leq m$, is called a *linear expression*. Note that if $t_i = 0$ for all $1 \leq i \leq m$, then a linear expression is an integer.

We extend the definition of a guard introduced earlier as follows. Let $x_i, x_j \in \mathcal{K}$, $x_i \neq x_j$, and e be a linear expression. We call the expressions of the form $x_i - x_j \prec e$ and $x_i \prec e$, where $\prec \in \{\leq, <\}$, the *simple parametric guards*. The conjunctions of the simple parametric guards are called the *parametric guards*. By $\mathcal{G}'_{\mathcal{X}}$ we denote the set of all the simple parametric guards, $\mathcal{G}_{\mathcal{X}}$ is the set of all the parametric guards, and by $\mathcal{G}^u_{\mathcal{X}}$ we mean the conjunctions of the simple parametric guards of type $x_i \prec e$. We omit the word "parametric" when mentioning the guards if this does not lead to a confusion.

Definition 3 *(PTA)* A *Parametric Timed Automaton* $\mathcal{AU} = (Q, l_0, \mathcal{A}, \mathcal{K}, \mathcal{X}, \rightarrow, \mathcal{I})$, is a 7-tuple where:

- Q is a finite set of locations,
- $l_0 \in Q$ is the initial location,
- $\mathcal{A}$ is a finite set of actions,
- $\mathcal{K}$ is a finite sets of clocks,
- $\mathcal{X}$ is a finite sets of parameters,
- $\rightarrow \subseteq Q \times A \times \mathcal{G}_{\mathcal{X}} \times \mathcal{R} \times Q$ is a transition relation,
- $\mathcal{I} : Q \rightarrow \mathcal{G}_{\mathcal{X}}$ is an invariant function.

The following example illustrates the above definition.

Example 1 (Parametric Timed Automaton) Consider the PTA $\mathcal{AU}$ shown in Fig. 1, where $\mathcal{X} = \{p, q, r\}$. The single parametric guard $g = x_2 - x_1 \geq p \wedge x_1 \geq 2p - q$ labels the transition *forward*. The invariant on *Sink* is parametric and we have $\mathcal{I}(Sink) = x_1 \leq r$.

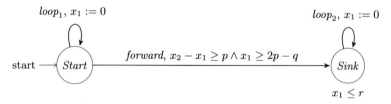

Fig. 1 Simple PTA

Our aim is to analyse how the properties of PTA depend on the values of the parameters. To this end we define a parameter valuation function $\upsilon : \mathcal{X} \to \mathbb{N}$. The set of all the parameter valuations is denoted by *ParVals*.

Let $\upsilon \in \textit{ParVals}$ and e be a linear expression. By $e[\upsilon]$ we denote the value obtained by substituting in e each parameter $p \in \mathcal{X}$ with $\upsilon(p)$. Formally, if $e = \sum_{i=1}^{m} t_i \cdot p_i + t_0$, then we have $e[\upsilon] = \sum_{i=1}^{m} t_i \cdot \upsilon(p_i) + t_0$.

If $g' = x_i - x_j \prec e \in \mathcal{G}'_{\mathcal{X}}$ is a parametric simple guard, then let $g'[\upsilon] = x_i - x_j \prec e[\upsilon]$. This extends in a natural way to the general case of $g \in \mathcal{G}_{\mathcal{X}}$ such that $g = \bigwedge_{i=0}^{k} g_i$, where $k \in \mathbb{N}$ and $\forall_{1 \le i \le k} g_i \in \mathcal{G}'$, as follows: $g[\upsilon] = \bigwedge_{i=0}^{k} g_i[\upsilon]$.

Let $\omega \in \Omega$. We write $\omega \models_{\upsilon} g$ iff $\omega \models g[\upsilon]$.

Finally, let $\mathcal{AU}$ be a parametric timed automaton and $\upsilon \in \textit{ParVals}$. By $\mathcal{AU}_{\upsilon}$ we denote the timed automaton obtained be replacing in each transition and each invariant of $\mathcal{AU}$ every parametric guard $g \in \mathcal{G}_{\mathcal{X}}$ with $g[\upsilon]$. Formally, if $\mathcal{AU} = (\mathcal{Q}, l_0, \mathcal{A}, \mathcal{K}, \mathcal{X}, \to, \mathcal{I})$ is a PTA, then $\mathcal{AU}_{\upsilon} = (\mathcal{Q}, l_0, \mathcal{A}, \mathcal{K}, \to', \mathcal{I}')$ is a TA such that: $\to' = \{(l, a, g[\upsilon], r, l') \mid (l, a, g, r, l') \in \to\}$ and $\mathcal{I}'(l) = \mathcal{I}(l)[\upsilon]$ for all $l \in \mathcal{Q}$.

2.3 Parametric Reachability for PTA

Let us formalise the problem of the synthesis for the parametric reachability, starting with the syntax and semantics of state formulae.

Definition 4 (*State Formulae*) Let $\mathcal{AU} = (\mathcal{Q}, l_0, \mathcal{A}, \mathcal{K}, \to, \mathcal{I})$ be a PTA. The *state formulae* are defined by the following grammar:

$$\psi ::= l \mid x_i \prec b \mid x_i - x_j \prec b \mid \psi \wedge \psi \mid \neg \psi,$$

where $l \in \mathcal{Q}$, $x_i, x_j \in \mathcal{K}$, $x_i \ne x_j$, $b \in \mathbb{N}$, and $\prec \in \{\le, <\}$.

We denote the set of all the state formulae by $\mathcal{SF}$.

Definition 5 (*State Formulae Semantics*) Let $l \in \mathcal{Q}$ and $\omega \in \Omega$. We recursively define the relation of satisfaction for $\mathcal{SF}$ as follows:

- $(l, \omega) \models l$,
- $(l, \omega) \models x_i \prec b$ iff $\omega(x_i) \prec b$,
- $(l, \omega) \models x_i - x_j \prec b$ iff $\omega(x_i) - \omega(x_j) \prec b$,

- $(l, \omega) \models \neg\phi$ iff $(l, \omega) \not\models \phi$,
- $(l, \omega) \models \phi \wedge \psi$ iff $(l, \omega) \models \phi$ and $(l, \omega) \models \psi$,

for all $\psi, \phi \in \mathcal{SF}$, $x_i, x_j \in \mathcal{K}$, $x_i \neq x_j$, $b \in \mathbb{N}$, and $\prec \in \{\leq, <\}$.

Observe that if $l \in \mathcal{Q}$, then $\neg l \equiv \bigvee_{l' \in \mathcal{Q}\setminus\{l\}} l'$. It is also not difficult to see that for all $l, l' \in \mathcal{Q}$ if $l \neq l'$, then $l \wedge l'$ is non-satisfiable. Additionally, a negated simple guard is also a simple guard. Each state formula can therefore be rewritten as a disjunction of terms of the form $l \wedge g$, where $l \in \mathcal{Q}$ and $g \in \mathcal{G}$.

Definition 6 (*Parametric Reachability*) Let $\mathcal{AU} = (\mathcal{Q}, l_0, \mathcal{A}, \mathcal{K}, \rightarrow, \mathcal{I})$ be a PTA, $\phi \in \mathcal{SF}$, and $\upsilon \in ParVals$. If there exists a path $\pi \in \Pi(\mathcal{AU}_\upsilon, l_0)$ such that $\pi_i \models \phi$ for some $i \in \mathbb{N}$, then ϕ is *reachable* in $\mathcal{AU}$ under υ.

We denote the reachability of ϕ in $\mathcal{AU}$ under υ by $\mathcal{AU} \models_\upsilon EF\phi$. As previously, we sometimes omit the PTA symbol if it is clear from the context.

By $\Gamma(\mathcal{AU}, \phi)$ we denote the set of all the valuations under which $\phi \in \mathcal{SF}$ is reachable in the parametric timed automaton $\mathcal{AU}$. Formally we have: $\Gamma(\mathcal{AU}, \phi) = \{\upsilon \in ParVals \mid \mathcal{AU} \models_\upsilon EF\phi\}$.

Example 2 (*Parametric Reachability*) Consider the PTA $\mathcal{AU}$ shown in Fig. 1. Under the parameter valuation υ such that $\upsilon(p) = 3$, $\upsilon(q) = 5$, and $\upsilon(r) = 2$ we have $g[\upsilon] = x_2 - x_1 \geq 3 \wedge x_1 \geq 1$ and $\mathcal{I}(Sink) = x_1 \leq 2$, therefore the state *Sink* is reachable from *Start* in $\mathcal{AU}_\upsilon$, i.e., $\models_\upsilon EF Sink$. On the other hand, let υ' be the valuation such that $\upsilon'(p) = 3$, $\upsilon'(q) = 1$, and $\upsilon'(r) = 2$. In this case we have $g[\upsilon'] = x_2 - x_1 \geq 3 \wedge x_1 \geq 5$ and $\mathcal{I}(Sink) = x_1 \leq 2$. Observe that the guard of the transition *forward* enforces $x_1 \geq 5$ and the invariant of *Sink* expects $x_1 \leq 2$, therefore in $\mathcal{AU}_{\upsilon'}$ the state *Sink* is not reachable from *Start*, i.e., $\not\models_\upsilon EF Sink$.

3 Parametric Reachability for L/U PTA via SMT

The question whether for a given PTA $\mathcal{AU}$ and state formula ϕ exists a parameter valuation $\upsilon \in ParVals$ such that $\mathcal{AU} \models_\upsilon EF\phi$ is called the *emptiness problem*. It is known that the emptiness problem is decidable for PTA with a single clock and undecidable for more than two clocks [1]. The case of two clocks is an open problem.

In [7] Hune et al. identified an important class of PTA, called Lower/Upper Bound Parametric Timed Automata. In this class, a given parameter can appear in the guards and the invariants exclusively either as a lower or an upper bound on clock values. The structure of the set of parameters enables for reducing the emptiness problem for L/U PTA to the problem of reachability for TA. As we show in what follows, it also allows for implementing a simple algorithm for parameter synthesis, based on the encoding of the transition function in a language accepted by SMT (Satisfiability Modulo Theory) solvers.

Definition 7 (*L/U PTA*) A PTA $\mathcal{AU} = (\mathcal{Q}, l_0, \mathcal{A}, \mathcal{K}, \mathcal{X}, \rightarrow, \mathcal{I})$ is called a *Lower/ Upper Bound Parametric Timed Automaton* if the set of parameters $\mathcal{X}$ can be partitioned into two disjoint sets $\mathcal{X}_L, \mathcal{X}_U \subseteq \mathcal{X}$ such that $\mathcal{X}_L = \{\lambda_1, \ldots, \lambda_l\}$, $\mathcal{X}_U = \{\mu_1, \ldots, \mu_u\}$, $\mathcal{X} = \mathcal{X}_L \cup \mathcal{X}_U$, and each linear expression present in the guards of the transitions and the invariants of $\mathcal{AU}$ can be written in the form:

$$\sum_{i=1}^{l} l_i \cdot \lambda_i + \sum_{j=1}^{u} u_j \cdot \mu_j + b_0,$$

where $l_i, u_j, b_0 \in \mathbb{Z}, l_i \leq 0, u_j \geq 0$ for all $1 \leq i \leq l$ and $1 \leq j \leq u$.

The elements of $\mathcal{X}_L$ ($\mathcal{X}_U$) are called the lower (upper, respectively) parameters.

For the remainder of this section we fix a L/U PTA $\mathcal{AU} = (\mathcal{Q}, l_0, \mathcal{A}, \mathcal{K}, \mathcal{X}, \rightarrow, \mathcal{I})$ with the lower parameters $\mathcal{X}_L = \{\lambda_1, \ldots, \lambda_l\}$ and the upper parameters $\mathcal{X}_U = \{\mu_1, \ldots, \mu_u\}$. Let us recall some basic notions concerning Lower/Upper Bound Parametric Timed Automata. If $\upsilon, \upsilon' \in ParVals$ are parameter valuations and we have $\forall_{\lambda \in \mathcal{X}_L} \upsilon'(\lambda) \leq \upsilon(\lambda)$ and $\forall_{\mu \in \mathcal{X}_U} \upsilon(\mu) \leq \upsilon'(\mu)$, then we write $\upsilon \leq \upsilon'$. Observe that if $g \in \mathcal{G}_\mathcal{X}$ is a parametric guard present in $\mathcal{AU}$ and $\upsilon \leq \upsilon'$ then for each clock valuation $\omega \in \Omega$ we have: $\omega \models_\upsilon g \implies \omega \models_{\upsilon'} g$.

Intuitively, this means that the constraints in L/U PTA can be uniformly relaxed by decreasing the values assigned to the lower parameters and increasing the values assigned to the upper parameters, without altering the enabledness of the invariants and the transitions. This operation also preserves the parametric reachability, as stated in the following proposition.

Proposition 1 (Corollary to Proposition 3 from [7]) *Let $\mathcal{AU}$ be a L/U PTA and $\phi \in \mathcal{SF}$. If $\upsilon, \upsilon' \in ParVals$, $\upsilon \leq \upsilon'$, and $\mathcal{AU} \models_\upsilon EF\phi$, then $\mathcal{AU} \models_{\upsilon'} EF\phi$.*

3.1 Basic Satisfiability Modulo Theories

Advances in the development of the techniques for deciding the satisfiability of propositional formulae (SAT) made the verification of large models possible. The formulae containing hundreds of thousands of variables became well within reach of modern SAT-solvers. Satisfiability Modulo Theories (SMT, for short) comprise a range of background theories for interpreting certain predicates. The language of SMT instances is a first order logic over these predicates. As with SAT, the task of an SMT-solver is to check whether a given formula is satisfiable; usually a model for the formula is returned as a by-product [6].

In what follows we present a simple translation from the problem of parametric reachability to the language of Quantifier-Free Linear Arithmetics (QF_LRA, for short). It should be noted that SMT-solvers use specialized algorithms for checking the satisfiability of linear constraints. These algorithms are usually based on the Simplex method. This means that in practice, despite the exponential worst-case complexity, one can expect quite an efficient average-case performance. Let us start with introducing the basic variables and relations used in building the formulae of QF_LRA.

Sorts and Predicates We use variables that belong to two *sorts*, i.e., types: Booleans and Reals. The locations Q of $\mathcal{AU}$ are encoded by means of enumerating them using propositional expressions. To this end we define for each $i \in \mathbb{N}$ the following set of propositional variables: $\mathcal{BV}^i = \{bv_1^i, bv_2^i, \ldots, bv_{\lceil log(|Q|) \rceil}^i\}$. We assume that $\mathcal{BV}^i \cap \mathcal{BV}^j = \emptyset$ if $i \neq j$, where $i, j \in \mathbb{N}$. The set of all the propositional formulae over $\mathcal{BV}^i$ is denoted by $\mathcal{BVE}^i$. Intuitively, we use the variables from $\mathcal{BV}^i$ to encode the locations found in the states of the ith position of any path. We define $BVars^k \stackrel{def}{=} \bigcup_{i \leq k} \mathcal{BV}^i$ for all $k \in \mathbb{N}$; we also let $BVars = \bigcup_{i \leq \infty} \mathcal{BV}^i$.

Recall that $\mathcal{K} = \{x_1, \ldots, x_n\}$, where $n \in \mathbb{N}$, is the set of the clocks. For each $i \in \mathbb{N}$ let $\mathcal{K}^i = \{x_1^i, x_2^i, \ldots, x_n^i\}$ be a set of real variables, where $\mathcal{K}^i \cap \mathcal{K}^j = \emptyset$ for $i \neq j$. The variables from $\mathcal{K}^i$ are used to encode the guards of the transitions leaving the ith position of any path. Similarly, let $T = \{t_0, t_1, \ldots\}$ be an infinite set of real variables. We use the members of T to represent the ith delay along a path. With a slight notational abuse we also treat as real variables the parameters from the set $\mathcal{X} = \{p_1, p_2, \ldots, p_m\}$, where $m \in \mathbb{N}$. For each $k \in \mathbb{N}$ we denote $Vars^k = \bigcup_{i=1}^{k} \mathcal{K}^i \cup T \cup \mathcal{X}$ and $Vars = \bigcup_{i=1}^{\infty} \mathcal{K}^i \cup T \cup \mathcal{X}$.

We also use two binary arithmetic predicates $\{\leq, <\}$, interpreted over reals in the usual way, and a single unary predicate *is_int* interpreted such that $is\_int(v)$ is true iff v is a natural number.

Quantifier-Free Linear Arithmetic We define two types of *atoms*. Firstly, all the real and propositional variables defined above are atoms. Secondly, so are all the expressions of the form $\sum_{i=1}^{m} d_i \cdot z_i + d_0 \prec 0$, where $d_i \in \mathbb{Q}$ and $z_i \in Vars$ for all $0 \leq i \leq m$, and $\prec \in \{\leq, <\}$.

Definition 8 (*QF_LRA Syntax*) The set of the formulae of Quantifier-Free Linear Arithmetics is defined by the following grammar:

$$\phi := \alpha \mid \neg\phi \mid \phi \vee \phi,$$

where α is an atom.

Intuitively, the language of QF_LRA consists of the propositional variables and linear expressions over clock and parameter variables, joined by the boolean connectives. We evaluate the truth of the QF_LRA formulae with respect to the valuations of the variables, i.e., functions $V : BVars \cup Vars \rightarrow \mathbb{B} \cup \mathbb{R}$ such that $V(v) \in \mathbb{B}$ for each $v \in BVars$ and $V(v) \in \mathbb{R}$ for all $v \in Vars$. The set of all such valuations is denoted by $QVals$.

For our convenience we define also for each $k \in \mathbb{N}$ the following formula:

$$TypeCut^k = \bigwedge_{v \in Vars^k} v \geq 0 \wedge \bigwedge_{p \in \mathcal{X}} is\_int(p).$$

Intuitively, k will correspond to the length of the considered paths. It is easy to notice that $V \models TypeCut^k$ ensures that the valuation V assigns non-negative values to the real variables and integer values to the parameters.

3.2 Encoding Models

To encode the locations of PTA we assume that there exists a family of functions $loc\_enc^i : Q \to \mathcal{BVE}^i$, where $i \in \mathbb{N}$, such that $loc\_enc^i(l) \wedge loc\_enc^j(l')$ is false for all $i, j \in \mathbb{N}$ and $l, l' \in Q$ such that $i \neq j$ or $l \neq l'$.

In our encoding we represent the paths in a unified manner, as interleaving sequences of time steps and actions. Let $k \in \mathbb{N}$ and $\upsilon \in ParVals$ be a valuation of parameters. By a k-path we mean a finite path $\pi^k \in \Pi(\mathcal{AU}_\upsilon, s_0)$ such that $\pi^k = (s_0, s_0', s_1, s_1', \ldots, s_k, s_k')$ and for each $0 \leq i \leq k$ we have $s_i \overset{d_i}{\hookrightarrow} s_i'$ for some $d_i \in \mathbb{R}_{\geq 0}$ and, if $0 \leq i < k$, then $s_i' \overset{a_i}{\hookrightarrow} s_{i+1}$ for some $a_i \in \mathcal{A}$. Intuitively, π^k is of the following form:

$$\pi^k = s_0 \overset{d_0}{\hookrightarrow} s_0' \overset{a_1}{\hookrightarrow} s_1 \overset{d_1}{\hookrightarrow} s_1' \overset{a_2}{\hookrightarrow} s_2 \overset{d_2}{\hookrightarrow} \ldots \overset{a_{k-1}}{\hookrightarrow} s_k \overset{d_k}{\hookrightarrow} s_k'.$$

If $\phi \in \mathcal{SF}$ is a state formula such that $s_k' \models_\upsilon \phi$, then we say that ϕ is k-reachable in $\mathcal{AU}_\upsilon$.

It is not difficult to see that every finite path can be represented as a k-path for some $k \in \mathbb{N}$. To this end it suffices to merge all the subsequent time steps and add zero-valued delays between any consecutive actions. In the analysis of the parametric reachability we can therefore focus on the k-paths. In what follows, we assume that in the considered transition relation there are no two transitions that share the same label. This assumption is not essential for the translation,[1] and it is used only to simplify the presentation of the results and the associated proofs.

Let $tr \in \to$ be a transition such that $tr = l \overset{a,g,r}{\hookrightarrow} l'$, where l, l' are respectively the source and the target location, a is the action label, g is the guard, and r is the reset. It is convenient to use the following notations: $source(tr) = l$, $target(tr) = l'$, $guard(tr) = g$, and $reset(tr) = r$.

We now present the building blocks of our encoding of tr. If α is a formula with free variables $a_1, a_2, \ldots, a_n$, where $n \in \mathbb{N}$, then by $\alpha[a_1'/a_1, a_2'/a_2, \ldots, a_n'/a_n]$ we denote the formula obtained by substituting in α each a_i with a_i', for all $1 \leq i \leq n$. Let $i \in \mathbb{N}$. Intuitively, for a given path, the value of i denotes the currently considered position. We define the encoding of $guard(tr)$ as $guard^i(tr) = guard(tr)[x_1^i/x_1, \ldots, x_n^i/x_n]$. The encoding $reset^i(tr)$ of $reset(tr)$ is defined as the smallest subset of QF_LRA such that for each $1 \leq j \leq n$ we have:

- $(x_j^i := b + t_i) \in reset^i(tr)$ if $(x_j := b) \in reset(tr)$,
- $(x_j^i := x_j^{i-1} + t_i) \in reset^i(tr)$ otherwise.

---

[1] We can always relabel the labels.

Intuitively, $reset^i(tr)$ models the new value of each clock after a consecutive reset and delay. For any location $l \in \mathcal{Q}$, we define the encoding of the invariant in l as $\mathcal{I}^i(l) = \mathcal{I}[x_1^i/x_1, \ldots, x_n^i/x_n]$.

We combine the above to define the encoding of the transition tr as follows:

$$tr\_enc^i(tr) = loc\_enc^i(source(tr)) \wedge guard^i(tr) \wedge reset^{i+1}(tr)$$
$$\wedge \; \mathcal{I}^{i+1}(target(tr)) \wedge loc\_enc^{i+1}(target(tr)).$$

Observe that $tr\_enc^i(tr) \in$ QF_LRA for each $i \in \mathbb{N}$. The meaning of the above construction is stated in the following lemma.

Lemma 1 *Let* $tr = l \overset{a,g,r}{\to} l'$ *be a transition,* $\upsilon \in ParVals$ *a parameter valuation,* $i \in \mathbb{N}$, $d \in \mathbb{R}_{\geq 0}$, *and* $(l, \omega) \in \mathcal{S}$ *a concrete state. We have:*

$$(l, \omega) \overset{a}{\hookrightarrow} (l', \omega[r]) \overset{d}{\hookrightarrow} (l', \omega[r] + d)$$

iff there exists a valuation $V \in QVals$ *s.t.* $V \models tr\_enc^i(tr) \wedge TypeCut^{i+1}$ *and:*

1. $\upsilon = V_{|\mathcal{X}}$,
2. $\omega = V_{|\mathcal{K}^i}[x_1/x_1^i, \ldots, x_n/x_n^i]$,
3. $d = V(t_{i+1})$,
4. $\omega[r] + d = V_{|\mathcal{K}^{i+1}}[x_1/x_1^{i+1}, \ldots, x_n/x_n^{i+1}]$.

Proof Observe that the locations are uniquely represented by their encodings, thus we can focus on nonboolean variables.

Let V be a valuation of the variables such that $V \models tr\_enc^i(tr) \wedge TypeCut^{i+1}$. Let $\omega = V_{|\mathcal{K}^i}[x_1/x_1^i, \ldots, x_n/x_n^i]$ and $\upsilon = V_{|\mathcal{X}}$. Denote $\omega^i = V_{|\mathcal{K}^i}$, then from $V \models guard^i(tr)$ we obtain $\omega^i \models_\upsilon guard^i(tr)$, which in turn yields that $\omega \models_\upsilon guard(tr)$.

Let $d = V(t_{i+1})$, denote $\omega^{i+1} = V_{|\mathcal{K}^{i+1}}$ and notice that from $V \models reset^{i+1}(tr)$ it follows that $\omega^{i+1}(x_j^{i+1}) = \omega^i[r](x_j^i) + d$ for all $1 \leq j \leq n$. Thus, if we denote $\omega' = V_{|\mathcal{K}^{i+1}}[x_1/x_1^{i+1}, \ldots, x_n/x_n^{i+1}]$, then $\omega' = \omega[r] + d$.

Now, observe that from $V \models inv^{i+1}(target(tr))$ we can infer that we have $\omega^{i+1} \models_\upsilon \mathcal{I}^{i+1}(target(tr))$, from which $\omega' \models_\upsilon \mathcal{I}(target(tr))$, i.e., $\omega[r] + d \models_\upsilon \mathcal{I}(target(tr))$.

As $d \geq 0$ and in view of the assumption that the invariants admit only upper bounds on clocks, we have also that $\omega[r] \models_\upsilon \mathcal{I}(target(tr))$. This, together with the fact that $V \models TypeCut^{i+1}$ assures that the used variables range over the correct sets, concludes this part of the proof.

The other part of the proof follows easily from the basic definitions and the construction of the encoding. $\square$

3.3 Encoding k-Paths and Reachability Testing

Recall that $n \in \mathbb{N}$ is the number of clocks and let $k \in \mathbb{N}$. We employ the formulae introduced earlier to obtain the following encoding of all the k-paths that start from the initial state:

$$model\_enc^k(\mathcal{AU}) = TypeCut^k \wedge \left(\bigwedge_{i=1}^{n} (x_i^0 = t_0) \wedge loc\_enc^0(l_0) \wedge \mathcal{I}^0(l_0) \right)$$

$$\wedge \bigwedge_{i=0}^{k-1} \bigvee_{tr \in \,\rightarrow} tr\_enc^i(tr).$$

Let $\phi \in \mathcal{SF}$ be a state formula. Assume that the set of the locations is $\mathcal{Q} = \{l_1, \ldots, l_m\}$, for some $m \in \mathbb{N}$. We define the encoding of ϕ as follows:

$$prop\_enc^k(\phi) = \phi[x_1^k/x_1, \ldots, x_n^k/x_n, loc\_enc^k(l_1)/l_1, \ldots, loc\_enc^k(l_m)/l_m].$$

In the above formula we substitute in ϕ each clock with its kth variable counterpart and each location with its encoding using boolean variables from $\mathcal{BV}^k$.

The formulae $model\_enc^k(\mathcal{AU})$ and $prop\_enc^k(\phi)$ combined together allow for the parameter synthesis for k-reachability, as shown in the following lemma.

Lemma 2 *Let $\phi \in \mathcal{SF}$ be a state formula, $\upsilon \in ParVals$ a parameter valuation, and $k \in \mathbb{N}$. A concrete state satisfying ϕ is k-reachable in $\mathcal{AU}_\upsilon$ iff there exists a valuation $V \in QVals$ such that $V \models model\_enc^k(\mathcal{AU}) \wedge prop\_enc^k(\phi)$ and $\upsilon = V_{|\mathcal{X}}$.*

Proof Let us focus on $model\_enc^k(\mathcal{AU})$. Observe that the first component of the formula ensures that all the variables range over the proper values. The second component of the formula sets all the initial clocks to some arbitrary common value, encodes the initial state, and makes sure that its invariant is satisfied (recall that the invariants represent the upper bounds on the clocks). Finally, notice that by applying Lemma 1 k times we obtain that the last component of the formula encodes all the possible k-paths. The formula $model\_enc^k(\mathcal{AU})$ therefore encodes all the k-paths that start in the initial state. To conclude, observe that by computing its conjunction with $prop\_enc^k(\phi)$ we select the k-paths whose last state satisfies ϕ. □

3.4 Approximative Synthesis of Parameters

From Lemma 2 we already know how to represent all the k-reachable states for which a given property holds. It might be beneficial to verify this formula as it is, as we can rely on the SMT-solver to obtain an exemplary witness, i.e., a correct valuation of the parameters. Our task is, however, to systematically explore the space of the

---

Algorithm 1 *SynthReachApprox* ($\mathcal{AU}$, ϕ, *depth*)

Input: PTA $\mathcal{AU}$, $\phi \in \mathcal{SF}$, *depth* $\in \mathbb{N}$
Output: *result* $\subseteq \Gamma(\mathcal{AU}, \phi)$

1: *result* := $\emptyset$
2: *reachApprox* := *model_enc*^{depth}($\mathcal{AU}$) $\wedge$ *prop_enc*^{depth}(ϕ)
3: **while** user requests to expand *result* and *reachApprox* is satisfiable **do**
4:    let V be such that $V \models$ *reachApprox* and $\upsilon := V_{|\mathcal{X}}$
5:    *result* := *result* $\cup \{\upsilon\}$
6:    *reachApprox* := *reachApprox* $\wedge$ *ComplClause*(υ)
7: **end while**
8: **return** *result*

---

admissible parameters, with a hope for "painting a part of the picture" from which an analyst can make further generalizations.

Let $\phi \in \mathcal{SF}$ and $\upsilon \in ParVals$ be a valuation of the parameters such that $\mathcal{AU} \models_\upsilon EF\phi$. Recall from Proposition 1 that in the class of the L/U PTA this means that a state satisfying ϕ is also reachable in $\mathcal{AU}_{\upsilon'}$ for each $\upsilon' \in ParVals$ such that $\upsilon \leq \upsilon'$. We define the *complementing clause* with respect to υ as follows:

$$ComplClause(\upsilon) = \bigvee_{i=1}^{l} (\lambda_i > \upsilon(\lambda_i)) \vee \bigvee_{i=1}^{u} (\mu_i < \upsilon(\mu_i)).$$

Observe that $\upsilon' \models ComplClause(\upsilon)$ iff $\upsilon \leq \upsilon'$ is not true. In Algorithm 1 we employ *ComplClause*(υ) to block the SMT-solver from seeking for those parameter valuations that can be inferred from the set of the already synthesized parameters using the properties of L/U PTA. The algorithm attemps to synthesise parameter valuations for which there exists a k-reachable state satisfying the property ϕ. If the search is successful, the user is presented with a newly synthesised parameter valuation υ and asked whether the procedure should be continued. If so, a new blocking *ComplClause*(υ) is added to the main formula and the loop takes another turn. Note that in the above algorithm testing for satisfiability (Line 3) and extraction of a witness valuation υ (Line 4) are performed by means of a call to an external SMT-solver.

The following lemma states that Algorithm 1 synthesises the correct parameter valuations. Moreover, all the possible executions of the routine cover every correct parameter valuation for parametric reachability.

Lemma 3 (Correctness and Completeness of Parameter Synthesis for L/U PTA) *If* $\mathcal{AU}$ *is a* L/U PTA, $\phi \in \mathcal{SF}$ *a state formula, then we have:*

$$\Gamma(\mathcal{AU}, \phi) = \bigcup_{i \in \mathbb{N}} SynthReachApprox(\mathcal{AU}, \phi, i).$$

Proof Follows easily from Lemma 2 and the properties of *ComplClause*. $\square$

Algorithm 1 can be used as the main building block of an interactive or guided parameter synthesis process. In the interactive process, the procedure *SynthReachApprox* ($\mathcal{AU}$, ϕ, *depth*) is called, starting from *depth* = 0. The user is presented with the synthesised results and decides whether *depth* should be incremented. In the guided approach, the user can provide a plan that helps the automated tool in resolving whether the already found values are sufficient.

4   Application and Evaluation

We have implemented the SMT-based approach presented in this section in a freely available open source tool PTA2SMT [9]. In the process of the synthesis the user is expected to supply a model file and a state formula together with an *experiment plan*. The plan consists of a sequence of pairs (k, No) of natural numbers, where k is the length of the runs to be considered, and No is the maximal number of parameter valuations to be synthesised should the verified property be found satisfiable. In practice this means that for each (k, No) in the experiment plan we call *SynthReachApprox* ($\mathcal{AU}$, ϕ, k) and request to synthesise further valuations until No are found or there are no more (Line 2 of Algorithm 1). We use parallel compositions of PTA as models.

Definition 9   Given $k \in \mathbb{N}$, let $I = \{1, \ldots, k\}$ be a finite set of indices and for each $i \in I$ let $\mathcal{AU}^i = (\mathcal{Q}^i, l_0^i, \mathcal{A}^i, \mathcal{K}^i, \mathcal{X}^i, \rightarrow^i, \mathcal{I}^i)$ be a L/U PTA such that $\mathcal{K}^i \cap \mathcal{K}^j = \emptyset$ if $i \neq j$ and $1 \leq i, j \leq k$. We define the *parallel product* of the network $\{\mathcal{AU}_i\}_{i \in I}$ as the PTA $\mathcal{AU} = (\mathcal{Q}, l_0, \mathcal{A}, \mathcal{K}, \mathcal{X}, \rightarrow, \mathcal{I})$ such that we have:

- $\mathcal{Q} = \prod_{i \in I} \mathcal{Q}^i$,
- $l_0 = (l_0^1, \ldots, l_0^k)$,
- $\mathcal{A} = \bigcup_{i \in I} \mathcal{A}^i$,
- $\mathcal{K} = \bigcup_{i \in I} \mathcal{K}^i$,
- $\mathcal{X} = \bigcup_{i \in I} \mathcal{X}^i$,
- $\mathcal{I}((l_1, \ldots, l_k)) = \bigwedge_{i \in I} \mathcal{I}(l_i)$, for each $(l_1, \ldots, l_k) \in \mathcal{Q}$,
- $(l_1, \ldots, l_k) \overset{a,g,r}{\rightarrow} (l_1', \ldots, l_k')$ iff for each $i \in I$ we have $l_i \overset{a,g,r}{\rightarrow} l_i'$ when $a \in \mathcal{A}^i$ and $l_i = l_i'$ otherwise, for each $(l_1, \ldots, l_k), (l_1', \ldots, l_k') \in \mathcal{Q}$.

In what follows we evaluate our approach on a parametric version of classical timed Fischer's mutual exclusion model. The experiments have been performed on an Intel P6200 dual core 2.13 GHz machine with 3.5GB RAM, running Linux operating system. In the evaluation we use CVC3 SMT-solver [4] that accepts the universal language proposed by the SMT-lib v2 initiative [3].

Fischer's Mutual Exclusion Protocol   In the problem of mutual exclusion we deal with multiple processes trying to access the critical section. Under a correct protocol a state where two or more processes are in this section is unreachable. In Fig. 2 we present the PTA network that models a timed version of the Fischer's solution to the

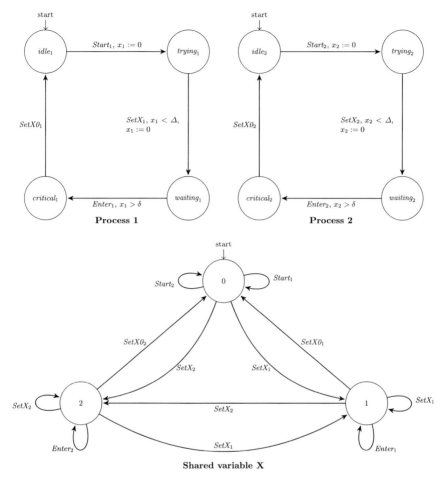

Fig. 2 Fischer's mutual exclusion protocol

mutual exclusion problem (this model originates from [17] and is a version of the one introduced earlier in [19]).

The system in question consists of $n \in \mathbb{N}$ independent processes synchronised via the shared variable X. There are n clocks, i.e., $\mathcal{K} = \{x_1, \ldots, x_n\}$, one lower bound parameter δ, and one upper bound parameter Δ.

Let $1 \leq i \leq n$. The ith process contains four locations, namely: $idle_i$, $trying_i$, $waiting_i$ and $critical_i$. The locations $idle_i$ and $trying_i$ are joined by the transition labeled with $Start_i$ and the reset $x_i := 0$. The transition labeled with $SetX_i$, guarded by $x_i < \Delta$, and reset by $x_i := 0$ joins $trying_i$ and $waiting_i$. The locations $waiting_i$ and $critical_i$ are connected by the transition labeled with $Enter_i$ and guarded by $x_i > \delta$. The return transition from $critical_i$ to $idle_i$ is labeled with $SetX0_i$.

Table 1 Fischer's Mutual Exclusion parameter synthesis results

n	k	Par. found	Form. size (MB)	Form. bdg. time (s)	Total CVC3 time (s)	Peak CVC3 mem. (MB)
7	1–5	–	1.54	1.7	1.8	20
7	6	10	3.51	2.25	37.2	41
8	1–5	–	2.95	3.95	3.6	30
8	6	10	7.13	5.45	75.3	70
9	1–5	–	5.48	7.1	6.5	48
9	6	10	13.71	11.86	187.8	119
10	1–5	–	9.43	13.1	11.33	69
10	6	10	24.62	24.06	245.75	198
11	1–5	–	15.25	23.29	20.71	108
11	6	10	41.84	46.23	504.35	331
12	1–5	–	24.94	40.13	25.11	170
12	6	10	71.17	85.07	726.51	560
13	1–5	–	38.89	66.1	40.78	256
13	6	10	115.8	149.24	1315.87	1000
14	1–5	–	57.76	105.98	67.50	384
14	6	10	180.71	253.99	3192.47	1600

We model X using the timed automaton with the locations $0, 1, \ldots, n$ that correspond to the values of the variable. The initial location 0 is joined with itself by the transitions labeled with $Start_1, \ldots, Start_n$. For each $1 \leq i \leq n$ there is a transition labeled with $SetX_i$ that joins 0 with i. Let $i, j \in \{1, \ldots, n\}$. Each location i is joined with the location 0 by the transition labeled with $SetX0_i$, with itself by the transition labeled with $Enter_i$, and with the location j by the transition labeled with $SetX_j$.

It is known (see [19]) that no two processes are able to simultaneously enter the critical sections iff $\Delta \leq \delta$. We have therefore chosen to analyse the parametric reachability of the state formula $\phi_1 = critical_1 \wedge critical_2$. Intuitively, this means that we aim to synthesise the values of the parameters δ and Δ under which the system behaves incorrectly, allowing two competing processes to jointly enter their critical sections.

Table 1 presents the evaluation of the performance of our approach. The first column corresponds to the number of the competing processes. The second column collects the lengths k of the k-paths; if no parameter valuation is found we present an interval, otherwise a single value. For simplicity of the presentation we treat the single value as a degenerate interval. In the third column we report the number of found parameter valuations. The fourth column contains the size of the file containing the largest formula generated in the interval. The fifth column collects the total time spent on the incremental construction of the formulae. The time spent on the verification of the formulae by CVC3 is reported in the sixth column. The last column presents the peak memory used during the verification.

In the experiment plan we have set *No* to 10. The set of the parameter valuations is the same in all the cases where at least one was found. Namely, we synthesise the parameter valuations that assign i to δ and $i + 1$ to Δ, for each $0 \leq i \leq 10$. Clearly, this is a subset of the full space of the solutions given by the inequality $\delta < \Delta$ (Fig. 3).

```
automaton Process_i {

locations:
initial location(idle_i)
        invariant: true, labelling: false;

        location(trying_i)
        invariant: true, labelling: false;

        location(waiting_i)
        invariant: true, labelling: false;

        location(critical_i)
        invariant: true, labelling: bad_i;

transitions:
                (idle_i, trying_i) action: start_i, guard: true, reset: x_i= 0;
                (trying_i, waiting_i) action: setx_i, guard: x_i < Delta, reset: x_i= 0;
                (waiting_i, critical_i) action: enter_i, guard: x_i > delta, reset: true;
                (critical_i, idle_i) action: setx0_i, guard: true, reset: true;
};

automaton VarX {

locations:
initial location(zero)
        invariant: true, labelling: false;

        location(one)
        invariant: true, labelling: false;

        location(two)
        invariant: true, labelling: false;

transitions:
                (zero, zero) action: start1, guard: true, reset: true;
                (zero, zero) action: start2, guard: true, reset: true;
                (zero, one) action: setx1, guard: true, reset: true;
                (zero, two) action: setx2, guard: true, reset: true;

                (one, zero) action: setx01, guard: true, reset: true;
                (one, one) action: enter1, guard: true, reset: true;
                (one, one) action: setx1, guard: true, reset: true;
                (one, two) action: setx2, guard: true, reset: true;

                (two, zero) action: setx02, guard: true, reset: true;
                (two, two) action: enter2, guard: true, reset: true;
                (two, two) action: setx2, guard: true, reset: true;
                (two, one) action: setx1, guard: true, reset: true;
};

property: bad_0 and bad_1;
```

Fig. 3 PTA2SMT template for Fischer's timed mutual exclusion protocol, $i \in \{1, 2\}$

5 Conclusions

In this paper we have presented a simple framework for experimental exploration of the space of solutions of parametric reachability problem for Parametric Timed Automata. The approach is based on a limited unfolding of the transition relation of PTA and encoding it in a form of an SMT instance. We have introduced a freely available tool PTA2SMT, together with the example of its application to the synthesis of valuations of parameters under which a timed mutual exclusion protocol fails. In the essence, the presented framework is a version of bounded model checking. Hence, in the future we plan to explore methods for efficient encoding of behaviours of PTA, similarly to how it is done for TA in [21].

Acknowledgments Michał Knapik is supported by the Foundation for Polish Science under Int. Ph.D. Projects in Intelligent Computing. Project financed from the EU within the Innovative Economy OP 2007–2013 and ERDF.

References

1. Alur, R., Henzinger, T., Vardi, M.: Parametric real-time reasoning. In: Proceedings of the 25th Annual Symposium on Theory of Computing, pp. 592–601. ACM (1993)
2. André, É., Chatain, T., Fribourg, L., Encrenaz, E.: An inverse method for parametric timed automata. Int. J. Found. Comput. Sci. **20**(5), 819–836 (2009)
3. Barrett, C., Stump, A., Tinelli, C.: The SMT-LIB Standard: Version 2.0. Technical report. Department of Computer Science, The University of Iowa, available at www.SMT-LIB.org (2010)
4. Barrett, C., Tinelli, C.: CVC3. In: Proceedings of the 19th International Conference on Computer Aided Verification (CAV). Lecture Notes in Computer Science, vol. 4590, pp. 298–302. Springer, Berlin (2007)
5. Bozzelli, L., La Torre, S.: Decision problems for lower/upper bound parametric timed automata. Form. Methods Syst. Des. **35**(2), 121–151 (2009). http://dx.doi.org/10.1007/s10703-009-0074-0
6. de Moura, L.M., Bjørner, N.: Satisfiability modulo theories: an appetizer. Formal Methods: Foundations and Applications. In: 12th Brazilian Symposium on Formal Methods, (SBMF), Revised Selected Papers. Lecture Notes in Computer Science, vol. 5902, pp. 23–36. Springer (2009)
7. Hune, T., Romijn, J., Stoelinga, M., Vaandrager, F.: Linear parametric model checking of timed automata. J. Log. Algebr. Program. **52–53**, 183–220 (2002)
8. Jovanovic, A., Lime, D., Roux, O.H.: Integer parameter synthesis for timed automata. In: Proceedings of Tools and Algorithms for the Construction and Analysis of Systems (TACAS). Lecture Notes in Computer Science, vol. 7795, pp. 401–415. Springer (2013)
9. Knapik, M.: https://michalknapik.github.io/PTA2SMT
10. Knapik, M., Penczek, W.: Bounded model checking for parametric timed automata. Trans. Petri Nets Models Concurr. **5**, 141–159 (2012)
11. Knapik, M., Penczek, W.: SMT-based parameter synthesis for L/U automata. In: Proceedings of the International Workshop on Petri Nets and Software Engineering, vol. 851, pp. 77–92. CEUR, Hamburg, 25–26 June 2012
12. Knapik, M., Penczek, W.: Parameter synthesis for timed Kripke structures. Fundam. Inform. **133**(2–3), 211–226 (2014). http://dx.doi.org/10.3233/FI-2014-1072

13. Knapik, M., Penczek, W., Szreter, M., Pólrola, A.: Bounded parametric verification for distributed time petri nets with discrete-time semantics. Fundam. Inform. **101**(1–2), 9–27 (2010)
14. Knapik, M., Szreter, M., Penczek, W.: Bounded parametric model checking for elementary net systems. Trans. Petri Nets Models Concurr. **4**, 42–71 (2010)
15. Larsen, K.G., Pettersson, P., Yi, W.: UPPAAL in a nutshell. Int. J. Softw. Tools Technol. Transf. **1**(1–2), 134–152 (1997)
16. Penczek, W., Pólrola, A., Zbrzezny, A.: SAT-based (parametric) reachability for a class of distributed time petri nets. Trans. Petri Nets Models Concurr. **4**, 72–97 (2010)
17. Penczek, W., Szreter, M.: SAT-based unbounded model checking of timed automata. Fundam. Inform. **85**(1–4), 425–440 (2008)
18. Spelberg, R.F.L., Toetenel, W.J.: Splitting trees and partition refinement in real-time model checking. In: 35th Hawaii International Conference on System Sciences (HICSS-35 2002), CD-ROM/Abstracts Proceedings, Big Island, HI, USA. p. 278, IEEE Computer Society, 7–10 January 2002. http://dx.doi.org/10.1109/HICSS.2002.994478
19. Tripakis, S., Yovine, S.: Analysis of timed systems using time-abstracting bisimulations. Form. Methods Syst. Design **18**(1), 25–68 (2001)
20. Wang, F.: Parametric analysis of computer systems. Form. Methods Syst. Design **17**(1), 39–60 (2000)
21. Zbrzezny, A.: SAT-based reachability checking for timed automata with diagonal constraints. Fundam. Inform. **67**(1–3), 303–322 (2005)

On the Identification of α-Asynchronous Cellular Automata in the Case of Partial Observations with Spatially Separated Gaps

Witold Bołt, Barbara Wolnik, Jan M. Baetens and Bernard De Baets

Abstract In this paper we present a statistical method, based on frequencies, for identifying so-called α-asynchronous Cellular Automata from partial observations, i.e. pre-recorded configurations of the system with some cells having an unknown (missing) state. The presented method, in addition to finding the unknown Cellular Automaton, is able to unveil the missing state values with high accuracy.

Keywords Asynchronous cellular automata · Identification · Parameter estimation

1 Introduction

Cellular Automata (CAs) are commonly used modelling constructs for addressing a variety of problems [12]. In order to use CAs for a practical modelling task, one needs to understand the underlying mechanisms of the phenomenon at stake, and translate them into a CA rule. Additionally, the state space, tessellation and neighborhood structure need to be pinned down beforehand. This hampers the use of CAs, since there are problems for which it is hard to manually design a proper local rule. In some cases, only the initial and final states of the system are known (e.g. [2, 20, 21]).

W. Bołt (✉)
Systems Research Institute, Polish Academy of Sciences,
Newelska 6, 01-447 Warsaw, Poland
e-mail: Witold.Bolt@ibspan.waw.pl

W. Bołt · J.M. Baetens · B. De Baets
KERMIT, Department of Mathematical Modelling, Statistics and Bioinformatics,
Ghent University, Coupure Links 653, 9000 Gent, Belgium
e-mail: jan.baetens@ugent.be

B. De Baets
e-mail: bernard.debaets@ugent.be

B. Wolnik
Institute of Mathematics, University of Gdańsk, Wita Stwosza 57,
80-952 Gdańsk, Poland
e-mail: Barbara.Wolnik@mat.ug.edu.pl

© Springer International Publishing Switzerland 2016
G. De Tré et al. (eds.), *Challenging Problems and Solutions in Intelligent Systems*, Studies in Computational Intelligence 634,
DOI 10.1007/978-3-319-30165-5_2

Besides classical deterministic CAs, Stochastic CAs (SCAs) are frequently used. Many efforts have been made in the direction of developing automated methods for constructing CAs and SCAs based on observed space-time diagrams [1, 3, 5, 6, 10, 11, 15–19, 22, 24, 26, 27]. Yet, there is only very limited literature for the case of incomplete observations in the deterministic case [7, 8]. To the best of our knowledge, the identification of CAs in the context of incomplete observations and stochastic rules has not yet been tackled by other authors.

The main goal of the research presented in this paper is to develop a method for the automated identification of a relatively simple class of SCAs, namely α-asynchronous CAs (α-ACAs) [14], based on partial observations. The presented method is based on statistical principles for estimating the parameters of binomial distributions based on frequencies of observed events. Moreover, we also present a method for completing those observations, i.e. filling the missing gaps into the observations. In addition to serving as a useful tool for building and analyzing models based on α-ACAs, the presented method is a first step towards an effective identification of SCAs based on partial observations.

The performance of the presented method is verified with computational experiments, for the class of α-ACAs corresponding to Elementary CAs (ECAs). The results show that the accuracy of the identification algorithm, when it comes to estimating the value of the synchrony rate α, finding the underlying CA and filling the missing states in the observation is very high.

This paper is organized as follows. In Sect. 2, we introduce definitions and present some well-known facts on CAs and SCAs. In Sect. 3, α-ACAs are formally introduced, while the formal definition of the identification problem is given in Sect. 4. Section 5 holds the description of the identification algorithm. The paper is concluded with Sect. 6, which presents the results of computational experiments.

2 Preliminaries

In this paper, we will concentrate on 1D, deterministic, two-state CAs with a symmetric neighborhood and a finite number of cells. Let $r \in \mathbb{N}_0$, $R = 2r + 1$ and let $f : \{0, 1\}^R \to \{0, 1\}$ be a function, then for $N > 0$, we define the N-cell global CA rule $A_N : \{0, 1\}^N \to \{0, 1\}^N$ as:

$$A_N(\ldots, s_i, \ldots) = (\ldots, f(s_{i-r}, \ldots, s_{i+r}), \ldots), \tag{1}$$

where periodic boundary conditions are assumed, i.e. for any $i \in \mathbb{Z}$ it holds that $s_{i+N} = s_i$. The function f used in (1) will be referred to as a local rule, and the integer r will be referred to as the radius of the neighborhood. Any local rule can be uniquely defined by a lookup table (LUT), which lists all the possible arguments of the local rule together with the corresponding function values. It is assumed that the arguments are listed in lexicographic order. The general form of such a LUT for $r = 1$ is shown in Table 1.

Table 1 The LUT of the local rule $n = (l_7, l_6, l_5, l_4, l_3, l_2, l_1, l_0)_2$

(1,1,1)	(1,1,0)	(1,0,1)	(1,0,0)	(0,1,1)	(0,1,0)	(0,0,1)	(0,0,0)
l_7	l_6	l_5	l_4	l_3	l_2	l_1	l_0

Table 2 The LUTs of ECAs 51 and 204

	(1,1,1)	(1,1,0)	(1,0,1)	(1,0,0)	(0,1,1)	(0,1,0)	(0,0,1)	(0,0,0)
51 negation	0	0	1	1	0	0	1	1
204 identity	1	1	0	0	1	1	0	0

The LUT can be used to enumerate local rules, as the coefficients l_i can be treated as digits in the binary representation of an integer n, i.e. in the case of $r = 1$ the number of a local rule is $n = \sum_{i=0}^{7} l_i \, 2^i$. Clearly, this reasoning extends to larger radii. Given that the ordering of the arguments in the LUT is fixed, only the second row needs to be known in order to uniquely define a CA, such that a LUT may be represented as a binary vector of length 2^R.

CAs for which there exists a local rule with radius $r = 1$ will be referred to as ECAs [25]. Due to their simplistic definition and rich dynamics, ECAs form a well-studied class of CAs. For that reason, the examples and experiments presented in this paper are based on ECAs and their asynchronous counterparts.

Example 1 In Table 2 the LUTs of ECA rules 51 and 204 are shown, which are known as negation and identity CAs, respectively. Both of these CAs can be expressed with a local rule of radius zero, but the ECA description is more commonly used.

With $\{0, 1\}^*$ we will denote the set of all binary sequences of finite length, i.e. $\{0, 1\}^* = \bigcup_{M=1}^{\infty} \{0, 1\}^M$. The function $A \colon \{0, 1\}^* \to \{0, 1\}^*$, defined by $A(X) = A_M(X)$ if $X \in \{0, 1\}^M$, with every global rule A_M being defined with the same local rule f, will be referred to as a generalized global CA rule. We will simply refer to such functions as global rules or rules. In this paper, a CA will be identified by its global rule, and by referring to a CA, we therefore always refer to its global rule in this generalized sense.

Every CA A can be uniquely defined by its local rule f with neighborhood radius $r \geq 0$. Every local rule can be uniquely described with a set of neighborhood configurations $\mathcal{C}(f)$, for which the local rule agrees with identity CA, i.e. $(x_1, \ldots, x_R) \in \mathcal{C}(f)$ if and only if $f(x_1, \ldots, x_R) = x_{r+1}$. As a consequence of the binary nature of the state set, it further holds that $f(x_1, x_2, \ldots, x_R) = 1 - x_{r+1}$ when $(x_1, \ldots, x_R) \notin \mathcal{C}(f)$.

Let A be a CA, $X \in \{0, 1\}^M$ for some M and $T > 0$. The finite sequence of vectors given by:

$$(X, A(X), A^2(X), \ldots, A^{T-1}(X)),$$

Table 3 pLUT of a stochastic ECA local rule

(1,1,1)	(1,1,0)	(1,0,1)	(1,0,0)	(0,1,1)	(0,1,0)	(0,0,1)	(0,0,0)
p_7	p_6	p_5	p_4	p_3	p_2	p_1	p_0

where A^t denotes the result of applying the rule A to $A^{t-1}(X)$, will be referred to as the space-time diagram covering T time steps. Each of its elements will be referred to as a configuration of the CA A, while the first element will be referred to as the initial configuration. For any $t = 0, 1, \ldots, T - 1$ and $m = 1, \ldots, M$, $A^t(X)[m]$ refers to the state of the mth cell in the tth row of the space-time diagram.

The CAs defined above are deterministic and are fully governed by their local rule. However, there are also Stochastic CAs (SCAs), for which the local rule is a random function that can be uniquely defined by a probability lookup table (pLUT). The pLUT lists all possible neighborhood configurations and maps them to the probabilities of transition to state 1. The general form of a pLUT for the stochastic counterparts of ECAs is shown in Table 3.

Formally, the meaning of the pLUT is the following. Let $\tilde{f}$ be the local rule of an SCA with unit neighborhood radius, and let $(x_1, x_2, x_3) \in \{0, 1\}^3$. Let i be an integer such that the vector (x_1, x_2, x_3) is its binary representation, then the entries in Table 3 are given by:

$$\mathbb{P}(\tilde{f}(x_1, x_2, x_3) = 1) = p_i .$$

Obviously, from this it follows that:

$$\mathbb{P}(\tilde{f}(x_1, x_2, x_3) = 0) = 1 - p_i .$$

In the case of SCAs, it is hard to define the space-time diagram in a strict, formal way. So if A is an SCA, any sequence of configurations, which can be obtained by simulating A, starting from some given initial configuration, is a space-time diagram. Formally it means that if $p_i \in]0, 1[$ for $i = 0, \ldots, 7$, any sequence of binary configurations makes up a space-time diagram. Yet, the likelihood of observing a given space-time diagram is uniform only in case $p_i = 0.5$ for all $i = 0, \ldots, 7$. In other cases, the probability distribution over the space of space-time diagrams might be more complex.

3 α-Asynchronous CAs

Classically, states in CAs are updated synchronously, i.e. a new state is assigned to all cells simultaneously at every time step according to the local rule. Yet, different approaches of breaking the synchronicity of CAs have been proposed [23]. Interestingly, the choice of the update scheme, which defines the order or timing of cell state updates, has very important repercussions on the dynamical properties of CAs [4]. Here, we focus on one of such schemes, namely α-asynchronous CAs (α-ACAs). A detailed definition of α-ACAs is presented below, while the description of their most important properties and applications can be found in [13].

Any α-ACA can be defined by a deterministic CA A and a probability α, called the synchrony rate, which controls whether or not its cells are updated. More precisely, α is the probability of applying the local rule f of A. Let $\tilde{f}$ be the random function (local rule) corresponding to an α-ACA, then for any $x_1, \ldots, x_R, y \in \{0, 1\}$ it holds that:

$$
\mathbb{P}(\tilde{f}(x_1, \ldots, x_R) = y) = \begin{cases} 0, & \text{if } (x_1, \ldots, x_R) \in \mathcal{C}(f) \wedge y \neq x_{r+1}, \\ 1, & \text{if } (x_1, \ldots, x_R) \in \mathcal{C}(f) \wedge y = x_{r+1}, \\ \alpha, & \text{if } (x_1, \ldots, x_R) \notin \mathcal{C}(f) \wedge y \neq x_{r+1}, \\ 1 - \alpha, & \text{if } (x_1, \ldots, x_R) \notin \mathcal{C}(f) \wedge y = x_{r+1}. \end{cases}
$$

Note that if $\alpha = 0$, such a system stays at its initial configuration, whereas the system is equivalent to a deterministic CA A if $\alpha = 1$.

The essential property of α-ACAs as defined here is that they may equivalently be considered as SCAs for which the local rule f is selected with probability α, while the identity rule is selected with probability $1 - \alpha$. Hence, we may say that CA A becomes stochastically mixed with the identity rule. In the remainder we will write A_α to denote the α-ACA which is defined with the use of CA A and synchrony rate α.

Let us assume that a local rule of an ECA A is defined by the LUT $(l_i)_{i=0}^7$. Clearly, α-ACAs form a special class of SCAs, therefore, we can represent A_α in terms of a pLUT. If $\alpha \in [0, 1]$ and $\bar{\alpha} = 1 - \alpha$, the pLUT of A_α is given by Table 4.

As can be inferred from the LUT shown in Table 4, A_α is deterministic on those neighborhood configurations belonging to the set $\mathcal{C}(f)$, where f is the local rule of A. For remaining neighborhood configurations, where f agrees with the negation rule, A_α is stochastic. This simple property is important in the construction of the identification algorithm.

Example 2 Let A be the ECA 150. The pLUT of A_α is given by Table 5. The space-time diagrams evolved for the same initial configuration for: (a) $\alpha = 0.1$, (b) $\alpha = 0.5$, (c) $\alpha = 0.9$ and (d) $\alpha = 1$ are shown in Fig. 1. As can be inferred from the plots, the behavior of the dynamical system is greatly affected by α.

Table 4 pLUT of an α-ACA local rule

(1,1,1)	(1,1,0)	(1,0,1)	(1,0,0)	(0,1,1)	(0,1,0)	(0,0,1)	(0,0,0)
$\alpha l_7 + \bar{\alpha}$	$\alpha l_6 + \bar{\alpha}$	αl_5	αl_4	$\alpha l_3 + \bar{\alpha}$	$\alpha l_2 + \bar{\alpha}$	αl_1	αl_0

Table 5 pLUT of A_α for A being ECA 150

(1,1,1)	(1,1,0)	(1,0,1)	(1,0,0)	(0,1,1)	(0,1,0)	(0,0,1)	(0,0,0)
1	$\bar{\alpha}$	0	α	$\bar{\alpha}$	1	α	0

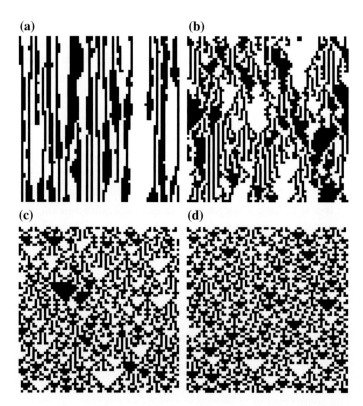

Fig. 1 Space-time diagrams of A_α evolved for the same initial configuration, for different synchrony rates, where A is ECA 150. **a** $\alpha = 0.1$. **b** $\alpha = 0.5$. **c** $\alpha = 0.9$. **d** $\alpha = 1$

4 Identification Problem

In this section we define the identification problem. Our formulation is based on the concept of an observation of a space-time diagram, which is assumed to be incomplete, i.e. it contains only partial information on the states of the underlying α-ACA.

Let I be an $N \times M$ array containing symbols belonging to the set $\{0, 1, ?\}$, where the symbols 0 and 1 denote valid states of an unknown α-ACA, while the symbol ? denotes an unknown state belonging to the set $\{0, 1\}$. Additionally, let the first row $I[1] \in \{0, 1\}^M$ represent the initial configuration of an α-ACA. Such an array I will be referred to as an observation. If an observation I does not contain the symbol ?, we refer to it as complete, while it will be referred to as partial otherwise. Entries in the observations occupied by the symbol ? will be denoted as gaps.

Note that the assumption of a completely observed initial configuration, i.e. $I[1] \in \{0, 1\}^M$ is crucial for the construction of the presented method and cannot be relaxed easily. Yet, this condition can be met in many practical applications as there are often means of controlling the initial state of the system in question.

Let $r \in \mathbb{N}_0$, then the vector $(I[n, m - r], \ldots, I[n, m + r])$ will be denoted by $I[n, m|r]$, for any $n, m \in \mathbb{Z}$, where periodic boundary conditions are assumed. Furthermore, we assume that there are strict limitations for what concerns the occurrence of gaps in observations. If for some n, m it holds that $I[n, m] = ?$, then $I[n - 1, m|r] \in \{0, 1\}$, $I[n + 1, m|r] \in \{0, 1\}$ and in $I[n, m|2r]$ only for the pair (n, m) it hols that $I[n, m] = ?$, i.e. for $m' = m - 2r, \ldots, m - 1, m + 1, \ldots, m + 2r$ it holds that $I[n, m'] \in \{0, 1\}$. Such a condition of spatial separation of gaps allows to consider each of the gaps separately in the gap filling process.

We consider the identification problem with the assumption that I was generated by some unknown α-ACA, denoted by A_α. Solving the identification problem means finding the CA A and $\widehat{\alpha}$, as an approximation of α, such that $\widehat{\alpha} \in]\alpha_1, \alpha_2[$, where $\alpha_2 - \alpha_1$ is as small as possible, and it is very likely that $\alpha \in]\alpha_1, \alpha_2[$. More formally, we select $L \in]0, 1[$ and assume a confidence level $1 - L$. We build an estimate for A, α_1 and α_2, based on observation I, such that observation I is a space-time diagram of A_α for some $\alpha \in]\alpha_1, \alpha_2[$ with probability $1 - L$. In those cases where the observation I is incomplete, the identification algorithm should yield the most likely values of the missing states.

We will consider the identification problem in the context of observation sets $\mathcal{I}$ containing one or more observation I_j for $j = 1, \ldots, |\mathcal{I}|$, of the behavior of some α-ACA, where $I_j \in \mathcal{I}$ contains M_{I_j} columns and N_{I_j} rows.

It should be mentioned that the identification problem presented here, has limited applicability when it comes to real-world modeling tasks. This is mostly due to the fact that binary CAs are typically too simple to mimic real-world processes as a consequence of their limited state set. Yet, the identification problem becomes of more direct importance in the case of multi-state and multi-dimensional CAs. Although the presented solution algorithm is tailored towards the binary case, it is possible to generalize it. This is one of the topics of future research in this area.

5 Identification Algorithm

5.1 Complete Observations

In this section we describe the algorithm for solving the identification problem in the case of complete observations. Besides, we assume that α is bounded between known bounds $a > 0$ and $b < 1$.

Based on a set of complete observations $\mathcal{I}$, we define frequency tables $N = (N_0, \ldots, N_{2^R-1})$ and $K = (K_0, \ldots, K_{2^R-1})$, where N_i denotes the number of occurrences of the ith neighborhood configuration in all of the observations $I \in \mathcal{I}$, where the last row of each observation is discarded, i.e. we count the occurrences of the neighborhoods in every row except for the last one, and store the results in N. To build table K, we additionally check the state of the central cell in the next row for each of the visited neighborhoods, and we count the number of cases where the value of the central cell changed, i.e. cases where the unknown α-ACA changed the value of a cell to its complement, and thus acted like the negation CA. It is obvious that for all i it holds that $K_i \leq N_i$.

Proposition 1 *Let* $N^* = \sum_{i:K_i>0} N_i$, $K^* = \sum_{i=0}^{2^R-1} K_i$ *and* $\widehat{\alpha} = \frac{K^*}{N^*}$. *The proportion* $\widehat{\alpha}$ *is a random variable following a binomial distribution with success probability equal to* α.

Following [9] there are various methods to estimate the confidence interval for α using $\widehat{\alpha}$. Here, we choose the normal distribution approximation, even though the authors of [9] advice against it. This choice is motivated by the fact that this method leads to an algorithm with a reasonable accuracy and at the same time its implementation is straightforward. Assuming that $1 - L$ is the selected confidence level, then the following holds with probability $1 - L$:

$$\widehat{\alpha} - z_L \sqrt{\frac{\widehat{\alpha}(1 - \widehat{\alpha})}{N^*}} \leq \alpha \leq \widehat{\alpha} + z_L \sqrt{\frac{\widehat{\alpha}(1 - \widehat{\alpha})}{N^*}}, \tag{2}$$

where z_L is the argument at which the cumulative standard normal distribution function takes the value of $1 - \frac{L}{2}$. The above holds if N^* is large enough, for example if both $N^*\alpha$ and $N^*(1 - \alpha)$ are greater than five [9]. Since α is unknown, we impose a bit stronger condition $N^* > \max(\frac{5}{a}, \frac{5}{1-b})$, which can be easily verified.

The estimated interval given by (2) can be adjusted, taking into account the assumption that $\alpha \in [a, b]$. For that purpose, let:

$$\alpha_1 = \max\left(a, \widehat{\alpha} - z_L \sqrt{\frac{\widehat{\alpha}(1 - \widehat{\alpha})}{N^*}}\right), \quad \alpha_2 = \min\left(b, \widehat{\alpha} + z_L \sqrt{\frac{\widehat{\alpha}(1 - \widehat{\alpha})}{N^*}}\right),$$

then it holds that $\alpha \in [\alpha_1, \alpha_2]$ with probability $1 - L$. Note that $\alpha_2 - \alpha_1 \leq \frac{z_L}{\sqrt{N^*}}$ and for commonly used confidence levels it holds that $z_L < 3$. Thus, if N^* is sufficiently large, we are sure that the interval $[\alpha_1, \alpha_2]$ narrows.

Summing up, we have formulated the estimation method for the confidence interval of α. If a point estimation is desired, the value of $\frac{\alpha_1 + \alpha_2}{2}$, which in most cases is equal to $\widehat{\alpha}$, shall be used. We now turn to the method for constructing the local rule of CA A.

Let f be the unknown local rule given by a LUT $(l_i)_{i=0}^{2^R-1} \in \{0, 1\}^{2^R}$, where the l_i's are unknown. Deciding on these might be seen as picking the same value as the identity rule or taking the opposite value. If $K_i > 0$ and $N_i > 0$ we pick the value opposite to identity, i.e. we select the value from the LUT of the negation rule. When $K_i = 0$ and $N_i > 0$, we are not sure if f agrees with the identity CA on the ith neighborhood, or too few samples were observed. Yet, if it holds that:

$$(1 - a)^{N_i} \le \frac{L}{2^{|\{i : K_i = 0\}|}}, \tag{3}$$

we may assume that f agrees with the identity. This assures that the total probability of picking the wrong value, over all the neighborhood configurations for which $K_i = 0$, is not higher than L. Otherwise if Eq. (3) is not fulfilled or $K_i = N_i = 0$, we are not able to select the value of l_i. Still, if we can simulate the unknown α-ACA starting from an arbitrary configuration, this bottleneck is eliminated. In the remainder we assume that such cases do not happen.

The computational complexity of the identification algorithm is linear, in the sense that it is proportional to the number of observed cells. Consequently, the algorithm is applicable even for relatively big observation sets.

5.2 Gap Filling Procedure

Having defined the estimates $\alpha_1, \alpha_2, \widehat{\alpha}$ and A in the case of complete observations, we now turn to the case of partial observations obeying the spatial separation condition formulated in Sect. 4, which guarantees that we can treat each of the gaps separately. We can find the estimates for α and A following the method described in Sect. 5.1, with the only change that K_i and N_i are calculated discarding those entries that contain the symbol ?. In other words, we ignore entries with gaps in the fist step of estimating the parameters of α-ACA. Then, for every $I \in \mathcal{I}$ and (n, m) such that $I[n, m] = ?$, we follow the procedure outlined below to find the missing state. Let the function $f : \{0, 1\}^R \to \{0, 1\}$ be the local rule of CA A, and $\tilde{f}$ be the corresponding random local rule of A_α.

If $f(I[n - 1, m|r]) = I[n - 1, m]$, then f agrees with the identity CA on the neighborhood configuration $I[n - 1, m|r]$, and we may replace the ? at (n, m) by $I[n - 1, m] \in \{0, 1\}$. Otherwise, we need to inspect values in the $n + 1$th row of observation I to find the most likely value for $I[n, m]$. In this case, from the definition of α-ACAs, we know that:

$$p_y = \mathbb{P}(I[n, m] = y) = \begin{cases} 1 - \alpha, & \text{if } I[n - 1, m] = y, \\ \alpha, & \text{if } I[n - 1, m] = 1 - y. \end{cases} \tag{4}$$

The informal meaning of p_y is that it is the probability of $I[n, m]$ being y, as calculated by only examining the $n - 1$th row of observation. For $h \in \{-r, \ldots, r\}$, let F_h denote the random event that, starting from the neighborhood configuration

$I[n-1, m|r]$, the α-ACA evolution leads to the state $I[n+1, m+h]$ after two time steps. For any $y \in \{0, 1\}$, let $p_{h,y}$ be defined as:

$$p_{h,y} = \mathbb{P}(F_h \mid I[n, m] = y). \tag{5}$$

We may calculate $p_{h,y}$ according to the following formula, where it is assumed that $I[n, m] = y$, and thus $I[n, m+h|r]$ depends on y:

$$p_{h,y} = \begin{cases} 0, & \text{if } I[n, m+h|r] \in \mathcal{C}(f) \wedge I[n, m+h] \neq I[n+1, m+h], \\ 1, & \text{if } I[n, m+h|r] \in \mathcal{C}(f) \wedge I[n, m+h] = I[n+1, m+h], \\ \alpha, & \text{if } I[n, m+h|r] \notin \mathcal{C}(f) \wedge I[n, m+h] \neq I[n+1, m+h], \\ 1-\alpha, & \text{if } I[n, m+h|r] \notin \mathcal{C}(f) \wedge I[n, m+h] = I[n+1, m+h]. \end{cases} \tag{6}$$

Since α is not known, we can only get an approximation of $p_{h,y}$. Yet it suffices for our purposes. The value $p_{h,y}$ is the probability of obtaining the $n + 1$th row, assuming that y is the missing value in the nth row. By combining those probabilities, we will find the most likely value for $I[n, m]$. More formally, according to the Total Probability theorem, it holds that:

$$\mathbb{P}\left(\bigcap_{h=-r}^{r} F_h\right) = \sum_{y=0}^{1} \mathbb{P}(I[n, m] = y) \, \mathbb{P}\left(\bigcap_{h=-r}^{r} F_h \mid I[n, m] = y\right)$$

$$= \sum_{y=0}^{1} p_y \, \mathbb{P}\left(\bigcap_{h=-r}^{r} F_h \mid I[n, m] = y\right)$$

$$= \sum_{y=0}^{1} p_y \prod_{h=-r}^{r} \mathbb{P}(F_h \mid I[n, m] = y)$$

$$= \sum_{y=0}^{1} p_y \prod_{h=-r}^{r} p_{h,y}.$$

This is justified as for $h_1 \neq h_2$, the events F_{h_1} and F_{h_2} are independent if $I[n, m]$ is known. The probability $\mathbb{P}\left(I[n, m] = y \mid \bigcap_{h=-r}^{r} F_h\right)$ is the probability of $I[n, m] = y$ assuming that all of the transitions from $I[n-1, m|r]$ to $n + 1$th row happened according to values recorded in observation I. Due to Bayes' theorem it holds that:

$$\mathbb{P}\left(I[n, m] = y \mid \bigcap_{h=-r}^{r} F_h\right) = \frac{p_y \prod_{h=-r}^{r} p_{h,y}}{\mathbb{P}\left(\bigcap_{h=-r}^{r} F_h\right)}. \tag{7}$$

Therefore, the most likely value for $I[n, m]$ is the one that maximizes the probability $\mathbb{P}\left(I[n, m] = y \mid \bigcap_{h=-r}^{r} F_h\right)$, and to find it, we only need to examine the numerators of the fractions in Eq. (7) for different values of y since the denominator does not depend on y.

Finally, the method for filling the gap in $I[n, m]$ works as follows. Firstly, we check whether the value can be selected deterministically according to the identity rule, which happens when $I[n-1, m|r] \in C(f)$. Otherwise, we calculate the two numerators from Eq. (7), for $y = 0$ and $y = 1$, using $\widehat{\alpha}$ instead of α. If the results differ, we pick the y for which the numerator is the largest. If both numerators are equal, we compare the probabilities p_0 and p_1, and pick this y for which the probability is greater. Finally, if $p_y = p_{1-y}$, we randomly assign a value to $I[n, m]$.

6 Experimental Results

For assessing the performance of the identification algorithm we evaluated 255 ECAs (all but ECA 204 which is the identity CA) for synchrony rates equal to $\alpha = 0.05, 0.01, \ldots, 0.95$. For each ECA and α, a set of 100 observations, each consisting of 49 time steps and 49 cells, was constructed by simulating the α-ACA and storing the resulting space-time diagrams. A common set of 100 randomly generated initial configurations was used. The 95 % confidence level was set, i.e. $L = 0.05$. The bounds for α were defined as $a = 0.05$ and $b = 0.95$. In each of the observation sets, 2500 gaps were introduced at random positions in randomly selected observations, but still such that the separation condition was fulfilled.

In all of the considered cases, the unknown ECA was discovered. To verify whether the estimation of α was reliable, we measured the relative error E defined as:

$$E = E(A_\alpha) = \frac{|\widehat{\alpha} - \alpha|}{\alpha} \times 100 \% , \tag{8}$$

where $\widehat{\alpha}$ is the estimate obtained for A_α. We obtained the following statistics of $E(A_\alpha)$ across the ECAs and synchrony rates (where the values were truncated to two significant digits):

$$\min(E) = 0.00 \% , \ \langle E \rangle = 0.51 \% , \ \max(E) = 8.37 \% , \ \sigma(E) = 0.68 \% ,$$

where $\langle E \rangle$ denotes the mean error, and $\sigma(E)$ is the standard deviation of the error. The histogram of E values with a bin width of 0.1 % is shown in Fig. 2. As can be inferred from this figure, the identification algorithm is able to find very good estimates of the synchrony rate. Not only was the maximum relative error 8.37 %, but more importantly the relative error was below 1 % in 85 % of the cases.

We now assess the performance of the second step of the identification algorithm, namely gap filling. We measured this as the percentage of correctly filled gaps in observations for a given case. For each ECA we averaged the success rate over the considered synchrony rates and the resulting quantity is denoted as S. The overall

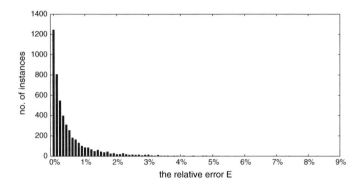

Fig. 2 Histogram of the relative error E with a bin width 0.1 %

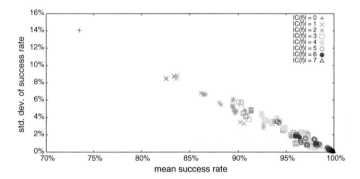

Fig. 3 Mean success rate versus the standard deviation for each of the ECAs. Different plot markers are used to indicate different sizes of the set $C(f)$

statistics of S are shown below (where the values were truncated to two significant digits):

$$\min(S) = 73.50\,\%, \ \langle S \rangle = 96.02\,\%, \ \max(S) = 99.95\,\%, \ \sigma(S) = 4.10\,\% .$$

Figure 3 shows both the mean success rate S and the standard deviation of the success rate for each ECA. The points on the plot close to the lower-right corner $(1, 0)$ correspond to the most successful cases. The results are labeled according to the size of the set $C(f)$, where f is the local rule of the considered ECA. As can be inferred from the plot, ECAs for which the set $C(f)$ is bigger, give rise to slightly better results, possible due to the fact that more transitions happened deterministically. Moreover, the outlier with a relatively low value of S originated from ECA 51 which is the negation CA, for which the α-ACA reduces to a weighted, random bit flip. Yet, even in this case we were able to correctly fill in more than 70 % of the gaps. The presented results indicate that the algorithm is very effective and accurate when it comes to gap filling.

7 Summary

In this paper the identification of α-ACAs was discussed. An algorithm for identifying the underlying CA and estimating the synchrony rate α was presented. Moreover, a method for gap filling was put forward. The experimental results presented in Sect. 6 for ECAs are very promising. In all cases the algorithm was able to find the correct CA, and a good estimate for the synchrony rate was obtained. Also the rate of correctly filled gaps was very high. The algorithm will be extended to a more general setting where the observations have less separated gaps and to richer classes of SCAs.

Acknowledgments Witold Bołt is supported by the Foundation for Polish Science under International PhD Projects in Intelligent Computing. This project is financed by the European Union within the Innovative Economy Operational Program 2007–2013 and the European Regional Development Fund.

References

1. Adamatzky, A.: Identification of Cellular Automata. Taylor & Francis Group, London (1994)
2. Al-Kheder, S., Wang, J., Shan, J.: Cellular automata urban growth model calibration with genetic algorithms. In: Urban Remote Sensing Joint Event, pp. 1–5. IEEE Press (2007)
3. Andre, D., Bennett, III, F.H., Koza, J.R.: Discovery by genetic programming of a cellular automata rule that is better than any known rule for the majority classification problem. In: Proceedings of the 1st Annual Conference on Genetic Programming, pp. 3–11. MIT Press, Cambridge (1996)
4. Baetens, J.M., Van der Weeën, P., De Baets, B.: Effect of asynchronous updating on the stability of cellular automata. Chaos Solitons Fractals **45**, 383–394 (2012)
5. Bandini, S., Manzoni, S., Vanneschi, L.: Evolving robust cellular automata rules with genetic programming. In: Adamatzky, A., Alonso-Sanz, R., Lawniczak, A.T., Martínez, G.J., Morita, K., Worsch, T. (eds.) Automata, pp. 542–556. Luniver Press, Frome (2008)
6. Billings, S.A., Yang, Y.: Identification of probabilistic cellular automata. IEEE Trans. Syst. Man Cybern. Part B Cybern. **33**, 225–236 (2003)
7. Bołt, W., Baetens, J.M., De Baets, B.: Identifying CAs with evolutionary algorithms. In: Proceedings of the 19th International Workshop on Cellular Automata and Discrete Complex Systems (AUTOMATA 2013) - Exploratory Papers, pp. 11–20 (2013)
8. Bołt, W., Baetens, J.M., De Baets, B.: An evolutionary approach to the identification of cellular automata based on partial observations. In: Proceedings of the 2015 Congress of Evolutionary Computation (CEC 2015). IEEE Press (2015)
9. Brown, L.D., Cai, T.T., DasGupta, A.: Interval estimation for a binomial proportion. Stat. Sci. **16**(2), 101–133 (2001)
10. Bull, L., Adamatzky, A.: A learning classifier system approach to the identification of cellular automata. J. Cell. Autom. **2**, 21–38 (2007)
11. Bäck, T., Breukelaar, R., Willmes, L.: Inverse design of cellular automata by genetic algorithms: an unconventional programming paradigm. In: Banâtre, J.P., Fradet, P., Giavitto, J.L., Michel, O. (eds.) Unconventional Programming Paradigms. Lecture Notes in Computer Science, vol. 3566, pp. 161–172. Springer, Berlin (2005)
12. Das, D.: A survey on cellular automata and its applications. In: Krishna, P., Babu, M., Ariwa, E. (eds.) Global Trends in Computing and Communication Systems, Communications in Computer and Information Science, vol. 269, pp. 753–762. Springer, Berlin (2012)

13. Fatès, N.: A guided tour of asynchronous cellular automata. In: Kari, J., Kutrib, M., Malcher, A. (eds.) Cellular Automata and Discrete Complex Systems. Lecture Notes in Computer Science, vol. 8155, pp. 15–30. Springer, Berlin (2013)
14. Fatès, N., Morvan, M.: An experimental study of robustness to asynchronism for elementary cellular automata. Complex Syst. **16**, 1–27 (2005)
15. Ferreira, C.: Gene Expression Programming: Mathematical Modeling by an Artificial Intelligence, Studies in Computational Intelligence, vol. 21. Springer, Berlin (2006)
16. Liu, X., Li, X., Liu, L., He, J., Ai, B.: A bottom-up approach to discover transition rules of cellular automata using ant intelligence. Int. J. Geogr. Inf. Sci. **22**, 1247–1269 (2008)
17. Maeda, K., Sakama, C.: Identifying cellular automata rules. J. Cell. Autom. **2**, 1–20 (2007)
18. Mitchell, M., Crutchfield, J.P., Das, R.: Evolving cellular automata with genetic algorithms: a review of recent work. In: Proceedings of the First International Conference on Evolutionary Computation and its Applications (EvCA'96) (1996)
19. Richards, F.C., Meyer, T.P., Packard, N.H.: Extracting cellular automaton rules directly from experimental data. Phys. D: Nonlinear Phenom. **45**, 189–202 (1990)
20. Rosin, P.L.: Image processing using 3-state cellular automata. Comput. Vis. Image Underst. **114**, 790–802 (2010)
21. Sapin, E., Bull, L., Adamatzky, A.: Genetic approaches to search for computing patterns in cellular automata. Comput. Intell. Mag. **4**, 20–28 (2009)
22. Sapin, E., Bailleux, O., Chabrier, J.J.: Research of a cellular automaton simulating logic gates by evolutionary algorithms. In: Proceedings of the 6th European Conference on Genetic Programming, EuroGP'03, pp. 414–423. Springer, Berlin (2003)
23. Schönfisch, B., de Roos, A.: Synchronous and asynchronous updating in cellular automata. Biosyst. **51**, 123–143 (1999)
24. Sun, X., Rosin, P.L., Martin, R.R.: Fast rule identification and neighborhood selection for cellular automata. IEEE Trans. Syst. Man Cybern. Part B: Cybern **41**, 749–760 (2011)
25. Wolfram, S.: Statistical mechanics of cellular automata. Rev. Mod. Phys. **55**, 601–644 (1983)
26. Yang, Y., Billings, S.A.: Neighborhood detection and rule selection from cellular automata patterns. IEEE Trans. Syst. Man Cybern. Part A: Syst. Hum. **30**, 840–847 (2000)
27. Yang, Y., Billings, S.A.: Extracting Boolean rules from CA patterns. IEEE Trans. Syst. Man Cybern. Part B: Cybern. **30**, 573–580 (2000)

k-Arithmetic Sequences—Theory and Applications

Adam Kołacz

Abstract The notion of an arithmetic progression was extended to embrace the class of polynomials of degree $k > 1$. Some properties of difference sequences are analyzed and their connections with some number-theory problems are studied. In particular, a certain aspects of Fermat's Last Theorem and the Fibonacci numbers are revisited.

Keywords Arithmetic progression · Difference sequence · Polynomials · Fibonacci numbers

2010 Mathematics Subject Classification: 11B83

1 Introduction

Arithmetic sequences (arithmetic progressions), both from the theoretical and applied points of view, are of interest in many areas including number theory, combinatorics, computer science, etc. Their diverse applications and generalizations are still inspiring and intriguing for researchers and practitioners (see [7]).

In this paper we propose the concepts of a *k*-arithmetic sequence and its characterizing sequence. The idea of such sequences is following: instead of the first (as in an arithmetic progression), let the *k*th difference sequence [2] be constant. A formula for a general term of a *k*-arithmetic sequence is presented which is similar to the one proposed by Newton [3].

The second part concerns the relations between the coefficients of a polynomial in two bases. Some connections between the coefficients and the terms of characterizing sequence of a polynomial are investigated. As a result, the basis numbers of the degree *k* are defined. Properties of these numbers lead, in particular, to another simple proof

A. Kołacz (✉)
Faculty of Mathematics and Information Science,
Warsaw University of Technology, Warsaw, Poland
e-mail: a.kolacz@mini.pw.edu.pl

© Springer International Publishing Switzerland 2016
G. De Tré et al. (eds.), *Challenging Problems and Solutions
in Intelligent Systems*, Studies in Computational Intelligence 634,
DOI 10.1007/978-3-319-30165-5_3

of Fermat's little theorem and an algorithm for calculation the coefficients of the sum of n initial terms of a polynomial of degree k.

The last section embraces some applications of the results presented in this paper. An interesting example of numbers that do not satisfy the Fermat's equation [5] is shown. After that, the formula for the general term of a k-arithmetic sequence is slightly modified to get the expression for the nth term of an arbitrary sequence of real numbers from which a certain generalization of binomial identity and an elegant recursive representation of the Fibonacci number are derived. This relationship between k-arithmetic sequences and the Fibonacci numbers is worth noting especially because of the increasing number of practical applications. Besides such areas like coding, optional compression, pseudorandom number generators, the Fibonacci numbers are recently used to predict stocks (see, e.g., [4]) and dynamic optimization problems (see, e.g., [1]).

2 k-Arithmetic Sequences

2.1 Basic Concepts

Consider an arbitrary, real-valued sequence a_n and the difference operator Δ defined as follows

$$\Delta a_n = a_{n+1} - a_n \quad \text{and} \quad \Delta^{k+1} a_n = \Delta(\Delta^k a_n),$$

for any natural k. The sequence $\Delta^{k+1} a_n$ is called the kth difference sequence of a_n.

Definition 1 We say that a sequence a_n is k-*arithmetic*, if it consists of at least $k + 2$ terms and its kth difference sequence is constant.

In other words, the $(k - 1)$th difference sequence of a k-arithmetic sequence is an arithmetic sequence. We will call the term of the constant sequence r—the *difference* of the sequence a_n and the number k—its *folds*. To simplify the notation let us denote the first difference sequence of a k-arithmetic sequence by a_{1n}, the second by a_{2n}, etc. and the mth—a_{mn}. The first of two indices of a term will be called the *difference index*. One may show that the mth term of the jth difference sequence is of the form

$$a_{jm} = \sum_{i=0}^{j} (-1)^i \binom{j}{i} a_{m+j-i}. \tag{1}$$

It can also be easily proven that every polynomial of the kth degree $f(n) = \sum_{i=0}^{k} c_i n^i$ is a k-arithmetic sequence with the difference $r = c_k k!$

2.2 Characterizing Sequence

Definition 2 The *characterizing sequence* of a k-arithmetic sequence a_n is the sequence of the first terms of consecutive difference sequences of a_n. It will be denoted by ψ_n, i.e. $(\psi_n)_{n=1}^{k+1} = (a_1, a_{11}, a_{21}, \ldots, a_{k-1,1}, r)$.

It is clear that the characterizing sequence of a k-arithmetic sequence consists of $k + 1$ terms and the last term is not equal to zero.

Theorem 1 *A k-arithmetic sequence is uniquely determined by its characterizing sequence.*

Proof Assume conversely that there exist two distinct k-arithmetic sequences having the same characterizing sequence. Denote these sequences by a_n^1 and a_n^2 and their consecutive difference sequences by $a_{1n}^1, a_{2n}^1, \ldots, a_{kn}^1$ and $a_{1n}^2, a_{2n}^2, \ldots, a_{kn}^2$ respectively. Since the sequences have the same characterizing sequences then, in particular, their differences are equal. Moreover, $a_1^1 = a_1^2, a_{21}^1 = a_{21}^2, \ldots, a_{k1}^1 = a_{k1}^2$. Because $a_{k-1,2}^1 - a_{k-1,1}^1 = r$ and $a_{k-1,2}^2 - a_{k-1,1}^2 = r$, we get that $a_{k-1,2}^1 = a_{k-1,2}^2$. Similarly, we check that $a_{km}^1 = a_{km}^2$ for any $m > 2$. By analogy $a_{j2}^1 - a_{j1}^1 = a_{j+1,1}^1$ and $a_{j2}^2 - a_{j1}^2 = a_{j+1,1}^2$ for $j < k$, hence $a_{j2}^1 = a_{j2}^2$ and $a_{jm}^1 = a_{jm}^2$ for any $m > 2$. Checking the consecutive difference sequences we finally get $a_m^1 = a_m^2$ for any m. It contradicts the assumption that the sequences have distinct terms, which proves the theorem. ∎

Consider now a k-arithmetic sequence a_n and the sequence $\Phi_n = \sum_{i=0}^{k} \binom{n-1}{i} \psi_{i+1}$, where k is the folds of a_n and ψ_i denotes the ith term of the characterizing sequence of a_n. Let us now investigate the characterizing sequence of Φ_n. Obviously, $\Phi_1 = \sum_{i=0}^{k} \binom{0}{i} \psi_{i+1} = \psi_1 = a_1$.

For the first difference sequence we have:

$$\Delta\Phi_n = \Phi_{n+1} - \Phi_n = \sum_{i=0}^{k}\left[\binom{n}{i} - \binom{n-1}{i}\right]\psi_{i+1} = \sum_{i=1}^{k}\binom{n-1}{i-1}\psi_{i+1}.$$

Therefore, we get $(\Delta^1\Phi)_1 = \sum_{i=1}^{k}\binom{0}{i-1}\psi_{i+1} = \psi_2 = a_{21}$. Analogously, let us examine the form of a term of the mth difference sequence

$$\Delta^m\Phi_n = \Delta^{m-1}\Phi_{n+1} - \Delta^{m-1}\Phi_n = \sum_{i=m-1}^{k}\left[\binom{n}{i-m+1} - \binom{n-1}{i-m+1}\right]\psi_{i+1}$$

$$= \sum_{i=m}^{k}\binom{n-1}{i-m}\psi_{i+1}.$$

Hence we get $(\Delta^m\Phi)_1 = \sum_{i=m}^{k}\binom{0}{i-m}\psi_{i+1} = \psi_{m+1} = a_{m1}$. Let us see what is the nth term of the kth difference sequence.

$$\Delta^k \Phi_n = \sum_{i=k}^{k} \binom{n-1}{i-k} \psi_{i+1} = \binom{n-1}{0} \psi_{k+1} = \psi_{k+1} = r.$$

It is the difference of a_n. The sequences a_n and Φ_n have the same characterizing sequences, therefore, by Theorem 3, we conclude that $a_n = \Phi_n$. Hence, the general term of a k-arithmetic sequence can be expressed as follows

$$a_n = \sum_{i=0}^{k} \binom{n-1}{i} \psi_{i+1}, \tag{2}$$

where (ψ_n) denotes the characterizing sequence of a_n.

Let us add that whenever $n \leqslant k$ one should remember the property of the binomial coefficients: $\forall m > n$ $\binom{n}{m} = 0$. Note that in particular for $k = 1$ we get $a_n = a_1 + (n-1)r$, which is the well-known formula for the general term of an arithmetic progression.

Theorem 2 *For any k-arithmetic sequence a_n let*

$$B(n) = \sum_{i=0}^{n-1} (-1)^i \binom{n-1}{i} a_{n-i}.$$

Then

(a) $B(n) = 0$, if $n \geqslant k + 2$,
(b) $B(n) = \psi_n$, if $n < k + 2$.

Proof Ad (a) We use the induction with respect to the number of terms. For $n = k + 2$, the kth difference sequence is constant and consists of exactly two terms equal to the difference of a_n. By (1) we get

$$r = a_{k1} = \sum_{i=0}^{k} (-1)^i \binom{k}{i} a_{k-i+1}, \tag{3}$$

$$r = a_{k2} = \sum_{i=0}^{k} (-1)^i \binom{k}{i} a_{k-i+2}.$$

Renumbering (3) we obtain

$$r = a_{k1} = \sum_{i=1}^{k+1} (-1)^{i+1} \binom{k}{i-1} a_{k-i+2}$$

and then subtracting the terms we have

$$0 = \sum_{i=1}^{k+1}(-1)^{i+1}\binom{k}{i-1}a_{k-i+2} - \sum_{i=0}^{k}(-1)^{i}\binom{k}{i}a_{k-i+2}$$

$$= (-1)^{k+2}a_1 - \sum_{i=1}^{k}(-1)^{i}\left[\binom{k}{i-1}+\binom{k}{i}\right]a_{k-i+2} - a_{k+2}$$

$$= (-1)^{k+2} - \sum_{i=1}^{k}(-1)^{i}\binom{k+1}{i}a_{k-i+2} - a_{k+2} = -\sum_{i=0}^{k+1}(-1)^{i}\binom{k+1}{i}a_{k-i+2}.$$

Assume now that the formula holds for any *k*-arithmetic sequence having $n > k + 2$ terms. Let a_n be a $(k + 1)$-arithmetic sequence having $n + 1$ terms. For a_n we get the following formula

$$B(n+1) = \sum_{i=0}^{n}(-1)^{i}\binom{n}{i}a_{n+1-i}$$

$$= (-1)^{n}a_1 + \sum_{i=1}^{n-1}(-1)^{i}\binom{n-1}{i-1}a_{n+1-i} + \sum_{i=0}^{n-1}(-1)^{i}\binom{n-1}{i}a_{n+1-i}$$

$$= (-1)^{n}a_1 + \sum_{i=0}^{n-2}(-1)^{i+1}\binom{n-1}{i}a_{n-i} + \sum_{i=0}^{n-1}(-1)^{i}\binom{n-1}{i}a_{n+1-i}$$

$$= \sum_{i=0}^{n-1}(-1)^{i+1}\binom{n-1}{i}a_{n-i} + \sum_{i=0}^{n-1}(-1)^{i}\binom{n-1}{i}a_{n+1-i}$$

$$= \sum_{i=0}^{n-1}(-1)^{i}\binom{n-1}{i}(a_{n+1-i} - a_{n-i}) = \sum_{i=0}^{n-1}(-1)^{i}\binom{n-1}{i}a_{1(n-i)}.$$

The sequence a_{1n} as the first difference sequence of a_n is *k*-arithmetic and it has n terms. Therefore, by the assumption the sum $\sum_{i=0}^{n-1}(-1)^{i}\binom{n-1}{i}a_{1(n-i)}$ equals to zero. That leads to $\sum_{i=0}^{n}(-1)^{i}\binom{n}{i}a_{n+1-i} = 0$.

Ad (b) The proof is a simple conclusion from (1) and the fact that a *k*-arithmetic sequence has exactly $k + 1$ non-zero-term levels. ■

Theorem 3 *The sum of n initial terms of a k-arithmetic sequence can be expressed as follows:*

$$S_n = \begin{cases} \sum_{i=0}^{n-1}\binom{n}{i+1}\psi_{i+1}, & \text{dla } n \leqslant k+1, \\ \\ \sum_{i=0}^{k}\binom{n}{i+1}\psi_{i+1}, & \text{dla } n > k+1. \end{cases} \tag{4}$$

Proof We use the induction with respect to the number of the summands. For $n = 1$ the theorem obviously holds. For $n = 2$ we have $S_2 = a_1 + a_2 = 2a_1 + a_{11}$, and for $n = 3$, $S_3 = a_1 + a_2 + a_3 = 2a_1 + a_{11} + a_3 = 2a_1 + a_{11} + a_2 + a_{12} = 3a_1 + 3a_{11} + a_{21}$. Now we show that if the formula holds for a certain $n - 1 < k + 2$ then it also holds for n. Let us consider

$$S_n = \sum_{i=0}^{n-1} \binom{n}{i+1} \psi_{i+1} = \psi_n + \sum_{i=0}^{n-2} \left[\binom{n-1}{i+1} + \binom{n-1}{i} \right] \psi_{i+1}$$

$$= \psi_n + \sum_{i=0}^{n-2} \binom{n-1}{i+1} \psi_{i+1} + \sum_{i=0}^{n-2} \binom{n-1}{i} \psi_{i+1}$$

$$= \sum_{i=0}^{n-2} \binom{n-1}{i+1} \psi_{i+1} + \sum_{i=0}^{n-1} \binom{n-1}{i} \psi_{i+1}.$$

The second summand is, by (2), equal to a_n. Therefore,

$$S_n = \sum_{i=0}^{n-2} \binom{n-1}{i+1} \psi_{i+1} + \sum_{i=0}^{n-1} \binom{n-1}{i} \psi_{i+1} = S_{n-1} + a_n.$$

Now consider the following sum

$$S_{k+2} = S_{k+1} + a_{k+2} = \sum_{i=0}^{k} \binom{k+1}{i+1} \psi_{i+1} + \sum_{i=0}^{k} \binom{k+1}{i} \psi_{i+1}$$

$$= \sum_{i=0}^{k} \left[\binom{k+1}{i+1} + \binom{k+1}{i} \right] \psi_{i+1} = \sum_{i=0}^{k} \binom{k+2}{i+1} \psi_{i+1}.$$

It is seen clearly that adding another element to the sum comes down only to increasing the binomial coefficient's upper parameter by 1. Then, by analogy, we derive the sum S_{k+p} for any natural p and then reformulate it to the form stated in the theorem. ∎

In particular, for $k = 1$ we get

$$S_n = \sum_{i=0}^{1} \binom{n}{i+1} \psi_{i+1} = na_1 + \frac{n(n-1)}{2} \cdot r = \frac{n}{2} [2a_1 + (n-1)r]$$

$$= \frac{n}{2} [a_1 + a_1 + (n-1)r],$$

which is the well-know formula for the sum of n initial terms of an arithmetic progression. Note that (4) allows to compute the sum of initial n terms of any polynomial knowing only its characterizing sequence.

3 Another View on Polynomials

3.1 Characterizing Sequence and Polynomial Coefficients

We know that every polynomial of degree k, say $P_n = A_k n^k + A_{k-1} n^{k-1} + \cdots + A_1 n + A_0$, is a k-arithmetic sequence. On the other hand, formula (2) gives the general form of every k-arithmetic sequence having a characterizing sequence (ψ_n). Therefore, we can write

$$\sum_{i=0}^{k} \binom{n-1}{i} \psi_{i+1} = A_k n^k + A_{k-1} n^{k-1} + \cdots + A_1 n + A_0.$$

Please, notice that in the space of all k-arithmetic sequences we can distinguish two bases: $\mathscr{B}_1 = [1, n, \ldots, n^k]$ and $\mathscr{B}_2 = \left[\binom{n-1}{0}, \binom{n-1}{1}, \ldots, \binom{n-1}{k} \right]$.

This section is dedicated to finding the relationships between coefficients $A_0, \ldots, A_k$ of a polynomial and the terms $\psi_1, \ldots, \psi_{k+1}$ of the characterizing sequence. We will also determine the transition matrices between $\mathscr{B}_1$ and $\mathscr{B}_2$ bases.

Theorem 4 *Let (ψ_n) denote a characterizing sequence of a k-arithmetic sequence a_n. Then*

$$A_0 = \sum_{i=0}^{k} (-1)^i \psi_{i+1}.$$

Proof By (2) we get

$$\sum_{i=0}^{k} \binom{n-1}{i} \psi_{i+1} = \psi_1 + (n-1)\psi_2 + \frac{(n-1)(n-2)}{2} \psi_3 + \cdots + \frac{(n-1)\ldots(n-k)}{k!} \psi_{k+1}.$$

It is easily seen that the zero-coefficient of the ith element is

$$\frac{-1(-2)\ldots(-i)}{i!} \psi_{i+1} = (-1)^i \psi_{i+1}.$$

Hence, the theorem follows from adding the elements. ∎

Theorem 5 *For any k-arithmetic sequence a_n the following formula holds*

$$\sum_{i=1}^{k} (-1)^i A_i = \sum_{i=1}^{k} (-1)^i i \psi_{i+1}.$$

First, we will prove the following

Lemma 1 *For a k-arithmetic sequence a_n, each coefficient A_i for $i = 0, 1, \ldots, k$ is a linear combination of $\psi_1, \ldots, \psi_{k+1}$ and*

$$A_i = \frac{\alpha_i^i}{i!}\psi_{i+1} + \cdots + \frac{\alpha_k^i}{k!}\psi_{k+1},$$

where α_i^j denotes a certain number, constant for the above expansion of A_j, standing by ψ_{i+1} and such that

$$\alpha_i^j = -\alpha_{i-1}^j \cdot j + \alpha_{i-1}^{j-1}. \tag{5}$$

Proof Let us express (2) in the following form

$$\sum_{i=0}^k \binom{n-1}{i}\psi_{i+1} = \psi_1 + (n-1)\psi_2 + \cdots + \frac{1}{k!}(n-1)\ldots(n-k)\psi_{k+1}$$

$$= \psi_1 + (n-1)\psi_2 + \cdots + \frac{1}{k!}(n^k + \cdots + (-1)^k k!)\psi_{k+1}.$$

We see that every element n^j is found in the above sum exactly $k + 1 - j$ times, hence we conclude that every coefficient A_j is a sum of elements of the form $\frac{\alpha_i^j}{i!}\psi_{i+1}$, where α_i^j is a certain number and constant for A_j. Because n^j appears only in ψ_{j+1}, then it is justified that the first element of the expansion for A_j is $\frac{\alpha_j^j}{j!}\psi_{j+1}$. Hence, for $j = 1, 2, \ldots, k$, we obtain

$$A_j = \sum_{l=j}^k \frac{\alpha_l^j}{l!}\psi_{l+1}.$$

For instance, a coefficient $A_k = \frac{\alpha_k^k}{k!}\psi_{k+1}$ stands by the term n^k and it is easy to note that $\alpha_k^k = 1$. In general $\alpha_i^i = 1$. Assume now that we know the first m elements of the expansion of A_j. They were obtained by summing the proper coefficients standing by the term n^j of polynomial $\frac{1}{j!}(n-1)\ldots(n-j)\psi_{j+1} + \cdots + \frac{1}{(j+m-1)!}(n-1)\ldots(n-j-m+1)\psi_{j+m}$. Let $\frac{\alpha_{j+m-1}^j}{(j+m-1)!}\psi_{j+m}$ be the coefficient from the element $\frac{1}{(j+m-1)!}(n-1)\ldots(n-j-m+1)\psi_{j+m}$ and $\frac{\alpha_{j+m-1}^{j-1}}{(j+m-1)!}\psi_{j+m}$ be the coefficient standing by n^{j-1} from the same element. In order to get the next, $(m+1)$th element, we need to consider $\frac{1}{(j+m)!}(n-1)\ldots(n-j-m)\psi_{j+m+1}$. Then, $\frac{\alpha_{j+m}^j}{(j+m)!}\psi_{j+m+1}$ stands by n^j, with α_{j+m}^j given by

$$\alpha_{j+m-1}^j(-j-m) + \alpha_{j+m-1}^{j-1} = -\alpha_{j+m-1}^j(j+m) + \alpha_{j+m-1}^{j-1}.$$

∎

Proof (Proof of Theorem 5) First, using induction we show that

$$\sum_{i=1}^{m}(-1)^i\alpha_m^i = (-1)^m m \cdot m!. \tag{6}$$

For $m = 1$ we have $-\alpha_1^1 = -1 = (-1)^1 1 \cdot 1!$.
For $m = 2$: $-\alpha_2^1 + \alpha_2^2 = -(\alpha_1^1 \cdot 2 + \alpha_1^0) + \alpha_2^2 = -(-2-1) + 1 = 4 = (-1)^2 2 \cdot 2!$.

Assume that (6) holds for a certain $m - 1$. Consider a sum $\sum_{i=1}^{m}(-1)^i\alpha_m^i$. Using (5) rewrite it as follows

$$\sum_{i=1}^{m}(-1)^i\alpha_m^i = \sum_{i=1}^{m}(-1)^i\left[-\alpha_{m-1}^i \cdot m + \alpha_{m-1}^{i-1}\right]$$

$$= \sum_{i=1}^{m}(-1)^{i+1}(m+1)\alpha_{m-1}^i + (-1)^{m+1}\alpha_{m-1}^m - \alpha_{m-1}^0$$

$$= -(m+1)\sum_{i=1}^{m}(-1)^i\alpha_{m-1}^i + (-1)^{m+1}\alpha_{m-1}^m - \alpha_{m-1}^0.$$

Because the coefficient A_m does not contain the term ψ_m, we consider the element $(-1)^{m+1}\alpha_{m-1}^m$ to be equal to 0. Furthermore, by Theorem 2 we conclude that

$$A_0 = \sum_{i=0}^{k}(-1)^i\psi_{i+1} = \sum_{i=0}^{k}(-1)^i\frac{\psi_{i+1} \cdot i!}{i!}.$$

Hence the element $-\alpha_{m-1}^0$ is equal to $(-1)^m(m-1)!$, we get

$$\sum_{i=1}^{m}(-1)^i\alpha_m^i = -(m+1)\sum_{i=1}^{m-1}(-1)^i\alpha_{m-1}^i + (-1)^m(m-1)!$$

$$= -(m+1)(-1)^{m-1}(m-1)(m-1)! + (-1)^m(m-1)!$$

$$= (-1)^m(m-1)!m^2 = (-1)^m m!m,$$

which proves (6). Because in the expansion of coefficients A_i, $\frac{1}{m!}\psi_{m+1}$ stands by α_m^i, then

$$\frac{1}{m!}\psi_{m+1}\sum_{i=1}^{m}(-1)^i\alpha_m^i = (-1)^m m \cdot \psi_{m+1}.$$

Summing-up over m we eventually get

Table 1 Table of the numbers α_i^j for $k = 5$

	0	1	2	3	4	5
0	1	−1	2	−6	24	−120
1	0	1	−3	11	−50	274
2	0	0	1	−6	35	−225
3	0	0	0	1	−10	85
4	0	0	0	0	1	−15
5	0	0	0	0	0	1

$$\sum_{m=1}^{k}\left[\sum_{i=1}^{m}(-1)^i\frac{\alpha_m^i}{m!}\psi_{m+1}\right] = \sum_{m=1}^{k}(-1)^m A_m = \sum_{m=1}^{k}(-1)^m m \cdot \psi_{m+1}.$$

∎

As an example, the numbers α_i^j for $k = 5$ are given in Table 1.

3.2   Basis Numbers

It follows from Lemma 1 that the matrix $\left(\frac{\alpha_i^j}{i!}\right)_{0\leqslant i,j\leqslant k}$ is a transition matrix between $\mathcal{B}_1$ and $\mathcal{B}_2$ bases. The following theorem shows its inverse.

Theorem 6 *For a k-arithmetic sequence a_n each term of its characterizing sequence ψ_i, $i = 1,\ldots,k+1$ is a linear combination of $A_0, A_1, \ldots, A_k$ and*

$$\psi_j = A_{j-1}(j-1)! - (j-1)!\sum_{i=0}^{k-j}\frac{\alpha_{k-i}^{j-1}}{(k-i)!}\psi_{k-i+1}, \quad 1 \leqslant j \leqslant k+1. \tag{7}$$

Proof (Sketch of the proof) It was shown in Lemma 1 that the coefficients of a polynomial are of the form

$$A_i = \frac{\alpha_i^i}{i!}\psi_{i+1} + \cdots + \frac{\alpha_k^i}{k!}\psi_{k+1}, \quad i = 0, 1, \ldots, k.$$

Notice that this representation forms a system of $k+1$ equations with $k+1$ variables $\psi_1, \psi_2, \ldots, \psi_{k+1}$. The augmented matrix of this system is of the form

$$\Lambda = \left(\begin{array}{ccccc|c} \frac{1}{0!} & \frac{\alpha_1^0}{1!} & \frac{\alpha_2^0}{2!} & \cdots & \frac{\alpha_k^0}{k!} & A_0 \\ 0 & \frac{1}{1!} & \frac{\alpha_2^1}{2!} & \cdots & \frac{\alpha_k^1}{k!} & A_1 \\ \multicolumn{6}{c}{\cdots\cdots\cdots\cdots\cdots\cdots} \\ \multicolumn{6}{c}{\cdots\cdots\cdots\cdots\cdots\cdots} \\ 0 & 0 & 0 & \cdots & \frac{1}{k!} & A_k \end{array} \right).$$

This matrix is nonsingular, hence the system has a solution which, after performing the basic row operations, is given by

$$\psi_j = A_{j-1}(j-1)! - (j-1)! \sum_{i=0}^{k-j} \frac{\alpha_{k-i}^{j-1}}{(k-i)!} \psi_{k-i+1}.$$

■

The solution (7) is given in a recursive form and ψ_j is a linear combination of $A_{j-1}, A_j, \ldots, A_k$. For $j = k + 1$ and $j = k$ we have

$$\psi_{k+1} = A_k k!, \quad \psi_k = A_{k-1}k! - A_k \alpha_k^{k-1}(k-1)!.$$

Looking on the form of the solution we conclude that, if only coefficients $A_0, \ldots, A_k$ are integers, then every element $\frac{\alpha_{k-i}^j \psi_{k-i+1}}{(k-i)!}$ is also an integer and each of the numbers ψ_j is divisible by $(j-1)!$. Then, we can write

$$\psi_j = \beta_j^{j-1} A_{j-1} + \beta_j^j A_j + \cdots + \beta_j^k A_k,$$

where β_j^i denotes a certain number, constant for the expansion of ψ_j and standing by the coefficient A_i. It is constant and the same for any k-arithmetic sequence. Moreover, it is easy to find out that $\beta_{i+1}^i = i!$ for any $i \geqslant 0$ and $\beta_j^i = 0$ for $i > j - 1$. Because of their properties, we will call the numbers β_j^k the **basis numbers of the kth degree**.

Theorem 7 *For any k, all the numbers β_i^k, $i = 1, \ldots, k + 1$ are natural.*

Let us first state two lemmas.

Lemma 2 *For a fixed k, $(\beta_i^k)_{i=1,\ldots,k+1}$ is a characterizing sequence of the sequence $K_n = n^k$.*

Proof We know that for any k-arithmetic sequence

$$\sum_{i=0}^{k} \binom{n-1}{i} \psi_{i+1} = A_k n^k + A_{k-1} n^{k-1} + \cdots + A_1 n + A_0.$$

Since $\psi_j = \beta_j^{j-1} A_{j-1} + \beta_j^j A_j + \cdots + \beta_j^k A_k$, then

$$\sum_{i=0}^{k} \binom{n-1}{i} \left(\beta_{i+1}^i A_i + \beta_{i+1}^{i+1} A_{i+1} + \cdots + \beta_j^k A_k \right) = A_k n^k + A_{k-1} n^{k-1} + \cdots + A_1 n + A_0.$$

After splitting the sum on the left-hand side of the equation into elements with respect to the coefficients A_i we obtain

$$A_0 \beta_1^0 + A_1 \left[\beta_1^1 + (n-1)\beta_2^1 \right] + \cdots + A_j \sum_{i=0}^{j} \binom{n-1}{i} \beta_{i+1}^j + \cdots + A_k \sum_{i=0}^{k} \binom{n-1}{i} \beta_{i+1}^k.$$

Since coefficients β_j^i are constant and equal for any k-arithmetic sequence, then the following condition

$$\sum_{i=0}^{j} \binom{n-1}{i} \beta_{i+1}^j = n^j, \quad \text{for } j = 0, 1, \ldots, k$$

is needed for the equality to hold. ∎

Lemma 3 *For any natural k and $1 \leqslant j \leqslant k$ the following identity holds*

$$\sum_{i=0}^{k} \binom{k+1}{i} \beta_j^i = \beta_{j+1}^{k+1}.$$

Proof Let $K_n = n^{k+1}$. By Lemma 2, its characterizing sequence is $(\beta_j^{k+1})_{j=1,\ldots,k+2}$. Consider a sequence $K_{1n} = \Delta^1 n^{k+1} = (n+1)^{k+1} - n^{k+1} = \sum_{i=0}^{k} \binom{k+1}{i} n^i$.

Because for each i the characterizing sequence of n^i is $(\beta_j^i)_{j=1,\ldots,i+1}$, then the sequence K_{1n} has the characterizing sequence of the form $\left(\sum_{i=0}^{k} \binom{k+1}{i} \beta_j^i \right)_{j=1,\ldots,k+1}$. Since the $(j+1)$th term of the characterizing sequence of K_n is the jth term of the characterizing sequence of K_{1n}, the proof is complete. ∎

Proof (Proof of Theorem 5) By Lemma 2 we conclude that $\beta_1^i = 1^i = 1$ for any $i \geqslant 0$. Moreover, the non-zero terms of the characterizing sequence of $a_n = n$ are $\psi_1 = 1$ and $\psi_2 = 1$, hence $\beta_2^1 = 1$. By Lemma 3, each number β_i^j, for $i, j \geqslant 2$ is a sum of natural numbers and therefore is a natural number itself. ∎

Notice that for every $i > 0$, $\sum_j (-1)^{j+1} \beta_j^i = 0$. Indeed, by Theorem 2, $A_0 = \sum_{i=0}^{k} (-1)^i \psi_{i+1}$. Each element of this sum is equal to $(-1)^i \left(\beta_{i+1}^i A_i + \cdots + \beta_{i+1}^k A_k \right)$. Summing-up over i and sorting with respect to A_i yields

$$A_0 = A_0 + A_1 \left(\beta_1^1 - \beta_2^1 \right) + \cdots + A_i \left(\sum_{j=1}^{i+1} (-1)^{j+1} \beta_j^i \right) + \cdots + A_k \left(\sum_{j=1}^{k+1} (-1)^{j+1} \beta_j^k \right).$$

Table 2 Table of the basis numbers β_i^j for $k = 5$

	0	1	2	3	4	5
1	1	1	1	1	1	1
2	0	1	3	7	15	31
3	0	0	2	12	50	180
4	0	0	0	6	60	390
5	0	0	0	0	24	360
6	0	0	0	0	0	120

To show an example of the basis numbers of the kth degree the collection of such numbers for $k = 5$ is given in Table 2.

Theorem 8 *For a prime number p all the basis numbers of the degree p, except for β_1^p and β_2^p, are divisible by p.*

Proof Obviously, $\beta_1^p = 1$ is not divisible by p. Moreover, by Lemma 3, we know that

$$\beta_j^p = \sum_{i=0}^{p-1} \binom{p}{i} \beta_{j-1}^i.$$

Now we use the following property of binomial coefficients: *for a prime p, $\binom{p}{i}$ is divisible by p, for $i = 1, \ldots, p-1$.* Note that

$$\beta_2^p = 1 + \sum_{i=1}^{p-1} \binom{p}{i} \beta_1^i,$$

because $\beta_1^0 = 1$ and

$$\beta_j^p = \sum_{i=1}^{p-1} \binom{p}{i} \beta_{j-1}^i, \quad j = 3, 4, \ldots,$$

since $\beta_j^0 = 0$, for $j = 2, 3, \ldots$. ∎

Notice that Fermat's little theorem follows immediately from the above. Actually, let n, p be coprime numbers with prime p. Then

$$n^p = \sum_{i=0}^{p} \binom{n-1}{i} \beta_{i+1}^p \equiv \beta_1^p + (n-1)\beta_2^p = 1 + n - 1 \equiv n \mod p.$$

Let us now return to the issue of finding the sum of n initial terms of a k-arithmetic sequence. Assume that a polynomial a_n of the degree k having coefficients $A_0, A_1, \ldots, A_k$ is given. We are interested in finding the coefficients of polynomial $S_n = \sum_{i=1}^{n} a_i$. Theorem 3 states that $S_n = \sum_{i=0}^{k} \binom{n}{i+1} \psi_{i+1}$. We may rewrite this sum as follows

$$S_n = \sum_{i=0}^{k} \binom{n}{i+1} \psi_{i+1} = \sum_{i=0}^{k} \left[\binom{n-1}{i} + \binom{n-1}{i+1} \right] \psi_{i+1} = a_n + \sum_{i=1}^{k+1} \binom{n-1}{i} \psi_i$$

$$= a_n + \sum_{i=0}^{k+1} \binom{n-1}{i} \psi_{i+1}^*,$$

where $\psi_1^* = 0$ and $\psi_{i+1}^* = \psi_i$ for $i = 1, \ldots, k+1$. Notice, that the second summand is the nth term of some $(k+1)$-arithmetic sequence a_n^* having the characterizing sequence $(\psi_i^*)_{i=1,\ldots,k+2}$. We then search for the coefficients $A_0^*, \ldots, A_{k+1}^*$ of the polynomial a_n^*. Obviously we have

$$\begin{pmatrix} A_0^* \\ A_1^* \\ \cdots \\ A_{k+1}^* \end{pmatrix} = \left(\frac{\alpha_i^j}{i!} \right)_{i=0,\ldots,k+1}^{j=0,\ldots,k+1} \cdot \begin{pmatrix} \psi_1^* \\ \psi_2^* \\ \cdots \\ \psi_{k+2}^* \end{pmatrix} = \left(\frac{\alpha_i^j}{i!} \right)_{i=0,\ldots,k+1}^{j=0,\ldots,k+1} \cdot \begin{pmatrix} 0 \\ \psi_1 \\ \psi_2 \\ \cdots \\ \psi_{k+1} \end{pmatrix}$$

and

$$\begin{pmatrix} 0 \\ \psi_1 \\ \psi_2 \\ \cdots \\ \psi_{k+1} \end{pmatrix} = \left(\frac{\mathbf{0}}{\mathbf{B}_{k+1 \times k+1}} \right) \cdot \begin{pmatrix} A_0 \\ A_1 \\ \cdots \\ A_k \end{pmatrix},$$

where $\mathbf{0}$ is a $(k+1)$-element vector of zeros and $\mathbf{B}_{k+1 \times k+1} = \left(\beta_i^j \right)_{i=1,\ldots,k+1}^{j=0,\ldots,k}$. This yields

$$\begin{pmatrix} A_0^* \\ A_1^* \\ \cdots \\ A_{k+1}^* \end{pmatrix} = \left(\frac{\alpha_i^j}{i!} \right)_{i=0,\ldots,k+1}^{j=0,\ldots,k+1} \cdot \left(\frac{\mathbf{0}}{\mathbf{B}_{k+1 \times k+1}} \right) \cdot \begin{pmatrix} A_0 \\ A_1 \\ \cdots \\ A_k \end{pmatrix}.$$

Notice that matrix $\left(\frac{\alpha_i^j}{i!} \right)_{i=0,\ldots,k+1}^{j=0,\ldots,k+1} \cdot \left(\frac{\mathbf{0}}{\mathbf{B}_{k+1 \times k+1}} \right)$ is a $k+2 \times k+1$ matrix so multiplication by the vector of coefficients is allowed. Finally, the coefficients of the sum S_n are obtained from $A_0, \ldots, A_k$ as follows

$$\left[\left(\frac{\mathbf{I}_{k+1}}{\mathbf{0}} \right) + \left(\frac{\alpha_i^j}{i!} \right)_{i=0,\ldots,k+1}^{j=0,\ldots,k+1} \cdot \left(\frac{\mathbf{0}}{\mathbf{B}_{k+1 \times k+1}} \right) \right] \cdot \begin{pmatrix} A_0 \\ A_1 \\ \cdots \\ A_k \end{pmatrix} = \mathbf{T}_{k+1} \cdot \begin{pmatrix} A_0 \\ A_1 \\ \cdots \\ A_k \end{pmatrix}.$$

For $k = 5$, we obtain

$$\mathbf{T}_6 = \begin{pmatrix} 0 & 0 & 0 & 0 & 0 \\ 1 & 0.5 & 0.17 & 0 & -0.03 \\ 0 & 0.5 & 0.5 & 0.25 & 0 \\ 0 & 0 & 0.33 & 0.5 & 0.33 \\ 0 & 0 & 0 & 0.25 & 0.5 \\ 0 & 0 & 0 & 0 & 0.2 \end{pmatrix}.$$

Note that for a fixed k, the matrix $\mathbf{T}_{k+1}$ is constant for any polynomial of the degree k. $\mathbf{T}_{k+2}$ is obtained from $\mathbf{T}_{k+1}$ by extending it by one row and one column.

4 Applications

4.1 Function ξ in Fermat's Equation

Consider a family of functions $\xi : \mathbb{N}_+ \to \mathbb{Z}$ defined in the following manner

$$\xi(i) := \xi_{n,p,q}(i) := \binom{n+q}{i} - \binom{n+p}{i} - \binom{n}{i} + (-1)^i,$$

where $n, p, q \in \mathbb{N}_+$ such that $n \geqslant 3$ and $p < q$. Now we will prove some properties of this class.

Property 1 *For any $i \geqslant 1$,*

$$\xi_{n,1,2}(i) = \inf_{p,q} \xi_{n,p,q}(i).$$

Proof Notice that

$$\binom{n+2}{i} - \binom{n+1}{i} = \binom{n+1}{i-1} \leqslant \binom{n+p}{i-1} = \binom{n+p+1}{i} - \binom{n+p}{i}$$
$$< \binom{n+q}{i} - \binom{n+p}{i}.$$

Adding $(-1)^i - \binom{n}{i}$ to both sides of inequality completes the proof. ∎

Property 2 *If, for some j and fixed $n, p, q, \xi(j) \geqslant 0$, then starting with $i = j$ up to $i = n + q - 1$ the values of $\xi(i)$ are positive.*

Proof The mere form of the function ξ implies that $\xi(i) \geqslant 0$ for $i \geqslant n$. Assume now that $\xi(j) \geqslant 0$ for some $1 \leqslant j \leqslant n - 1$. Then, omitting the $(-1)^j$ term, we have

$$\xi(j+1) = \binom{n+q}{j+1} - \binom{n+p}{j+1} - \binom{n}{j+1}$$

$$= \binom{n+q}{j}\frac{n+q-j}{j+1} - \binom{n+p}{j}\frac{n+p-j}{j+1} - \binom{n}{j}\frac{n-j}{j+1}$$

$$= \frac{n+q-j}{j+1}\left[\binom{n+q}{j} - \binom{n+p}{j} - \binom{n}{j}\right] + \frac{q-p}{j+1}\binom{n+p}{j} + \frac{q}{j+1}\binom{n}{j}$$

$$= \frac{n+q-j}{j+1}\left[\xi(j) - (-1)^j\right] + (q-p)\frac{1}{j+1}\binom{n+p}{j} + q\frac{1}{j+1}\binom{n}{j}.$$

Because $\xi(j) \geqslant 0$, then $\xi(j) - (-1)^j \geqslant -1$. Hence,

$$\xi(j+1) \geqslant \frac{(q-p)\binom{n+p}{j} + q\binom{n}{j} - (n+q) + j}{j+1} - 1. \tag{8}$$

Since qn is the minimal value of $q\binom{n}{j}$ and the assumptions for ξ give $qn > q+n$, then $q\binom{n}{j} - (q+n) > 0$. Moreover, $(q-p)\binom{n+p}{j} \geqslant 4$. Inserting this into (6) we get

$$\xi(j+1) \geqslant \frac{5+j}{j+1} - 1 = \frac{4}{j+1} > 0.$$

For $n - 1 \leqslant i \leqslant n+q-1$ the proof is a direct consequence of the form of ξ. ∎

Property 3 *At most $\lfloor \frac{n}{2} \rfloor$ initial values of $\xi(i)$ are negative.*

Proof We will carry out the proof for an odd n. The case od even n is analogous. We will show that $\xi_{n,1,2}\left(\lfloor \frac{n}{2} \rfloor + 1\right) > 0$ which, by Property 1, implies what is needed. Substitute $n + 1 = 2m$.

$$\xi_{2m-1,1,2}(m) = \binom{2m+1}{m} - \binom{2m}{m} - \binom{2m-1}{m} + (-1)^m$$

$$= \binom{2m}{m-1} - \binom{2m-1}{m} + (-1)^m > \binom{2m}{m-1} - \binom{2m-1}{m-1} - 1$$

$$= \binom{2m-1}{m-2} - 1.$$

Let $C_m = \binom{2m-1}{m-2} - 1$. Because for $m \geqslant 3$ the sequence C_m is increasing and $C_2 = 0$, then $\xi(m) > C_m \geqslant 0$. As a result we get

$$\xi_{n,p,q}\left(\frac{n+1}{2}\right) > \xi_{n,1,2}\left(\frac{n+1}{2}\right) > 0,$$

for any n, p, q chosen according to the assumptions. By Property 2, the last possible negative value of $\xi_{n,p,q}$ is the one for $i = \frac{n+1}{2} - 1 = \lfloor \frac{n}{2} \rfloor$. ∎

Property 4 *For a fixed j, if $\xi(j) \leqslant 0$, then the sequence $a_n = \xi_{n,p,q}(j)$ is non-increasing.*

Proof Fix j and determine a_{1n}. Then

$$a_{1n} = \xi_{n+1,p,q}(j) - \xi_{n,p,q}(j) = \binom{n+q}{j-1} - \binom{n+p}{j-1} - \binom{n}{j-1}$$

$$= \xi_{n,p,q}(j-1) + (-1)^j.$$

Since $\xi(j) \leqslant 0$ and by Property 2, we obtain $\xi(j-1) < 0$. Therefore, $a_{1n} = \xi_{n,p,q}(j-1) + (-1)^j \leqslant 0$, which proves that a_n is non-increasing. ∎

Theorem 9 *Let $x_0 < y_0 < z_0 \in \mathbb{N}$ be such that $y_0 - x_0 = p$ and $z_0 - x_0 = q$. If for some prime $k > 2$, $\xi_{x_0--1,p,q}(k) \leqslant 0$ then the Fermat Last Theorem holds for the exponent k and any triple $(x, x + p, x + q)$, where $x \geqslant x_0$.*

Proof We will prove that the equation

$$x^k + y^k = z^k \tag{9}$$

has no solutions (x, y, z) as in the theorem. Assume the opposite. Let us denote $x = n + 1$, where $n \geqslant 3$ (we omit the proof for $x = 1, 2, 3$). As $z > y > x$, then

$$\exists_{p \in \mathbb{N}_+} : \quad y = n + 1 + p \quad \text{and} \quad \exists_{q > p \in \mathbb{N}_+} : \quad y = n + 1 + q.$$

Now (9) takes the form

$$(n+1)^k + (n+p+1)^k = (n+q+1)^k. \tag{10}$$

Notice that each element of (10) is a term of the *k*-arithmetic sequence $K_m = m^k$. Therefore, we get

$$(n+1)^k = \sum_{i=0}^{k} \binom{n}{i} \beta_{i+1}^k,$$

$$(n+p+1)^k = \sum_{i=0}^{k} \binom{n+p}{i} \beta_{i+1}^k, \tag{11}$$

$$(n+q+1)^k = \sum_{i=0}^{k} \binom{n+q}{i} \beta_{i+1}^k.$$

Inserting (11) into (10) and converting yields

$$\sum_{i=0}^{k} \binom{n+q}{i} \beta_{i+1}^k - \sum_{i=0}^{k} \binom{n+p}{i} \beta_{i+1}^k - \sum_{i=0}^{k} \binom{n}{i} \beta_{i+1}^k = 0,$$

2

1

or equivalently

$$\sum_{i=1}^{k}\left[\binom{n+q}{i}-\binom{n+p}{i}-\binom{n}{i}\right]\beta_{i+1}^{k}-1=0. \tag{12}$$

By Theorem 2, $1=\sum_{i=1}^{k}(-1)^{i+1}\beta_{i+1}^{k}$. Inserting this into (10) we obtain

$$\sum_{i=1}^{k}\left[\binom{n+q}{i}-\binom{n+p}{i}-\binom{n}{i}+(-1)^{i}\right]\beta_{i+1}^{k}=\sum_{i=1}^{k}\xi_{n,p,q}(i)\beta_{i+1}^{k}=0.$$

By Property 2, it is impossible that all $\xi(i)$ were zero and the assumptions of the numbers n, p, q imply that $\xi(1)<0$. Therefore, there exists such partition I_{+}, I_{-} of the set of indices that

$$\forall_{i\in I_{+}}\ \xi(i)\geqslant 0\ \ \text{and}\ \ \forall_{i\in I_{-}}\ \xi(i)<0,$$

where $\forall_{i\in I_{+},j\in I_{-}}\ i>j$. In particular, if for some n_{0}, p, q, $I_{+}=\emptyset$, i.e. $\xi_{n_{0},p,q}(k)\leqslant 0$, then (10) cannot be satisfied. But, by Property 4, also $\xi_{n,p,q}(k)\leqslant 0$ for all $n\geqslant n_{0}$. Hence the theorem holds. ∎

4.2 Generalization to Arbitrary Sequences

The previous part of this paper embraced a particular class of sequences whose folds is a natural number. As a matter of fact, this class is very "poor" when compared to the class of all sequences. It is then justified to propose the following generalization of the former concepts.

Let a_{n} be an arbitrary sequence of real numbers such that there does not exist $k\in\mathbb{N}$ for which $\Delta^{k}a_{n}=\text{const}$. We assume then that $k=\infty$. As the characterizing sequence of a_{n} is, in this case, infinite, then we can state Theorem 1 as follows:

Theorem 10 *The initial n terms of the characterizing sequence of a sequence a_{n} determine it uniquely up to the nth term.*

The proof is analogous. What follows immediately is the fact that the general term of a sequence a_{n} can be expressed as

$$a_{n}=\sum_{i=0}^{n-1}\binom{n-1}{i}\psi_{i+1}, \tag{13}$$

where ψ_{i} denotes the ith term of the characterizing sequence of a_{n}.

Example 1 Let $a_n = c^n$, with $c \in \mathbb{R}$ i $c \neq 0$. Then

$$a_{1n} = c^{n+1} - c^n = c^n(c-1), \quad a_{2n} = c^n(c-1)^2, \quad a_{in} = c^n(c-1)^i.$$

It gives the characterizing sequence of a_n. Namely, $\psi_i = c(c-1)^{i-1}$. We can then write

$$c^n = \sum_{i=0}^{n-1} \binom{n-1}{i} c(c-1)^i$$

or, equivalently,

$$c^{n-1} = \sum_{i=0}^{n-1} \binom{n-1}{i} (c-1)^i.$$

Notice that the above identity corresponds to the binomial identity, hence the conclusion that formula (13), applied to any sequence, can be thought of as a generalization of the binomial theorem.

Example 2 Let F_n denote the Fibonacci sequence, i.e. $F_1 = 1, F_2 = 1, F_3 = 2, \dots$, $F_n = F_{n-2} + F_{n-1}$. Then we have

$$(F_{1n}) = 0, 1, 1, 2, \dots, F_{n-1}, \dots, \quad (F_{2n}) = 1, 0, 1, 1, \dots, F_{n-2}, \dots,$$

$$(F_{3n}) = -1, 1, 0, 1, \dots, F_{n-3}, \dots, \quad (F_{4n}) = 2, -1, 1, 0, \dots, F_{n-4}, \dots,$$

$$\dots \dots$$

$$(F_{in}) = (-1)^i F_{i-1}, (-1)^{i-1} F_{i-2}, (-1)^{i-2} F_{i-3}, \dots, 0, 1, 1, \dots, F_{n-i}, \dots$$

Which gives its characterizing sequence

$$\psi_1 = 1, \quad \psi_2 = 0, \quad \psi_i = (-1)^{i-1} F_{i-2}, \quad \text{for } i > 2$$

and the formula

$$F_n = 1 + \sum_{i=2}^{n-1} (-1)^i \binom{n-1}{i} F_{i-1}.$$

Notice, that for a prime number p, we obtain the following congruence:

$$F_{p+1} - 1 \equiv -F_{p-1} \mod p.$$

You can find some similar formulae in [6].

In general, for an arbitrary sequence a_n of integers the formula (13) implies that:

$$a_{p+1} - a_1 \equiv \psi_{p+1} \mod p.$$

Example 3 For the sequence $\sin mx$, after simple trigonometric transformations, we get

$$\sin mx = \sum_{i=0}^{m-1} \binom{m-1}{i} 2^i \sin^i \frac{x}{2} \sin x \left[\frac{i}{2}(\pi + 1) + 1\right].$$

5 Conclusions

A generalization of the arithmetic progression, which assumes that more than one differencing must be performed to obtain a constant sequence was considered in this paper. A theorem that gives an expanded formula for a general term was proposed. Moreover, the basis numbers were defined, which allowed to explain the structure of polynomial sequences more deeply and therefore to prove some facts from the number theory. We concluded by giving the formula for the nth term of an arbitrary sequence of real numbers as a combination of its characterizing sequence, a generalization of the binomial identity and an elegant recursive formula for the nth Fibonacci number.

References

1. von Brasch, T., Bystrm, J., Lystad, L.P.: Optimal control and the fibonacci sequence. J Optim Theory Appl **154**, 857–878 (2012)
2. Jordan, C.: Calculus of Finite Differences. Chelsea Publishing, New York (1939/1965)
3. Newton, I.: J Principia, Book III, Lemma V, Case 1 (1687)
4. Nowakowski, J., Borowski K.: Zastosowanie teorii Carolana i Fischera na rynku kapitaowym. Difin (2006)
5. Stark, H.: An Introduction to Number Theory. MIT Press, Cambridge (1978)
6. Vajda, S.: Fibonacci and Lucas Numbers, and the Golden Section. p. 179. Halsted Press, Sydney (1989)
7. Wagstaff Jr., S.S.: Some questions about arithmetic progressions. Am. Math. Mon. **86**, 579–582 (1979)

Part II
Intelligent Techniques in
Data Mining

Forecasting of Short Time Series with Intelligent Computing

Katarzyna Kaczmarek and Olgierd Hryniewicz

Abstract Although time series analysis and forecasting have been studied since the seventeenth century and the literature related to its statistical foundations is extensive, the problem arises when the assumptions underlying statistical modeling are not fulfilled due to the shortness of available data. In such cases, additional expert knowledge is needed to support the forecasting process. The inclusion of prior knowledge may be easily formalized with the Bayesian approach. However, the proper formulation of prior probability distributions is still one of the main challenges for practitioners. Hopefully, intelligent computing can support the formulation of the prior knowledge. In this paper, we review recent trends and challenges of the interdisciplinary research on time series forecasting with the use of intelligent computing, especially fuzzy systems. Then, we propose a method that incorporates fuzzy trends and linguistic summaries for the forecasting of short time series. Experiments show that it is a very promising and human-consistent approach.

Keywords Time series · Forecasting · Fuzzy sets · Soft computing · Bayesian methods

1 Introduction

Recalling the 'No Free Lunch' theorem of Wolpert [1], there is no forecasting method that is best for any problem. However, some forecasting methods outperform the others for certain problems. Within this paper, we focus on forecasting of short time series, because short time series usually do not meet the assumptions underlying the statistical modeling, and therefore most methods seem infeasible and ineffective for

K. Kaczmarek (✉) · O. Hryniewicz
Systems Research Institute, Polish Academy of Sciences,
Newelska 6, 01-447 Warsaw, Poland
e-mail: K.Kaczmarek@ibspan.waw.pl

O. Hryniewicz
e-mail: Olgierd.Hryniewicz@ibspan.waw.pl

© Springer International Publishing Switzerland 2016
G. De Tré et al. (eds.), *Challenging Problems and Solutions in Intelligent Systems*, Studies in Computational Intelligence 634,
DOI 10.1007/978-3-319-30165-5_4

short data. Furthermore, they are not flexible enough to include human–computer interaction, which is usually needed in practice during the starting phase of the data collection process. Moreover, in the real-world environment, the data are usually imprecisely reported, missing, come from inhomogeneous sources and rarely can be described by precisely defined mathematical models.

Hopefully, the purpose of the soft computing methodologies is to formalize the human ability to reason in such uncertain and imprecise situations [2]. Zadeh's computing with words paradigm [3] has been continuously developed with the means of soft computing methodologies, like fuzzy logic, that aim at the analysis and design of the intelligent systems to solve complex real-life problems. At the same time, the human-consistent results of the knowledge discovery from time series datasets have been applied mostly for the descriptive or analytical purposes, e.g., [4–7], infrequently also for classification [8] or decision making and prediction [9, 10]. Within this contribution, we show that linguistic summaries [4, 11, 12] may successfully support the forecasting of short time series. One of the main advantages of the proposed method is its human-consistency.

This paper is a continuation of our previous works which deal with the problem of supporting time series analysis and forecasting with linguistic information [13, 14]. The proposed method is also in line with the granular computing perspective as introduced in [15]. The proposed approach assumes employing techniques from the time series analysis and forecasting, the fuzzy set theory and data mining (segmentation, summarization, supervised learning).

The structure of this paper is as follows. Next chapter reviews the recent trends and challenges of supporting time series forecasting with intelligent computing, especially fuzzy systems. In Chap. 3, the proposed approach for forecasting of short time series using fuzzy trends and linguistic summaries is presented. Chapter 4 provides illustrative examples. This paper concludes with general remarks and a description of further research opportunities gathered in Chap. 5.

2 Recent Trends in Forecasting with Intelligent Computing

The field of Computational Intelligence was formally initiated in 1994 during the IEEE World Congress on Computational Intelligence in Orlando. Computational Intelligence (CI) is defined as *a methodology involving computing that exhibits an ability to learn and/or deal with new situations such that the system is perceived to possess one or more attributes of reason, such as generalization, discovery, association, and abstraction* cf. [16]. Following Zadeh [17], the Soft Computing is based on the Computational Intelligence, and together with the Hard Computing (that is based on the Artificial Intelligence) form the Machine Intelligence. Due to Kruse et al. [18], the main stream methodologies developed within the Computational Intelligence are Fuzzy Systems, Artificial Neural Networks, Evolutionary Algorithms and Bayesian Networks.

In this chapter, we review the recent trends and challenges of the interdisciplinary research on the time series forecasting with the use of fuzzy systems.

2.1 Fuzzy Time Series

Common approach to address the imprecision and uncertainty related to time series data is based on the concept of Fuzzy Time Series introduced by Song and Chissom [19]. Fuzzy time series employ the linguistic variables and fuzzy relations. The approach assumes the split of the universum into intervals and defining the linguistic variables and relations over intervals. Then, the observations are fuzzified and the forecasts are generated for them. Finally, the outputs are defuzzified.

There have been proposed a lot of further modifications to the Song and Chissom's approach, e.g., [20–22]. For example, Chen and Chen [23] introduce the fuzzified variation and the fuzzy logical relationship groups.

The main advantage of the fuzzy time series is their simplicity and the direct linkage to the linguistic expressions. However, their scheme for representation of the fuzzy logical relationships may not be relevant for complex problems.

2.2 Fuzzification of Model Parameters

One of the first solutions reflecting the imprecision and uncertainty about the time series and model itself is the fuzzification of parameters in the predictive model. Instead of processing crisp parameters, fuzzy numbers are defined and then, the calculations may be performed for the α-cuts. For basic definitions of the fuzzy random variables, see e.g., [24, 25].

Various models with fuzzy parameters have been presented in the literature. Tanaka et al. [26] introduce the idea of fuzzy regression. In [13], Kaczmarek and Hryniewicz build the vector autoregression model basing on the imprecise segments derived from crisp time series. Helin and Koivisto [27] suggest employing fuzzy conditional distributions in the GARCH models. Tseng et al. [28, 29] propose to use the fuzzy triangular numbers instead of crisp parameters in the ARIMA models.

2.3 Fuzzy Rule Based Systems

Fuzzy rule based systems are yet another forecasting approach that addresses the imprecision related to time series data. In [30], Agrawal et al. propose an algorithm for discovery of rules in sets of items. The problem is to find interesting rules of the form: *IF P THEN Q with chance p*. An example of such a rule is as follows: '*IF exchange rate for currency X decreases slowly, THEN the number of international travels will increase rapidly'*. Association rules have been successfully applied to describe huge datasets of different disciplines e.g., [31–36]. In [8], Höppner et al. focus on enhancing patterns with a context information and operating on block constraints instead of Allen's relations e.g., *'IF A happens before B and in the meantime we do not observe C, THEN we have a failure of class X'*. An important challenge is the natural lack of precision when handling patterns and rules. Segmentation methods usually return interval bounds that are uncertain due to various reasons. There is uncertainty related, not only to the segment bounds, but also to the human perception of patterns and rules. To circumvent this, Chen et al. [35] describe the fuzzy association rules.

In [37], Höppner provides an example of the short-term weather prediction for sailors with the use of frequent patterns. Experienced sailors are able to predict strong wind or storm based on the simple rules including the local changes of air pressure and wind direction e.g., *'Windspeed decreasing segment occurs after air pressure highly increasing segment'*. Such local patterns could be formulated with the use of the Allen's temporal logic [38]. Schockaert and Cock [39] fuzzify Allen's temporal interval relations and propose a complete framework to represent, calculate and reason about the temporal relationships for fuzzy intervals about the time series and the temporal data in general. Another kind of the relations based on the temporal sequences is introduced in Wilbik and Kacprzyk [40]. This idea originated from linguistic summaries in the sense of Yager [41] developed by Kacprzyk et al. [42, 43]. Linguistic summaries describe general facts about evolution of time series with quasi natural language and may be exemplified by *'Most medium and constant trends are low'*. The concept of pattern recognition has been widely discussed also by Mörchen et al. [7] and Kempe et al. [6].

Another idea of combining forecasts by the fuzzy rule system has been proposed recently by Burda et al. [9], who suggest to combine the multiple forecasting methods by the fuzzy rule-based ensemble, and the weights for the linear combination are the result of the linguistic association mining for the time series features, like trend or seasonality. Such associations may be exemplified by *'IF Strength of Seasonality is Small AND Coefficient of Variation is Roughly Small THEN Weight of the j th method is Big'*. In [44], Petrovic et al. combine the autoregressive and moving average models with the fuzzy rule based approach using Mamdani inference mixed with some techniques of counting in fuzzy sets. Chen et al. [45] propose an approach for the automatic generation of the fuzzy rule base for the combination of forecasts.

In general, the association rules are an example of the information granule about time series data retrieved with the knowledge discovery techniques. In [15], Hryniewicz and Kaczmarek list the following information granules related to time series data: labeled intervals (trends), linguistic summaries, frequent patterns, association rules, linguistic descriptions and labeled time series. To conclude, although, the primary goal of data mining is the human-consistent description of huge datasets, such descriptions may be also useful for predictive purposes and may be included in the rule based systems.

2.4 Hybrid Systems

In the recent years, there have been proposed also various hybrid systems to support time series forecasting. First forecasting competition for the fuzzy time series models to predict multiple heterogeneous time series was held during IFSA-EUSFLAT 2015 conference [46]. The dataset of this competition includes 91 time series of different length, time frequencies and behaviour. The winner of the competition is the hybrid system by Afanasieva et al. [47], that combines time series decomposition and the F-transform technique. In [48, 49], Perfilieva et al. propose forecasting solutions basing on the F-transform combined with among others fuzzy relations, neural networks and decomposition techniques. The Perfilieva's system seems to be very successful for the analysis and forecasting of short time series. Recently, in [50, 51], Novak et al. further develop the concept of F-transform and trend extraction from time series using fuzzy natural logic.

Valenzuela et al. [52] propose yet another hybridization of intelligent techniques such as ANNs, fuzzy systems and evolutionary algorithms and the use of fuzzy rules to identify the appropriate ARMA model for prediction. Another expert system with fuzzy clustering and fuzzy rule interpolation techniques is proposed by Chen and Chang [53]. In [54], the authors propose the hybrid system that integrates the fuzzy time series representation with the neural network.

There is a wide range of other examples of the applications of the soft computing methods with the use of neural networks, e.g., [55–59]. The fast interval predictors for large-scale, nonlinear time series with noisy data using fuzzy granular support vector machines are presented by Ruan et al. [60]. For further application examples and empirical studies related to hybrid systems, see e.g., [61–66].

To conclude, in the opinion of the authors of this paper, the research on forecasting with intelligent computing shall still focus on discovery of meaningful and human-consistent information granules and the proper representation of the imprecision related to such granules. Furthermore, the dependencies between the information granules and probabilistic models describing the time series shall be further analyzed. Finally, to enable the inclusion of expert knowledge in the forecasting process, user-friendly interfaces shall be designed to amend the human–computer interaction.

3 Proposed Approach Using Linguistic Summaries

Finding an appropriate predictive model for short time series and formulating its assumptions may become very challenging tasks for practitioners. To facilitate the formulation of assumptions, we propose the following forecasting procedure. Its main objective is to combine various predictive models according to weights being a result of data mining algorithms and human–computer interaction. Algorithm 1 depicts a high-level description of the proposed method.

Algorithm 1 Forecasting of short time series using linguistic summaries (FTLS)

Input: y, M, S
Output: y_{n+1}

1: **procedure** FTLS(y, M, S)
2:   **for** $i = 1$ **to** J **do**   ▷ Generation of time series from template models
3:     **for** $x = 1$ **to** k **do**
4:       $Y_{i,x}, C_{i,x} \leftarrow$ generateTSFromModel(M_i, x)
5:   $d \leftarrow k * J$
6:   **for** $j = 1$ **to** L **do**   ▷ Defining of fuzzy numbers for trends
7:     $\mu_j \leftarrow$ defineFuzzyNumber(S_j)
8:   $TR^d, LS^d, V^d = \emptyset$ ▷ Discovery of linguistic summaries
9:   **for** $i = 1$ **to** d **do**
10:     $\tilde{Y}_i \leftarrow$ brokenLineSegmentation(Y_i)
11:     $TR_i \leftarrow$ extractFuzzyTrends(S, $\tilde{Y}_i$)
12:     $LS_i \leftarrow$ summarizeFuzzyTrends(S, TR_i)
13:     $V_i \leftarrow$ degreeOfTruth(LS_i)
14:   $\tilde{V}^d \leftarrow$ reduceDimension(V^d, C^d)   ▷ Supervised learning of predictive models
15:   $CL \leftarrow$ kNNclassification($\tilde{V}^d$, C^d)
16:   **if** crossValidation(CL, $\tilde{V}^d$, C^d) $\geq \min_{acc}$ **then**
17:     return to *Line 2*
18:   **else**
19:     $\tilde{y} \leftarrow$ brokenLineSegmentation(y)   ▷ Evaluation of linguistic summaries for y
20:     $TR_{pr} \leftarrow$ extractFuzzyTrends(S, $\tilde{y}$)
21:     $LS_{pr} \leftarrow$ summarizeFuzzyTrends(S, TR_{pr})
22:     $V_{pr} \leftarrow$ degreeOfTruth($LS_p r$)
23:     $V_e =$ expertEvaluation (LS_{pr}, V_{pr})
24:   **while** $i \in J$ **do**   ▷ Posterior simulation and forecasting
25:     $w^{M_i} =$ calculateWeights(V_e, CL)
26:   p=constructPriorProbabilities(M, w^M)
27:   y_{n+1}=MCMCPosteriorSimulation (y, p)
28:   **return** y_{n+1} ▷ Return forecast for y

The input for the algorithm is the short discrete time series for prediction y:

$$y = \{y_t\}_{t=1}^n \in \mathbb{Y}, n \in \{n_{min}, \ldots, n_{max}\} \subseteq \mathbb{N} \tag{1}$$

where $\mathbb{Y}$ is a space of discrete time series. Let us assume that $n_{min} = 10, n_{max} = 20$. Also, the following set of template probabilistic models needs to be defined a priori: $M = \{M_1, M_2, \ldots, M_J\} \subseteq \mathbb{M}$. Within this procedure for short time series, we adapt various ARMA processes as template models M. Additionally, the set S of linguistic expressions needs to be formulated a priori, e.g., $S = \{increasing, decreasing\}$. The definitions for expressions in S are defined and validated with a user during the operation of the whole forecasting procedure. It needs to be stated, that for the clarity reasons, we focus on the one-step-ahead forecast. For the h-step-ahead forecast, the procedure shall be iterated.

The proposed approach consists of the following steps:

1. The algorithm starts from the generation of the sample time series Y^d for the training database. Sample time series are generated from the template probabilistic models M. As a result, we obtain $[Y^d, C^d]$ where C^d consists of labels (classes) describing models that have produced the respective time series Y^d.

2. Secondly, the definitions of the linguistic expressions from S, describing the trends in time series, are established. They are represented and processed as fuzzy numbers.

3. Next, the discovery of fuzzy trends and linguistic summaries for time series of the training database is performed. Each time series in Y^d is segmented and summarized. Then, its linguistic summaries are evaluated. Within this approach, the degree of truth is adapted as quality measure, it is defined as follows:

$$T(LS) = \mu_Q \left(\frac{\sum_{i=1}^{n} (\mu_R(y_i) \wedge \mu_P(y_i))}{\sum_{i=1}^{n} \mu_R(y_i)} \right) \quad (2)$$

where $\mu_R, \mu_P, \mu_Q : \mathbb{R} \to [0, 1]$ are the membership functions of the fuzzy sets representing the qualifier R, summarizer P and quantifier Q, respectively. Vector V^d consists of the degrees of truth about the linguistic summaries.

4. Then, the supervised learning of the probabilistic models basing on the linguistic summaries is performed. The input for the supervised learning algorithm are $[V^d, C^d]$ where V^d consists of the degrees of truth for the linguistic summaries LS^d and C^d consists of labels (classes) describing models that have produced the respective time series. The features V^d are processed with conventional classification algorithm (k-NN is adapted within this approach). The accuracy of the learned classifier is validated using the 2-fold cross validation, which is one of the simplest variation of k-fold cross-validation and tough, useful in practice. Each data point is used for both training and validation on each fold. If the approximated accuracy of the classifier exceeds the desired threshold min_{acc}, the algorithm proceeds. Otherwise, the whole procedure (model selection, generation of sample time series, defining of fuzzy trends) is repeated.

5. In the next step, the provisional linguistic summaries about the evolution of time series considered for prediction are generated. They are provisional due to the shortness of data, therefore are validated with the user.

6. Finally, the validated evaluations of linguistic summaries for the time series considered for prediction are classified according to the classifier learned on the training database. The classification scores are applied as a priori weights in the Bayesian averaging and the Markov chain Monte Carlo Posterior Simulations are executed. Due to the Bayesian forecasting, the posterior density $p(\omega|y, M)$ is a weighted average of the posterior densities of models $\{M_1, M_2, \ldots, M_J\}$:

$$p(\omega|y, M) = \sum_{j=1}^{J} p(\omega|y, M_j) p(M_j|M) \quad (3)$$

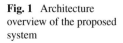 **Fig. 1** Architecture
overview of the proposed
system

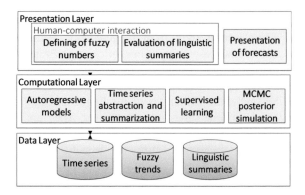

The illustration in Fig. 1 shows the architecture of the proposed approach. The proposed system is divided into the classical three tiers: *Presentation Layer*, which is the interface to the user; *Computational Layer*, which handles the data objects and modifies them; *Data Layer*, which consists of a database that loads and stores data.

As presented in Fig. 1, the *Data Layer* consists of the following 3 types of data: *Time Series*—real-valued discrete time series, *Fuzzy Trends*—the sequences of fuzzy numbers related to the linguistic expressions describing trends in data; *Linguistic Summaries*—the sets of linguistic summaries with their descriptions in natural language and the corresponding values of quality measures (e.g., degrees of truth).

Within the *Computation Layer*, the main time series data mining and forecasting algorithms are implemented. They modify the data objects aiming at the generation of forecasts. There are also implemented services responsible for the preprocessing of data and reporting. The *Autoregressive models* module implements methods that handle the multiple autoregressive and moving average processes. Additionally, it includes various functions for the time series analysis, especially model estimation and residual analysis. The *Time series abstraction and summarization* module handles the data imprecision and operations on fuzzy numbers, and performs communication with the *Trend Analysis System* [67] adapting the results about the linguistic summaries. The *Supervised learning* module consists of methods that perform the vector classification for the multi-class classification task. Among others, the k-NN algorithm is adapted. Additionally, the Principal Component Analysis (PCA) is implemented to reduce the dimensionality of the feature vectors. Finally, the *MCMC posterior simulation* module gathers methods for generating samples and approaching the posterior densities by the MCMC simulation. It includes also methods for goodness-of-fit, convergence diagnostics and the Bayesian averaging.

The *Presentation Layer* deals with the communication between the system and the user. The language summaries, e.g., *'Most increasing trends are short'* are presented to the decision maker who points his confidence that this summary is true about the considered time series.

As an output, the system delivers both, the crisp forecasts and the descriptions of linguistic summaries, that have been processed during the forecasting process.

4 Numerical Results

The empirical experiments have been performed on the sample time series from the M3-Competition dataset by Makridakis and Hibon [68]. The goal of these experiments is to provide an illustrative demonstration of the proposed *Forecasting of short time series using linguistic summaries* (FTLS) approach.

4.1 Illustrative Example

First, we choose one of the shortest time series (that has only 14 observations) and present the performance of the proposed approach in detail. The considered time series and its preprocessed version are presented in Fig. 2.

As described in Chap. 3, the proposed forecasting procedure consists of 6 steps. Now, we illustrate its performance in detail.

1. The algorithm starts from the generation of the learning time series. Within this experiment, the set of 3 template probabilistic models $M = \{M_1, M_2, M_3\}$ is considered. Sample time series are created from the following models:

$$\tilde{y}_t = \sum_{i=1}^{p} \phi_i \tilde{y}_{t-i} + a_t, \quad a_t \sim N(0, \sigma^2), \quad \tilde{y}_t = y_t - \mu \qquad (4)$$

   where $\sigma^2 = 0.1$ and the autoregressive coefficient $\phi_1 \in (-1.0, -0.4)$ for M_1, $\phi_1 \in [-0.4, 0.4]$ for M_2 and $\phi_1 \in (0.4, 1)$ for M_3.
2. Secondly, the definitions of fuzzy numbers that describe the trends and linguistic summaries are validated with the user of the system. Table 1 presents the exemplary set of attributes with their imprecise labels, where $f_1, \ldots, f_4$ denote the successive points defining the fuzzy trapezoidal number.

   For example, the duration of a trend may be characterized with label *short*, and trends that have length 2 or 3 are surely short. However, trends longer than 3 or shorter than 2 are short only to some degree, and trends that are longer than 6 are not short at all.

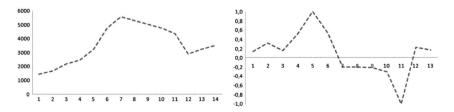

Fig. 2 Time series for prediction (N3) before and after preprocessing

Table 1 Interpretation for the set of attributes for trends with imprecise labels as fuzzy trapezoidal numbers

Attribute name	Imprecise label	f_1	f_2	f_3	f_4
Duration	*short*	**0.0**	**2.0**	**3.0**	**6.0**
Duration	*medium*	0.0	5.0	15.3	25.3
Duration	*long*	7.1	17.1	50.0	50.0
Dynamics	*decreasing*	−90.0	−90.0	−15.1	−10.1
Dynamics	*constant*	−15.1	−10.1	10.1	15.1
Dynamics	*increasing*	10.1	15.1	90.0	90.0
Variability	*low*	0.0	0.0	0.2	0.4
Variability	*moderate*	0.2	0.4	0.6	0.8
Variability	*high*	0.6	0.8	1.0	1.0

Table 2 Exemplary sample linguistic summaries from the training database

Description of linguistic summary	Degree of truth	Time series	Model
Most trends are constant	**1.0**	**001**	**1**
Most trends are constant	1.0	002	2
Most trends are constant	0.99	003	1
Most trends are constant	1.0	004	1
Most trends are constant	1.0	005	3
Most increasing trends are short	**0.40**	**001**	**1**
Most increasing trends are short	0.66	002	2
Most increasing trends are short	0.39	003	1
Most increasing trends are short	0.45	004	1
Most increasing trends are short	0.70	005	3

3. Then, the knowledge discovery techniques are run. Table 2 presents exemplary linguistic summaries for time series from the training database.
   For example, time series *001* (that has been generated from model 1) is described by the linguistic summary: *'Most trends are constant'* with the degree of truth 1, and with the summary: *'Most increasing trends are short'* only to some degree (0.40).
4. Next, the supervised learning of the probabilistic models basing on the linguistic summaries is performed. As presented in Table 2, there are multiple time series

generated from the same model. Each time series of the training database is linked to some class that determines the probabilistic model, including the probability distributions for all its parameters. For example, the following time series: *001, 003, 004* are generated from model *1*, and the degree of truth of the linguistic summary *'Most increasing trends are short'* amounts to 0.40, 0.66 and 0.39, respectively.

5. Then, the provisional linguistic summaries for the time series considered for prediction are generated. They are provisional because of the shortness data (14 observations), and therefore require user's validation. The user points his confidence about the expected evolution of the time series to be predicted. An exemplary set of linguistic summaries is presented in Table 3.

   Types of linguistic summaries considered for validation have been chosen basing on the analysis of most significant linguistic summaries in the training database. As observed in Table 3, for the considered time series, the following two types of linguistic summaries are most possible to be true: *'Most decreasing trends are medium'* and *'Most trends are constant'*.

6. Finally, the prior probability distributions of parameters for the probabilistic models M are retrieved. The prior probabilities $p(M_i|M)$ are set as uniform distributions basing on the classification scores. As a result of the classification, the following weights are calculated: 0.05, 0.43 and 0.52 for models M_1, M_2 and M_3 respectively, and the forecast is simulated with the MCMC algorithm according to formula (3). Finally, for the considered time series, the 1-step-ahead forecast results to be 3692.69. Whereas, the actual value of time series occurs to be 3698.17, so the calculated forecast is very close to the actual. Furthermore, to justify this crisp forecast, linguistic descriptions of most possible linguistic summaries may be provided. These summaries are related to generation of this forecast: *'Most decreasing trends are medium'* and *'Most trends are constant'*.

Table 3 Exemplary set of linguistic summaries of the time series to be predicted

Description of linguistic summary	Degree of truth
Most trends are constant	**0.6**
Most decreasing trends are medium	**0.7**
Most decreasing trends are moderate	0.5
Most trends are decreasing	0.5
Most trends are medium	0.4
Most trends are moderate	0.4
Most trends are short	0.3
Most trends are increasing	0.1
Most trends most are low	0.1

Table 4 Forecasting accuracy measured with Mean Squared Error (MSE) of the proposed FTLS approach and *Robust-Trend* method

Method	TS-N1	TS-N2	TS-N3	TS-N4	TS-N5
FTLS	80	246 376	394 345	130 208	118 134
Robust-Trend	30 776	79 637	458 519	180 132	211 876

4.2 Comparative Analysis

The proposed *Forecasting of short time series using linguistic summaries* (FTLS) approach has been evaluated for the benchmark datasets: the subset of the 5 first yearly time series (N1-N5) that have length 20 from the M3-Competition Datasets Repository [68]. We perform the in-sample evaluation of the method. Only 14 observations are used to learn the model and the remaining 6 observations are used to calculate the forecasting accuracy. The performance of the proposed FTLS approach is compared to the Robust-Trend method (a non-parametric version of Holt's linear model with median based estimate of trend). Table 4 shows the forecasting accuracy measured by the Mean Squared Error.

As demonstrated by results in Table 4, for most time series the proposed FTLS approach delivers more accurate forecasts than the Robust Trend method. Furthermore, although the human input is highly subjective, it helps to eliminate the need to manually express the assumptions as prior probability distributions, which may be difficult to handle by non-mathematician decision makers. At the same time, the procedure enables to effectively perform the posterior simulation for various predictive models.

5 Conclusion

When forecasting short time series, some additional knowledge is usually needed to select the predictive model and establish its prior assumptions. The inclusion of the prior knowledge may be easily formalized with the Bayesian methods. However, formulating prior assumptions may become very challenging task for practitioners. Hopefully, intelligent computing techniques can support the Bayesian forecasting of short time series.

Within this contribution, we have introduced the forecasting method *Forecasting of short time series using linguistic summaries* (FTLS). It incorporates fuzzy trends and linguistic summaries into the process of forecasting. Instead of providing definitions of prior probability distributions, users are asked to evaluate linguistic summaries (they are intuitive and easy for interpretation). The proposed forecasting method is human-consistent and easy for interpretation thanks to the use of linguistic expressions, that are processed with fuzzy logic. Finally, the numerical examples have been provided to illustrate the performance of the proposed approach. As confirmed by experiments, it is a very promising approach for forecasting of short time series.

In this paper, we have also reviewed the recent trends of the interdisciplinary research on the time series forecasting with the use of fuzzy systems. As described and advocated with examples, fuzzy logic may successfully address the imprecision related to real-life time series data.

In future research, the analysis of other forms of linguistic descriptions, fuzzy classification rules and frequent temporal patterns is considered. Also, further experiments including next predictive models are planned.

Acknowledgments Katarzyna Kaczmarek is supported by the Foundation for Polish Science under International Ph.D. Projects in Intelligent Computing financed from the European Union within the Innovative Economy Operational Programme 2007–2013 and European Regional Development Fund.

References

1. Wolpert, D.: The lack of a priori distinctions between learning algorithms. Neural Comput. **8**(7), 1341–1390 (1996)
2. Zadeh, L.: Fuzzy sets. Inf. Control **8**, 338–353 (1965)
3. Zadeh, L.A.: From computing with numbers to computing with words—from manipulation of measurements to manipulation of perceptions. Intelligent Systems and Soft Computing. Lecture Notes in Computer Science, pp. 3–40. Springer, New York (2000)
4. Kacprzyk, J., Wilbik, A.: Using fuzzy linguistic summaries for the comparison of time series: an application to the analysis of investment fund quotations. IFSA/EUSFLAT Conf. 1321–1326 (2009)
5. Conde-Clemente, P., Alonso, J., Trivino, G.: Interpretable fuzzy system allowing to be framed in a profile photo through linguistic expressions, In: Proceedings of 8th Conference of the European Society for Fuzzy Logic and Technology (EUSFLAT), pp. 463–468 (2013)
6. Kempe, S., Hipp, J., Lanquillon, C., Kruse, R.: Mining frequent temporal patterns in interval sequences. Fuzziness Knowl.-Based Syst. Int. J. Uncertain. **16**(5), 645–661 (2008)
7. Mörchen, F., Batal, I., Fradkin, D., Harrison, J., Hauskrecht, M.: Mining recent temporal patterns for event detection in multivariate time series data, KDD, pp. 280–288 (2012)
8. Höppner, F., Peter, S., Berthold, M.: Enriching multivariate temporal patterns with context information to support classification. Comput. Intell. Intell. Data Anal. Studies Comput. Intell. **445**, 195–206 (2013)
9. Burda, M., Štěpnička, M., Štěpničková, L.: Fuzzy rule-based ensamble for time series prediction: progresses with associations mining. Strengthening Links Between Data Analysis and Soft Computing, vol. 315, pp. 261–271, Springer, Berlin (2014)
10. Yarushkina, N., Perfilieva, I., Afanasieva, T., Igonin, A., Romanov, A., Shishkina, V.: Time series processing and forecasting using soft computing tools. In: RSFDGrC'11 Proceedings of the 13th International Conference on Rough Sets, Fuzzy Sets, Data Mining and Granular Computing, pp. 155–162, Springer, Berlin (2011)
11. Kacprzyk, J., Zadrożny, S.: Protoforms of linguistic data summaries: towards more general natural-language-based data mining tools. Soft Computing Systems. IOS Press, Amsterdam (2002)
12. Kacprzyk, J., Zadrożny, S.: Linguistic database summaries and their protoforms: towards natural language based knowledge discovery tools. Inf. Sci. **173**, 281–304 (2005)

13. Kaczmarek, K., Hryniewicz, O.: Linguistic knowledge about temporal data in bayesian linear regression model to support forecasting of time series. In: Proceedings of the Federated Conference on Computer Science and Information Systems, pp. 655–658 (2013)
14. Kaczmarek, K., Hryniewicz, O., Kruse, R.: Human input about linguistic summaries in time series forecasting, In: Proceedings of the Eighth International Conference on Advances in Computer-Human Interactions ACHI, pp. 9–13 (2015)
15. Hryniewicz, O., Kaczmarek, K.: Bayesian analysis of time series using granular computing approach. Appl. Soft Comput. **1**(3) (2015)
16. Eberhart, R., Simpson, P., Dobbins, R.: Computational Intelligence PC Tools. Academic Press, Boston (1996)
17. Zadeh, L.A.: Fuzzy logic, neural networks, and soft computing. Commun. ACM **37**, 77–84 (1994)
18. Kruse, R., Borgelt, C., Klawonn, F., Moewes, C., Steinbrecher, M., Held, P.: Computational intelligence. Texts in Computer Science. fuzzy Sets and fuzzy logic. Springer, London (2013)
19. Song, Q., Chissom, B.: Fuzzy time series and its models. Fuzzy Sets Syst. **54**(3), 269–277 (1993)
20. Liu, H.: An integrated fuzzy time series forecasting system. Expert Syst. Appl. **36**, 10045–10053 (2009)
21. Li, S., Kuo, S., Cheng, Y., Chen, C.: A vector forecasting model for fuzzy time series. Appl. Soft Comput. J. **11**(3), 3125–3134 (2011)
22. Chen, M., Chen, B.: Online fuzzy time series analysis based on entropy discretization and a fast fourier transform. Appl. Soft Comput. J. **14**, 156–166 (2014)
23. Chen, S., Chen, C.: Taiex forecasting based on fuzzy time series and fuzzy variation groups. IEEE Trans. Fuzzy Syst. **19**(1), 1–12 (2011)
24. Gil, M., Hryniewicz, O.: Statistics with imprecise data. Encyclopedia of Complexity and Systems Science, pp. 8679–8690 (2009)
25. Shapiro, A.: Fuzzy random variables. Insur.: Math. Econ. **44**, 307–314 (2009)
26. Tanaka, H., Uejima, S., Asai, K.: Linear regression analysis with fuzzy model. IEEE Trans. Syst. Man Cybern. **12**, 903–907 (1982)
27. Helin, T., Koivisto, H.: The garch-fuzzydensity method for density forecasting. Appl. Soft Comput. J. **11**(6), 4212–4225 (2011)
28. Tseng, F., Tzeng, G., Yu, H., Yuan, B.: Fuzzy arima model for forecasting the foreign exchange market. Fuzzy Sets Syst. **118**, 9–19 (2001)
29. Tseng, F., Tzeng, G.: A fuzzy seasonal arima model for forecasting. Fuzzy Sets Syst. **126**, 367–376 (2002)
30. Agrawal, R., Mannila, H., Srikant, R., Toivonen, H.: Fast discovery of association rules. Advances in Knowledge Discovery and Data mining, pp. 307–328. AAAI Press, Menlo Park Calif (1996)
31. Klawonn, F., Kruse, R.: Derivation of fuzzy classifcation rules from multidimensional data. Advances in Intelligent Data Analysis, pp. 90–94 (1995)
32. Mörchen, F., Ultsch, A.: Efficient mining of understandable patterns from multivariate interval time series. Data Min. Knowl. Discov. **15**(2), 181–215 (2007)
33. Peter, S., Hoeppner, F.: Finding temporal patterns using constraints on (partial) absence, presence and duration, Lecture Notes in Computer Science, pp. 442–451 (2010)
34. Mörchen, F., Thies, M., Ultsch, A.: Efficient mining of all margin-closed itemsets with applications in temporal knowledge discovery and classification by compression. Knowl. Inf. Syst. **29**(1), 55–80 (2011)
35. Chen, C., Hong, T., Tseng, V.: Fuzzy data mining for time-series data. Appl. Soft Comput. J. **12**(1), 536–542 (2012)
36. Alvarez, M., Felix, P., Carinena, P.: Discovering metric temporal constraint networks. Artif. Intell. Med. **58**(3), 139–154 (2013)
37. Höppner, F.: Knowledge Discovery from Sequential Data. Ph.D. thesis (2002)
38. Allen, J.: Maintaining knowledge about temporal intervals. Commun. ACM **26**(11), 832–843 (1983)

39. Schockaert, S., Cock, M.D.: Temporal reasoning about fuzzy intervals. Artif. Intell. **172**, 1158–1193 (2008)
40. Wilbik, A., Kacprzyk, J.: Temporal sequence related protoform in liguistic summarization of time series. In: Proceedings of WConSC, San Francisco, CA, USA (2011)
41. Yager, R.: A new approach to the summarization of data. Inf. Sci. **28**(1), 69–86 (1982)
42. Kacprzyk, J., Strykowski, P.: Linguistic summaries of sales data at a computer retailer: a case study. In: Proceedings of IFSA'99, pp. 29–33 (1999)
43. Kacprzyk, J., Yager, R.: Linguistic summaries of data using fuzzy logic. Int. J. Gen. Syst. **30**(2), 133–154 (2001)
44. Petrovic, D., Xie, Y., Burnham, K.: Fuzzy decision support system for demand forecasting with a learning mechanism. Fuzzy Sets Syst. **157**(12), 1713–1725 (2006)
45. Chen, D., Wang, J., Zou, F., Zhang, H., Hou, W.: Linguistic fuzzy model identification based on pso with different length of particles. Appl. Soft Comput. J. **12**(11), 3390–3400 (2012)
46. Štěpnička, M., Burda, M.: Computational intelligence in forecasting - the results of the time series forecasting competition. In: Proceedings of the 2015 Conference of the International Fuzzy Systems Association and the European Society for Fuzzy Logic and Technology (2015)
47. Afanasieva, T., Yarushkina, N., Toneryan, M., Zavarzin, D., Sapunkov, A., Sibirev, I.: Time series forecasting using fuzzy techniques. In: José, M.R., Alonso, M., Bustince, H. (eds.) Proceedings of the 2015 Conference of the International Fuzzy Systems Association and the European Society for Fuzzy Logic and Technology, pp. 1068–1075 (2015)
48. Perfilieva, I.: Fuzzy transforms: theory and applications. Fuzzy Sets. Syst. **157**, 993–1023 (2006)
49. Perfilieva, I., Yarushkina, N., Afanasieva, T., Romanov, A.: Time series analysis using soft computing methods. Int. J. Gen. Syst. **42**(6), 687–705 (2013)
50. Novak, V., Štěpnička, M., Dvořák, A., Perfilieva, I., Pavliska, V., Vavříčková, L.: Analysis of seasonal time series using fuzzy approach. Int. J. Gen. Syst. **39**, 305–328 (2010)
51. Novak, V., Pavliska, V., Štepnicka, M., Štepnicková, L.: Time series trend extraction and its linguistic evaluation using f-transform and fuzzy natural logic. Recent Developments and New Directions in Soft Computing, pp. 429–442 (2015)
52. Valenzuela, O., Rojas, I., Rojas, F., Pomares, H., Herrera, L., Guillen, A., Marquez, L., Pasadas, M.: Hybridization of intelligent techniques and arima models for time series prediction. Fuzzy Sets Syst. **159**, 821–845 (2008)
53. Chen, S., Chang, Y.: Multi-variable fuzzy forecasting based on fuzzy clustering and fuzzy rule interpolation techniques. Inf. Sci. **180**(24), 4772–4783 (2010)
54. Sharma, S., Chouhan, M.: A review: fuzzy time series model for forecasting. Int. J. Adv. Sci. Technol. (IJAST) **2**, 32–35 (2014)
55. Khashei, M., Bijari, M.: A novel hybridization of artificial neural networks and arima models for time series forecasting. Appl. Soft Comput. J. **11**(2), 2664–2675 (2011)
56. Jain, A., Kumar, A.: Hybrid neural network models for hydrologic time series forecasting. Appl. Soft Comput. J. **7**(2), 585–592 (2007)
57. Tewari, A., Macdonald, M.: Knowledge-based parameter identification of tsk fuzzy models. Appl. Soft Comput. J. **10**(2), 481–489 (2010)
58. Yap, W., Karri, V.: Comparative analysis of artificial neural networks and dynamic models as virtual sensors. Appl. Soft Comput. J. **13**(1), 181–188 (2013)
59. Toro, C., Gómez, M., Gálvez, J., Fdez-Riverola, F.: A hybrid artificial intelligence model for river flow forecasting. Appl. Soft Comput. J. **13**(8), 3449–3458 (2013)
60. Ruan, J., Wang, X., Shi, Y.: Developing fast predictors for large-scale time series using fuzzy granular support vector machines. Appl. Soft Comput. J. **13**(9), 3981–4000 (2013)
61. Höppner, F., Klawonn, F.: Finding informative rules in interval sequences. Intell. Data Anal. **6**(3), 237–256 (2002)
62. Das, G., Lin, K., Mannila, H., Renganathan, G., Smyth, P.: Rule discovery from time series. In: Proceedings of the 4th International Conference on Knowledge Discovery and Data Mining, pp. 16–22 (2008)

63. Froelich, W., Papageorgiou, E., Samarinas, M., Skriapas, K.: Application of evolutionary fuzzy cognitive maps to the long-term prediction of prostate cancer. Appl. Soft Comput. J. **1**, 3810–3817 (2012)
64. Yang, Y., Sun, T., Huo, C., Yu, Y., Liu, C., Tsai, C.: A novel self-constructing radial basis function neural-fuzzy system. Appl. Soft Comput. **13**(5), 2390–2404 (2013)
65. Mahmoud, S., Lotfi, A., Langensiepen, C.: Behavioural pattern identification and prediction in intelligent environments. Appl. Soft Comput. J. **13**(4), 1813–1822 (2013)
66. Bautu, E., Barbulescu, A.: Forecasting meteorological time series using soft computing methods: an empirical study. Appl. Math. Inf. Sci. **7**(4), 1297–1306 (2013)
67. Kacprzyk, J., Wilbik, A., Partyka, A., Ziółkowski, A.: Trend Analysis System. Systems Research Institute, Polish Academy of Sciences, Warsaw (2011)
68. Makridakis, S., Hibon, M.: The m3-competition: results, conclusions and implications. Int. J. Forecast. pp. 451–476 (2000)

An Improved Adaptive Self-Organizing Map

Dominik Olszewski, Janusz Kacprzyk and Sławomir Zadrożny

Abstract We propose a novel adaptive Self-Organizing Map (SOM). In the introduced approach, the SOM neurons' neighborhood widths are computed adaptively using the information about the frequencies of occurrences of input patterns in the input space. The neighborhood widths are determined independently for each neuron in the SOM grid. In this way, the proposed SOM properly visualizes the input data, especially, when there are significant differences in frequencies of occurrences of input patterns. The experimental study on real data, on three different datasets, verifies and confirms the effectiveness of the proposed adaptive SOM.

Keywords Self-organizing map · Adaptive self-organizing map · Neighborhood width · Gaussian kernel · Data visualization

1 Introduction

The Self-Organizing Map (SOM) [15, 16, 26, 30, 33] is an example of the artificial neural network architecture. It was introduced by Kohonen in [18] as a generalization and extension of the concepts proposed in [17]. This approach can be also interpreted as a visualization technique, since the algorithm may perform a projection from multidimensional space to 2-dimensional space, this way creating a map structure. The location of points in 2-dimensional grid aims to reflect the similarities between the corresponding patterns in multidimensional space. Therefore, the SOM algorithm allows for visualization of relationships between patterns in multidimensional space.

D. Olszewski (✉)
Faculty of Electrical Engineering, Warsaw University of Technology, Warsaw, Poland
e-mail: dominik.olszewski@ee.pw.edu.pl

J. Kacprzyk · S. Zadrożny
Systems Research Institute, Polish Academy of Sciences, Warsaw, Poland
e-mail: janusz.kacprzyk@ibspan.waw.pl

S. Zadrożny
e-mail: slawomir.zadrozny@ibspan.waw.pl

© Springer International Publishing Switzerland 2016
G. De Tré et al. (eds.), *Challenging Problems and Solutions
in Intelligent Systems*, Studies in Computational Intelligence 634,
DOI 10.1007/978-3-319-30165-5_5

75

The SOM technique is an unsupervised data analysis approach, i.e., there are no additional training data required. Although the method consists of two substantial phases, i.e., the training phase and the testing phase, both of the phases proceed using the same testing dataset. During the training phase, the weights corresponding to each neuron in the SOM grid are being computed. An important step during this process is updating of the neurons in the neighborhood of the Best Matching Unit (BMU)—the closest neuron to the currently matched input pattern. Usually, the neighborhood of the BMU is selected using the Gaussian kernel (see [15] for other choices of neighborhood functions). However, the choice of the neighborhood function parameters, and the choice of the function itself is always to some extent arbitrary, because there are no strict guidelines, and resulting optimal solutions in this matter. Therefore, any justified proposals regarding the neighborhood width of the BMU are desirable, because that choice strongly affects the quality of the final SOM visualization, and consequently, the performance of the entire analysis.

1.1 Our Proposal

In this paper, we propose a method for the SOM neurons' neighborhood widths adaptive computation. The neighborhood widths are determined independently for each neuron in the SOM grid. The introduced method is based on the measurement of the frequencies of occurrences of patterns in the input space. The Gaussian kernel is employed as the neurons' neighborhood function, and the radius of the Gaussian kernel determining the neurons' neighborhood width is calculated adaptively on the basis of the mentioned frequencies. Therefore, the whole considered SOM is an adaptive enhancement to the traditional approach. In case of input patterns appearing frequently in the input space (or certain groups of highly similar input patterns—for details, see Sect. 4), the corresponding BMU's neighborhood is wider than in case of input patterns occurring rarely in the input space. Consequently, the proposed adaptive SOM reserves larger area for frequent input patterns, and smaller area for rare input patterns. In this way, the novel SOM properly visualizes the input data, especially, when there are significant differences in frequencies of occurrences of input patterns in the input space. As a result, the entire visualization constituting the final result will reflect the input data more accurately.

1.2 Remainder of This Paper

The rest of this paper is organized as follows: in Sect. 2, the appropriate related work is presented as the background for the proposal of this paper, and a theoretical justification of the proposed approach is provided; in Sect. 3, the traditional version of the SOM algorithm is described; in Sect. 4, the main proposal of the paper, i.e., the novel adaptive SOM is introduced; in Sect. 5, our experimental results are reported; while Sect. 6 summarizes the whole paper, provides some concluding remarks, and points out certain directions of future research.

2   Related Work

The SOM visualization technique has been extensively studied, and numerous improvements and extensions have been developed, including the Growing Hierarchical SOM (GHSOM) [31], the asymmetric SOM [21, 24, 27], and the adaptive SOM [3, 5, 14, 22, 32, 34, 35], to name a few. Naturally, the adaptive SOM versions are of particular interest for the purposes of our research.

The paper [5] introduces a conscience mechanism in the self-organization process. The author of [5] does not yet use the term "Self-Organizing Map" and the corresponding "SOM" abbreviation, because it was published before the terms appeared in the literature, i.e., before the Kohonen's [18] publication. Therefore, the author uses the notions of "self-organizing neural network" with "Kohonen learning." The goal of the conscience mechanism is to bring all the processing elements (i.e., all neurons in the grid) available into the solution quickly, and to bias the competition process so that each neuron can win the competition with approximately equal probability. Compared to the work [5], our method goes one step further, and adjusts the winning probability to the frequencies of occurrences of input patterns in the input space, by adapting the neurons' neighborhood widths according to the mentioned frequencies. In this way, the proposed approach aims to assure the accurate mapping and visualizing of the input data by taking into account differences regarding the distribution of input patterns in the input space.

In the paper [22], a statistical iterative Gaussian kernel smoothing problem is considered. The authors propose a batch SOM algorithm consisting of two steps. In the first step, the training data are partitioned according to the Voronoi regions of the map unit locations. In the second step, the units are updated by taking weighted centroids of the data falling into the Voronoi regions, with the weighing function given by the neighborhood. The neighborhood width is decreased in each iteration of the algorithm. The difference between the approach from the work [22] and the method developed in our paper is that in [22], the neighborhood width is being constantly decreased exponentially according to the adaptation rule (3) introduced in [22], while in our work, the neighborhood width is adapted to a given dataset depending on the dataset's specific properties.

The authors of [13] introduce a Self Organizing with Adaptive Neighborhood (SOAN) neural network. The presented technique utilizes an interaction radius between neurons, which varies spatially and temporally, and an adaptive neighborhood function. The concept is closely related to our proposal, however, the main difference is that we utilize the relationship between the number of input patterns (frequency of occurrences of input patterns), and the widths of the neighborhood functions, while the authors of [13] use the correlation between the spatial variety (variance) of neurons in the grid, and the widths of the neighborhood functions. Consequently, in the proposal of the paper [13], only the topography of the SOAN grid is taken into account, when the neighborhood widths are determined, whereas we establish an association between the information derived from the original input space, and the widths of the neighborhood functions on the constructed SOM grid (i.e., in the output space).

The work [8] proposes an Auto-SOM, an algorithm that estimates the learning parameters during the training of SOMs automatically. Auto-SOM consists of a Kalman filter implementation of the SOM coupled with a recursive parameter estimation method. The Kalman filter trains the neurons' weights with estimated learning coefficients so as to minimize the variance of the estimation error. The recursive parameter estimation method estimates the width of the neighborhood function by minimizing the prediction error variance of the Kalman filter. This is another example of an adaptive SOM extension, in which the neighborhood width is one of the estimated parameters. Hence, there are noticeable similarities between the method from [8], and the method proposed in our paper. However, the significant difference is that in Auto-SOM, the prediction error variance of the Kalman filter must be computed in order to adapt the neighborhood widths. In case of our method, the sole information about the numbers of input patterns is sufficient to effectively adjust the widths of the neighborhood functions. Consequently, the solution proposed in our paper is significantly simpler conceptually, and therefore, it is also less computationally complex, and naturally, more efficient.

Further, in the paper [32], an Adaptive Double SOM (ADSOM) is introduced. The constructed map is designed for subsequent clustering analysis without requiring of a priori knowledge about the number of clusters. ADSOM updates its free parameters and allows convergence of its position vectors to a fairly consistent number of clusters provided its initial number of nodes is greater than the expected number of clusters.

The paper [34] proposes a Time Adaptive SOM (TASOM). The work, along with the papers [13, 22], and [8], is especially important in the context of our research, because it also introduces a method for neurons neighborhood width adaptive computation. In the approach proposed in [34], every neuron has its own learning rate and neighborhood width. The difference between the solution from [34] and our method is the following. In [34], the adaptation of the neighborhood width results from the "closed-loop" learning of the parameter, i.e., the neighborhood width is updated on the basis of the final quality of visualization (so as to minimize an appropriate error function). Consequently, a learning process is a necessary stage of that analysis. On the other hand, in case of our approach, the neighborhood width is computed in the "open-loop" system, only on the basis of the input dataset analysis (i.e., measurement of frequencies of occurrences of input patterns). No learning process is required, and the method does not rely on the final results of the visualization. Consequently, no additional error function is necessary.

The SOM adaptation process may also concern the number of neurons in the SOM lattice and the generated topological connections among neurons. Such an adaptive SOM version is introduced in the paper [3]. The authors of [3] propose a new self-organizing model with growing mechanism called Diffusion and Growing

Self-Organizing Map (DGSOM). The DGSOM model adds neurons through competition mechanism, updates the topology of network using a Competitive Hebbian Learning (CHL) fashion, and uses diffusion mechanism of Nitric Oxide (NO) as global coordinator of self-organizing learning. Although, the DGSOM model incorporates an adaptation SOM behavior, conceptually, it is not similar to our approach, because in our research, we focus on the neurons' neighborhood widths in SOM, as opposed to [3], where the number of neurons and the topological connections among them are studied.

Another way of gaining a control over the neurons' neighborhood widths in SOM is the magnification control approach. The issue is thoroughly studied in the work [39], where the three learning rule modifications for SOM are considered, namely, the localized learning, the winner-relaxing learning, and the concave–convex learning. The one closest to our research is the localized learning modification leading to inserting the local learning step size in the SOM weights update formula, in this way, affecting the neurons' neighborhood widths. The local learning step size depends on the stimulus density of the weight vectors (prototypes) of SOM. As it is noticed in [39], a major drawback of the approach is that one has to estimate the generally unknown data distribution corresponding to the mentioned stimulus density, which may lead to numerical instabilities of the control mechanism [12, 39]. Such a drawback does not concern the proposal of the present paper, because in our method, there is no necessity of any data distribution estimation. The second important difference between our technique and the localized learning is that our frequencies measurements refer to the input patterns in the SOM input space, whereas in case of the localized learning, the local learning step size is determined on the basis of the stimulus density of the weight vectors of SOM, i.e., on the basis of the intrinsic SOM structural information.

A different learning strategy is utilized in [1], where the authors propose a Parameter-Less Self-Organizing Map (PLSOM). The method eliminates the need for a learning rate and annealing schemes for learning rate and neighborhood width. Therefore, the entire approach essentially differs from the classical adaptive versions of SOM.

In the work [35], an adaptive hierarchical structure called "Binary Tree TASOM" (BTASOM) is introduced. The considered SOM enhancement resembles a binary natural tree having nodes composed of TASOM networks. The BTASOM is proposed to make TASOM fast and adaptive in the number of its neurons.

Finally, in the paper [14], an adaptive GHSOM-based approach (A-GHSOM) is introduced as an effective technique to deal with the anomaly detection problem. As the authors claim, their GHSOM enhancement can adapt on-line to the ever-changing anomaly detection. Consequently, according to the authors, A-GHSOM is superior over the standard GHSOM-based methods, and it provides higher accuracy in identifying intrusions, particularly "unknown" attacks.

2.1 Justification of the Proposed Approach

The idea behind the proposed approach is to provide a mechanism for accurate SOM visualization by taking into account the information about frequencies of occurrences of input patterns. The goal of the traditional SOM technique is to construct a 2-dimensional map structure reflecting all the data phenomena, properties, and relationships among the input data. Consequently, the input patterns appearing more frequently in the input space (or certain groups of highly similar input patterns) should occur more frequently also on the SOM grid so as to satisfy the natural and principal requirement of the SOM technique, i.e., the correct input data visualization. On the other hand, the rare input patterns should be treated in the opposite way. To achieve this goal, we propose an intuitive, straightforward, and computationally efficient mechanism of controlling the neurons' neighborhood widths via the input patterns' frequencies information. In this simple way, the frequent input patterns will receive a wider area on the SOM grid, in contrast to the rare input patterns being allocated in the small areas on the SOM lattice. The approach proposed in this paper supports the SOM method in preserving and maintaining the input patterns' frequencies in the generated grid, and properly visualizing input patterns of different frequencies of occurrences in the input space.

Taking into consideration the biological perspective of human's cognition ability, our proposal aims to follow the human's skill of distinguishing the significance of various external stimuli on the basis of their frequencies of occurrences. Dominating external stimuli and phenomena gain higher significance and importance during the knowledge self-organization process, and strongly affect the human's environment perception. The mechanism of the frequency distinction in the input data follows the human's ability to successfully acquire the diverse external information.

Notice that the proposed adaptive method does not use any feedback information in a "closed loop" system, and consequently, no error computation is necessary in order to determine the neurons' neighborhood widths. Also, there is no need of any additional data distribution estimation. As a result, our approach is not flawed by the aforementioned drawbacks and constraints, which makes it robust, mathematically simple, and consequently, computationally efficient. Furthermore, the proposed solution is not endangered by numerical instability as it is in case of the localized learning discussed in the work [39].

3 Traditional Self-Organizing Map

The SOM algorithm provides a non-linear mapping between a high-dimensional original data space and a 2-dimensional map of neurons. The neurons are arranged according to a regular grid, in such a way that the similar vectors in input space are represented by the neurons close in the grid. Therefore, the SOM technique visualizes the data associations in the high-dimensional input space.

Beside the classical algorithmic description of the SOM method, which is well-known in the existing literature (see, for example, [15]), an additional mathematical scaffolding has been provided in [21]. The authors of [21] utilized a specific error function reflecting the behavior of SOM in a purely analytical manner. In this way, a formalized optimization-framework-based description has been created, and it may be found beneficial and more convenient for certain goals of the SOM analysis.

It turns out that in case of our research, the error-function-based SOM description is of higher value in terms of the derivation of our proposal, and in terms of the clear and comprehensive explanation and justification of the novelty, introduced in this paper.

According to [21], the results obtained by the SOM method are equivalent to the results obtained by minimizing the following error function with respect to the prototypes w_r and w_s:

$$
\begin{aligned}
e &= \sum_r \sum_{x_i \in V_r} \sum_s h_{rs} d_{\text{Euc}}^2 (x_i, \ w_s) \\
&\approx \sum_r \sum_{x_i \in V_r} d_{\text{Euc}}^2 (x_i, \ w_r) \\
&\quad + K \sum_r \sum_{s \neq r} h_{rs} d_{\text{Euc}}^2 (w_r, \ w_s),
\end{aligned}
\tag{1}
$$

where x_i, $i = 1, \ldots, n$ is the ith input pattern in high-dimensional input space, n is the total number of input patterns; w_r, $r = 1, \ldots, M$ and w_s, $s = 1, \ldots, M$ are the prototypes of input patterns in the grid (the different indices r and s are used in order to compute the sum of distances between neurons within the SOM grid, including the values of the function h_{rs}); M is the total number of prototypes/neurons in the grid; h_{rs} is a neighborhood function (for example, the Gaussian kernel) that transforms non-linearly the neuron distances (see [15] for other choices of neighborhood functions); $d_{\text{Euc}} (\cdot, \ \cdot)$ is the Euclidean distance; and V_r is the Voronoi region corresponding to prototype w_r. The number of prototypes is assumed to be sufficiently large so that $d_{\text{Euc}}^2 (x_i, \ w_s) \approx d_{\text{Euc}}^2 (x_i, \ w_r) + d_{\text{Euc}}^2 (w_r, \ w_s)$.

According to (1), the SOM error function can be decomposed as the sum of the quantization error and the topological error. The first one minimizes the loss of information, when the input patterns are represented by a set of prototypes. By minimizing the second one, we assure the maximal correlation between the prototype dissimilarities and the corresponding neuron distances, this way assuring a proper visualization of the data relationships in the input space.

The neighborhood function h_{rs} is of particular importance in our study, because it determines the neighborhood of neurons in the SOM grid, i.e., the area around BMU, where the neurons' weights are updated together with BMU itself. Therefore, the neighborhood function h_{rs} may be used as a tool allowing for manipulating the neurons' neighborhood widths, which is the main purpose of our research.

The SOM version considered in our study is the one using the batch training algorithm.

The SOM error function can be optimized by an iterative algorithm consisting of two steps (discussed in [11]). First, a quantization algorithm is executed. This algorithm represents each input pattern by the nearest neighbor prototype. This operation minimizes the first component in (1). Next, the prototypes are arranged along the grid of neurons by minimizing the second component in the error function. This optimization problem can be solved explicitly using the following adaptation rule for each prototype [15]:

$$w_s = \frac{\sum_{r=1}^{M} \sum_{x_\mu \in V_r} h_{rs} x_\mu}{\sum_{r=1}^{M} \sum_{x_\mu \in V_r} h_{rs}}, \tag{2}$$

where M is the number of neurons, and h_{rs} is a neighborhood function (for example, the Gaussian kernel of width $\sigma(t)$). The width of the kernel is adapted in each iteration of the algorithm using the rule proposed by [22], i.e.,

$$\sigma(t) = \sigma_m \left(\sigma_f / \sigma_m\right)^{t/N_{\text{iter}}}, \tag{3}$$

where $\sigma_m \approx M/2$ is typically assumed in the literature (for example, in [15]), σ_f is the parameter that determines the smoothing degree of the principal curve generated by the SOM algorithm [22], and N_{iter} is the total number of iterations during the SOM training phase.

4 A Novel Adaptive Self-Organizing Map

In this paper, we propose a novel adaptation rule of the SOM neurons' neighborhood widths. The neighborhood widths are determined independently for each neuron in the SOM grid. The proposed rule employs the exponential update (3) from the work [22], includes the information about the frequencies of occurrences of all input patterns, and consequently, provides a more accurate and effective adaptation process than the rule (3) itself.

In case of input patterns represented by a large number of features, which might occur, for example, in case of the time series analysis (see our experimental research in Sects. 5.8 and 5.9), there is a very low probability that a certain input pattern will entirely (all features) appear in the input space more than once. However, certain input patterns may be very similar, and from the point of view of the correct mapping onto SOM (meant, in this case, as the correct adjustment of the neighborhood widths), and the proper representation in the SOM grid, these patterns should be treated as the same, because most likely, they will match the same BMU in the SOM grid. Such a goal might be achieved by establishing a small area in the input space, which will be gathering the input patterns of sufficiently high similarity to each other (determined by a given tolerance threshold).

We propose to accomplish this, by using a function returning the normalized number of input patterns, which are not farther (in the sense of the Euclidean distance) than a settled threshold from a pattern given as the argument to the function. The term "normalized" in this case means that the calculated number of input patterns will be divided by the total number of patterns in the input space. The function $\varphi(\cdot)$ is defined in the following way:

$$\varphi(x_i) = \left|\left\{x_j : d_{\text{Euc}}(x_i, x_j) \leq \tau\right\}\right|, \tag{4}$$

where $x_i, i = 1, \ldots, n$, is the ith input pattern represented by a vector of features, τ is the arbitrarily settled tolerance threshold, n is the total number of input patterns, $|\cdot|$ is the number of occurrences (in the input space) of an input pattern given as the argument, and $d_{\text{Euc}}(\cdot, \cdot)$ is the Euclidean distance.

The function $\varphi(\cdot)$ assumes values in the interval $\langle 0, 1 \rangle$, which is a consequence of the beforehand mentioned normalization.

Using the function $\varphi(\cdot)$, we may proceed to determination of the neighborhood widths.

The SOM neurons' neighborhood widths are adapted in our research using the Gaussian kernels of the following radius:

$$\sigma(x_i, t) = (1 + \varphi(x_i)) \sigma_m \left(\sigma_f / \sigma_m\right)^{t/N_{\text{iter}}}, \tag{5}$$

where $\sigma(\cdot, \cdot)$ is the function returning the radius of the Gaussian kernel around the BMU corresponding to a given input pattern in analyzed dataset, $x_i, i = 1, \ldots, n$, is the ith input pattern represented by an appropriate vector of features, n is the total number of input patterns, and the rest of the notation is explained in (3).

In (5), the values of the function $\varphi(\cdot)$ are not used directly, but instead, they are increased by one, which is necessary, in order to assure the proper association between the standard adaptation rule (3), and the rule introduced in our paper. The coefficient function $(1 + \varphi(x_i))$ returns values from the interval $\langle 1, 2 \rangle$, i.e., the value of 1 (or values close to 1) for the input patterns occurring rarely in the input space, and the value of 2 (or values close to 2) for the input patterns occurring in the input space frequently. In this way, it increases the kernel widths computed according to the adaptation rule (3) for the frequent input patterns, while for the rare input patterns, it changes the kernel widths calculated according to (3) very slightly, or it may even leave them unchanged. It should be noted that by inserting into (3) the function $\varphi(\cdot)$ only, we obtain too narrow Gaussian kernels. On the other hand, by applying the coefficient function $(1 + \varphi(x_i))$, we obtain the desirable adaptation effect, and the rule (5), proposed in this way in this paper, can be regarded as an enhancement to the classical SOM adaptation approach (3).

By utilizing the information about the frequencies of occurrences of input patterns, our method exploits the specific nature and character of a given dataset, and this way, it visualizes the dataset in the SOM grid more accurately by better adjusting to the dataset features and properties.

If the Gaussian kernels specifying the SOM neurons' neighborhood width are fitted to the frequencies of occurrences of input patterns, then the resulting SOM will assign the wider neighborhoods (i.e., the larger area in the SOM grid) to the neurons corresponding to the input patterns appearing more frequently in the input space, and likewise, the obtained SOM will assign the narrower neighborhoods (i.e., the smaller area in the SOM grid) to the neurons corresponding to the input patterns appearing less frequently in the input space.

The desirable consequence of this phenomenon is that the proposed improved adaptive SOM is dataset-dependent, and therefore, it reflects properly the relationships between input patterns, especially if the input dataset is highly diverse with respect to the input patterns' frequencies of occurrences.

The SOM enhancement proposed in this paper refers to the batch version of the discussed technique.

4.1 Computational Complexity

The estimated computational complexity of the proposed adaptive SOM extension comes down to the cost of computing the values of the function $\varphi(\cdot)$. As it can be easily derived from (4), the complexity of the function $\varphi(\cdot)$ is quadratic with respect to the number of input patterns, i.e., it can be denoted as $\mathcal{O}(n^2)$, where n is the number of input patterns. Notice that the values of the function $\varphi(\cdot)$ need to be computed only once, and the computation does not have to be repeated during the SOM training.

As it is shown in [37], a rough lower bound estimate of the amount of memory used by batch-trained SOM, implemented in the Matlab programming language, is given by: $8\left(5(M+n)d + 3M^2\right)$ bytes, where M is the number of neurons in the SOM lattice, and d is the number of dimensions in the input space. The majority of the memory use comes from the last term of the memory complexity formula. Therefore, the entire memory complexity of the classical SOM (regardless of the implementation), can be estimated as $\mathcal{O}(M^2)$.

The increase of the memory complexity of SOM associated with our extension is n only, consequently, the estimated memory complexity of our method is given by the same formula as in case of the standard technique, i.e., $\mathcal{O}(M^2)$.

On the other hand, the time complexity of the traditional SOM is linear with respect to both variables: M and n, i.e., $\mathcal{O}(Mn)$.

In case of the proposed SOM improvement, the application of the function $\varphi(\cdot)$ causes the increase in the time complexity regarding the quadratic dependency on n, i.e., $Mn + n^2$. However, notice that the term n^2 is added (not multiplied) to the term Mn. That is because the computation of the values of the function $\varphi(\cdot)$ is executed only once, and it does not slow down the SOM training process. Taking into account the property of the "big $\mathcal{O}$ notation," which hides constant factors and smaller terms, the resulting estimated time complexity of our entire approach can be described in the following way: $\mathcal{O}(n^2)$.

5   Experiments

In our experimental study, we have evaluated effectiveness of the proposed improved adaptive SOM technique on the basis of the comparison with 11 reference methods. The comparison has been performed by conducting the clustering process in the SOM grid obtained using the proposed approach, and the clustering process in the output spaces of 9 reference visualization methods (including 7 different adaptive SOMs and 2 different visualization concepts cooperating with 2 modern clustering approaches).

5.1   Reference Methods

As the reference adaptive SOMs, we have used TASOM from [34], Auto-SOM from [8], SOAN from [13], batch SOM from [22], ADSOM from [32], BTA-SOM from [35], and A-GHSOM from [14] techniques. As the other 2 visualization competitors, we have chosen the recent developments, i.e., t-distributed Stochastic Neighbor Embedding (t-SNE) technique [20] and the Neighbor Retrieval Visualizer (NeRV) method [36]. As the clustering methods, we have employed the standard well-known k-means clustering algorithm, and the two modern clustering techniques, i.e., topic-model-based clustering and Non-negative Matrix Factorization (NMF) [19, 29]. The topic-model-based clustering has been implemented using the Bayesian generative probabilistic model—Latent Dirichlet Allocation (LDA) proposed in [2]. In case of all the investigated clustering methods, the correct number of clusters has been provided a priori as the input data. Clustering process has been carried out in the space of the SOM grid, i.e., prototypes in the SOM grid have been the subject of clustering. In case of the adaptive SOM approaches, the standard k-means clustering algorithm has been used, whereas in case of the t-SNE and NeRV visualization techniques, the LDA-based and NMF clustering approaches have been utilized. The reason for this choice is that we want to test our method competing not only with other adaptive SOMs, but also with different modern visualization and clustering techniques combinations, i.e., t-SNE and LDA, t-SNE and NMF, NeRV and LDA, and NeRV and NMF. As a result, we test our approach versus 11 reference methods.

5.2   Datasets Overview

The experimental research aims to ascertain the superiority of the introduced adaptive SOM on the basis of the comparison of the clustering results obtained using the proposed SOM and the reference methods. The experiments have been conducted on real data in the three different research fields: in the field of words clustering, in the field of sound signals clustering, and in the field of human heart rhythm signals clustering. The first part of the experimental study has been carried out on the large dataset of high-dimensionality (Sect. 5.7), while the remaining two experimental parts have been conducted on smaller datasets, but also of high-dimensionality (Sects. 5.8 and 5.9). In this way, one can assess the performance of the investigated methods operating on datasets of different size and nature, and consequently, one can better evaluate the effectiveness of the proposed approach.

The words visualization and clustering experiment was conducted on the "Bag of Words" dataset from the UCI Machine Learning Repository [6].

The sound signals visualization and clustering was carried out on the piano music recordings, and the human heart rhythm signals analysis was conducted using the ECG recordings derived from the MIT-BIH ECG Databases [7].

In case of the piano music dataset and the ECG recordings dataset, a graphical illustration of the U-matrices generated in a single run of two chosen SOM methods and of the t-SNE and NeRV techniques is provided, while in case of the "Bag of Words" dataset no such illustration is given, because of the high number of instances in that dataset, which would make such images unreadable.

5.3   Evaluation Criteria

In case of all three parts of our experiments, we have compared the clustering results obtained using the investigated methods. As the basis of the comparisons, i.e., as the clustering evaluation criteria, we have used the total accuracy rate [24, 28], and the uncertainty degree [21, 24].

In general, the results of clustering can be assessed with use of two groups of evaluation criteria [9] (also called as the validity indices). First group are the external criteria, which are computed by using the ground knowledge about the clustered data, i.e., what should be the correct result of clustering. These criteria are much easier to formulate, and they allow for the precise assessment of the clustering results, however, they are useless in real-world clustering problems, where no additional information about the analyzed data is available. Second group are the internal criteria, which are computed without using the ground knowledge about the clustered data. Formulation of these criteria is, naturally, more difficult, however, they can be employed to assess the results of clustering in real-life problems. Therefore, their usefulness in data analysis is of much value. More information on the issue of clustering output assessment can be found, for example, in [9, 10].

In case of all three parts of our empirical study, the ground knowledge about the data was known. Therefore, an application of external evaluation criterion (the total accuracy rate) was possible. In order to provide more reliable assessment of the clustering results, we have used also the second—internal evaluation criterion (the uncertainty degree).

Hence, the following two evaluation criteria have been used:

1. **Total accuracy rate**. This evaluation criterion determines the number of correctly assigned patterns divided by the total number of patterns.

   Hence, for the ith cluster, the cluster accuracy rate is determined as follows:

   $$q_i = \frac{m_i}{n_i}, \tag{6}$$

   where m_i, $i = 1, \ldots, k$, is the number of patterns correctly assigned to the ith cluster, n_i, $i = 1, \ldots, k$, is the number of patterns in the ith cluster, and k is the number of clusters.

   And, for the entire dataset, the total accuracy rate is determined as follows:

   $$q_{\text{total}} = \frac{m}{n}, \tag{7}$$

   where m is the total number of correctly assigned patterns, and n is the total number of patterns in the entire dataset.

   The cluster accuracy rates q_i and the total accuracy rate q_{total} assume values in the interval $\langle 0, 1 \rangle$, and naturally, greater values are preferred.

   The total accuracy rate q_{total} was used in our experimental study as the main basis of the clustering accuracy comparison of all the investigated approaches.

2. **Uncertainty degree**. This evaluation criterion is formulated on the basis of the number of overlapping patterns divided by the total number of patterns in a dataset. This means, the number of patterns, which are in the overlapping area between clusters, divided by the total number of patterns. A pattern belonging to the overlapping area is determined on the basis of the ratio of the Euclidean distances between the pattern and the two nearest clusters' centroids. If this ratio is in the interval $\langle 0.9, 1.1 \rangle$, then the corresponding pattern is said to be in the overlapping area. In other words, the patterns in the overlapping area are more likely to be assigned to incorrect clusters than other patterns, and therefore, their number should be minimized. One can also say that the uncertainty degree determines the uncertainty of the assignments of patterns to correct clusters.

   The uncertainty degree is determined as follows:

   $$U_d = \frac{\mu}{n}, \tag{8}$$

   where μ is the number of overlapping patterns in the dataset, and n is the total number of patterns in the dataset.

   The uncertainty degree assumes values in the interval $\langle 0, 1 \rangle$, and, smaller values are desired.

5.4 Statistical Significance

In case of all comparisons between the proposed approach and the reference methods, the statistical significance has been verified on the basis of the statistical Student's t-test, this way confirming that the difference in the results produced by a pair of evaluated approaches is statistically significant. The p-values calculated in case of each comparison are reported in Tables 2, 3, and 4. Each p-value corresponding to a given comparison method should be referenced to the proposed method. The p-values computed in case of each comparison are lower than the significance level $\alpha = 0.001$, which indicates the high statistical significance of the obtained empirical results. Therefore, the Student's t-test confirmed the reliability and high statistical significance of the conducted experimental research.

5.5 Time Series Feature Extraction

Feature discovery process preceding the actual visualization and clustering is an important stage of data pre-processing. It has a strong impact on the final accuracy of clustering, and consequently, on the performance of the whole analysis. Feature discovery aims to form possibly smallest set of most relevant, informative, and discriminative features. A proper choice of the feature set results in higher visualization and clustering quality.

Features of the time series considered in Sects. 5.8 and 5.9 have been extracted using a method based on the Discrete Fourier Transform (DFT), which is described in details in [25]. In general, the DFT-based feature extraction is described, for example, in [4].

5.6 Parameters Tuning

In case of all the 11 reference methods, all their parameters have been tuned according to the suggestions and recommendations (oriented on the problem of visualization and clustering) from the papers introducing them, if such recommendations have been provided. Otherwise, the exact values applied in the experimental researches of the papers proposing the reference methods have been used in our experimental study. We decided to use the these values as the universal ones recommended by the inventors of the methods, and therefore, the reliable and convincing choices providing the high quality results and high performance analysis.

On the other hand, in case of the approach proposed in the present paper, there were four parameters involved in the method, namely, M—the number of neurons in the SOM lattice computed according to the recommendation from [38], i.e., $M \approx 5\sqrt{n}$, where n is the number of input patterns (generally, the size and shape of the SOM grid

Table 1 The values of the parameters of our method corresponding to the investigated datasets

Parameter	Words	Music	ECG
M	16,500 (110×150)	221 (17×13)	189 (27×7)
τ	0.064	0.066	0.07
σ_m	8250	110.5	94.5
σ_f	$3.2 * M = 52,800$	$2.3 * M = 508.3$	$2.4 * M = 453.6$

Fig. 1 The function of the total accuracy rate versus the parameter τ for the "Bag of Words" dataset

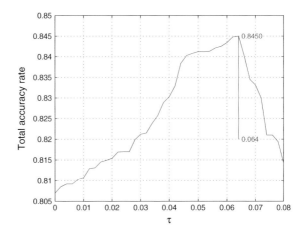

should fit the input data manifold, therefore the value of M may sometimes differ from $5\sqrt{n}$); τ—the tolerance threshold in (4), in the function $\varphi(\cdot)$; $\sigma_m = M/2$; and σ_f—the parameter that determines the smoothing degree of the principal curve generated by the SOM algorithm [22]. The parameters τ and σ_f were chosen empirically for each of the investigated datasets. The values of τ were changed in the range from 0.0 to 0.08 with the step 0.002, whereas the values of σ_f were changed in the range from 2.0 to $4.0M$ with the step $0.1M$. The values of the parameters τ and σ_f were chosen so as to maximize the total accuracy rate—the primary evaluation criterion. As a result, the values were obtained for each of the examined datasets are gathered in Table 1.

The respective graphs illustrating the function of the total accuracy rate versus the values of τ are depicted in Figs. 1, 6, and 11, while the graphs illustrating the function of the total accuracy rate versus the values of σ_f are depicted in Figs. 2, 7, and 12.

5.7 Words Visualization and Clustering

In the first part of our experimental study, we have evaluated the investigated methods on the basis of the clustering accuracy of the visualized words. All the methods have been forming five clusters (the correct number of clusters has been provided to the methods as the input data). The total accuracy rate and uncertainty degree have been used as the evaluation criteria.

We have utilized excerpts from the "Bag of Words" dataset from the UCI Machine Learning Repository [6]. It is a high-dimensional dataset of large volume, especially useful in case of our research. It is so, because of the significant differences in frequencies of occurrences of different words in the entire dataset. Therefore, the experimental investigation on the "Bag of Words" dataset clearly shows the superiority of the proposed adaptive approach over the other examined techniques.

Dataset Description The "Bag of Words" dataset consists of five text collections:

- **Enron E-mail Collection**. This collection was prepared by the CALO Project (A Cognitive Assistant that Learns and Organizes). It contains data from about 150 users, mostly senior management of Enron, organized into folders. The number of documents in this collection is 39,861, the number of words in the vocabulary is 28,102, and the total number of words is approximately 6,400,000.
- **Neural Information Processing Systems (NIPS) full papers**. The number of documents in this collection is 1500, the number of words in the vocabulary is 12,419, and the total number of words is approximately 1,900,000.
- **Daily KOS Blog Entries**. The number of documents in this collection is 3430, the number of words in the vocabulary is 6906, and the total number of words is approximately 468,000.
- **New York Times News Articles**. The number of documents in this collection is 300,000, the number of words in the vocabulary is 102,660, and the total number of words is approximately 100,000,000. We have utilized only an excerpt from this collection, i.e., 3000 documents, 11,203 words in the vocabulary, and approximately 2,000,000 of words in total.
- **PubMed Abstracts**. This is the collections of abstracts of the U.S. National Library of Medicine, National Institute of Health. The number of documents in this collection is 8,200,000, the number of words in the vocabulary is 141,043, and the total number of words is approximately 730,000,000. We have utilized only an excerpt from this collection, i.e., 1000 documents, 8520 words in the vocabulary, and approximately 100,000 of words in total.

The total number of analyzed words was approximately 10868,000.

On the visualizations generated by the investigated methods, five clusters representing those five text collections in the "Bag of Words" dataset were formed.

Text Keywords Extraction Keywords extraction of the textual data investigated in this part of our experimental study was carried out using the term frequency— inverse document frequency (tf-idf) approach. The Vector Space Model (VSM) constructed in this way is particularly useful in our research, because it implicitly captures the terms frequency (both: local—document-dependent and global— collection-dependent).

In order to obtain a unified, compact, and cohesive data model, we have utilized the common length of all vocabularies in the five considered text collections. As the basis length, we have chosen the length of the shortest vocabulary, i.e., the vocabulary

Table 2 Accuracy rates, standard deviations, p-values, and uncertainty degrees for the words clustering

Investigated method	q_{total}	s	p-value	U_d
TASOM	0.7719	2592.2	$< 10^{-4}$	0.2120
Auto-SOM	0.7479	3005.4	$< 10^{-4}$	0.2298
SOAN	0.7360	2767.2	$< 10^{-4}$	0.3301
Batch SOM	0.7416	2946.0	$< 10^{-4}$	0.1845
ADSOM	0.7285	3094.9	$< 10^{-4}$	0.2038
BTASOM	0.7798	3048.9	$< 10^{-4}$	0.1812
A-GHSOM	0.7381	2629.3	$< 10^{-4}$	0.2073
t-SNE and LDA	0.6955	2826.7	$< 10^{-4}$	0.2830
t-SNE and NMF	0.6929	2906.9	$< 10^{-4}$	0.2759
NeRV and LDA	0.6903	2446.9	$< 10^{-4}$	0.2578
NeRV and NMF	0.6894	2548.4	$< 10^{-4}$	0.3047
Proposed adaptive SOM	0.8450	2738.9	–	0.1402

of the Daily KOS Blog Entries. Consequently, the number of keywords utilized in this part of our experimental study was 6906. It was necessary to truncate the longer vocabularies in order to build the data matrix constituting the analyzed VSM model. The truncation has been executed in the lexicographic manner, because all the keywords in the five vocabularies have been arranged in alphabetical order. As a result, not all of the keywords in the remaining four text collections have been taken into account. Nevertheless, the considered experimental problem remains a high-dimensionality issue, and the number and variety of the keywords in the analyzed vocabularies makes the problem complex and challenging.

Experimental Results The results of this part of our experiments are reported in Table 2, where the total accuracy rates, standard deviations, p-values (of the statistical t-test), and uncertainty degrees corresponding to each investigated approach are presented.

Each of the examined methods was run 50 times, because all the methods are non-deterministic, and by repeating their runs, we obtain results, which may be considered as more reliable. The randomness of the methods lies in both stages of our data analysis, i.e., in visualization and in clustering.

Figure 1 illustrates the relationship between the total accuracy rate and the values of the parameter τ. The values of τ belong to the range from 0.0 to 0.08 (with the step 0.002).

Figure 2 shows the function of the total accuracy rate versus the values of the parameter σ_f. The values of σ_f belong to the range from 2.0 to 4.0M (with the step 0.1M).

The results of this part of our experimental study show that clustering of the SOM grid obtained using the introduced adaptive method outperforms clustering of the 2D

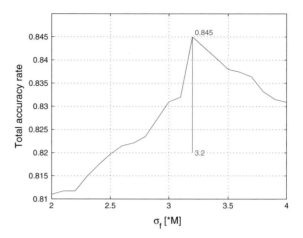

Fig. 2 The function of the total accuracy rate versus the parameter σ_f for the "Bag of Words" dataset

visualizations returned by the other investigated approaches. The proposed approach leads to the higher clustering quality measured on the basis of the total accuracy rate, and also to the lower clustering uncertainty measured on the basis of the uncertainty degree.

5.8 Piano Music Composer Visualization and Clustering

In this part of our experiments, we have evaluated the investigated methods on the basis of the clustering quality of the visualized music pieces. All the methods have been forming three clusters (the correct number of clusters has been provided to the methods as the input data) representing three piano music composers: Johann Sebastian Bach, Ludwig van Beethoven, and Fryderyk Chopin. The total accuracy rate and uncertainty degree have been used as the evaluation criteria.

Dataset Description Each music piece was represented by a 30 s sound signal sampled with the 44,100 Hz frequency. The entire dataset consisted of 160 sound signals. Some of the sound signals corresponded to the same music piece played by different pianists. In this way, we were able to increase the number of instances in this dataset. Feature extraction process was carried out according to the Discrete-Fourier-Transform-based (DFT-based) method, as it is described in Sect. 5.5.

Experimental Results The results of this part of our experiments are demonstrated in Fig. 3 and in Table 3. Figure 3 presents the maps (U-matrices) generated in a single run of the TASOM method (Fig. 3a) and of the proposed adaptive method (Fig. 3b). The U-matrix is a graphical presentation of SOM. Each entry of the U-matrix corresponds to a neuron in the SOM grid, while value of that entry is the average distance between the neuron and its neighbors. We provide the images of the U-matrices of the two chosen methods (not all investigated adaptive SOM versions), because this is only

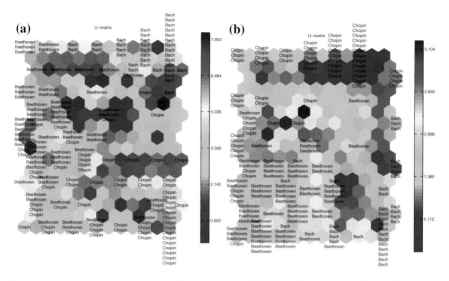

Fig. 3 Piano music composers maps (U-matrices). **a** TASOM. **b** Proposed adaptive SOM

Table 3 Accuracy rates, standard deviations, p-values, and uncertainty degrees for the piano music composer clustering

Investigated method	q_{total}	s	p-value	U_d
TASOM	0.8563	0.6987	$< 10^{-4}$	0.0688
Auto-SOM	0.8750	0.7730	$< 10^{-4}$	0.0938
SOAN	0.7625	0.7140	$< 10^{-4}$	0.0625
Batch SOM	0.8375	0.6652	$< 10^{-4}$	0.0750
ADSOM	0.7688	0.7273	$< 10^{-4}$	0.0563
BTASOM	0.8813	0.6047	$< 10^{-4}$	0.1063
A-GHSOM	0.8375	0.7354	$< 10^{-4}$	0.0563
t-SNE and LDA	0.7563	0.7384	$< 10^{-4}$	0.2375
t-SNE and NMF	0.7438	0.7117	$< 10^{-4}$	0.2500
NeRV and LDA	0.7000	0.6704	$< 10^{-4}$	0.3375
NeRV and NMF	0.6438	0.6713	$< 10^{-4}$	0.3563
Proposed adaptive SOM	0.9125	0.7656	–	0.0313

a graphical insight into a method, and the most important results allowing for the comparison and evaluation of the tested approaches are given in Table 3.

Table 3, in turn, presents the accuracy rates, standard deviations, p-values of the statistical t-test, and the uncertainty degrees corresponding to each of the examined approaches. Each of the examined methods was run 50 times.

Fig. 4 The t-SNE visualization of the music dataset

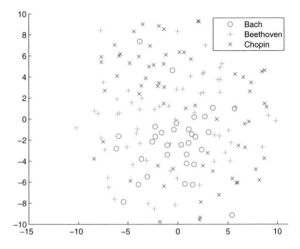

Fig. 5 The NeRV visualization of the music dataset

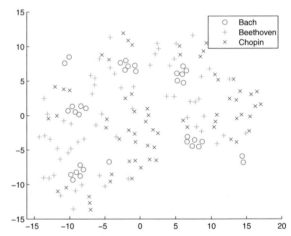

The visualizations provided by the two conceptually different than SOM dimensionality reduction methods, i.e., t-SNE and NeRV (single run of each of them), are depicted in Figs. 4 and 5, respectively.

Figure 6 illustrates the relationship between the total accuracy rate and the values of the parameter τ. The values of τ belong to the range from 0.0 to 0.08 (with the step 0.002).

Figure 7 shows the function of the total accuracy rate versus the values of the parameter σ_f. The values of σ_f belong to the range from 2.0 to 4.0M (with the step 0.1M).

Also in this part of our experiments, the proposal of this paper appeared to be superior over the other examined visualization and clustering techniques.

Fig. 6 The function of the total accuracy rate versus the parameter τ for the music dataset

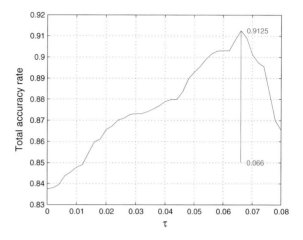

Fig. 7 The function of the total accuracy rate versus the parameter σ_f for the music dataset

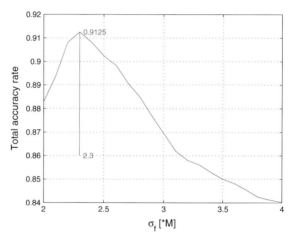

5.9 Human Heart Rhythms Visualization and Clustering

The human heart rhythm signals visualization and clustering experiment was carried out on the dataset of ECG recordings derived from the MIT-BIH ECG Databases [7].

In this part of our experiments, we have evaluated the investigated methods on the basis of the clustering quality of the visualized human heart rhythm (ECG) recordings. All the methods have been forming three clusters (the correct number of clusters has been provided to the methods as the input data) representing three types of human heart rhythms: normal sinus rhythm, atrial arrhythmia, and ventricular arrhythmia. This kind of clustering can be interpreted as the cardiac arrhythmia detection and recognition based on the ECG recordings. The total accuracy rate and uncertainty degree have been used as the evaluation criteria.

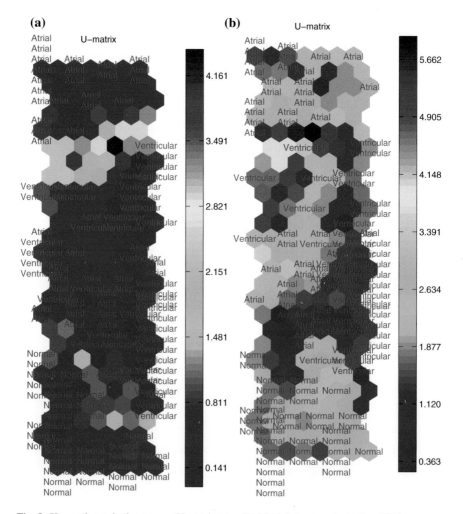

Fig. 8 Human heart rhythms maps (U-matrices). **a** TASOM. **b** Proposed adaptive SOM

Dataset Description In general, the cardiac arrhythmia disease may be classified either by rate (tachycardias—the heart beat is too fast, and bradycardias—the heart beat is too slow) or by site of origin (atrial arrhythmias—they begin in the atria, and ventricular arrhythmias—they begin in the ventricles). Our clustering recognizes the normal rhythm, and also, recognizes arrhythmias originating in the atria and in the ventricles.

We analyzed 20 min ECG holter recordings sampled with the 250 Hz frequency. The entire dataset consisted of 126 ECG signals. Feature extraction was carried out according to the DFT-based method, as it is described in Sect. 5.5.

Table 4 Accuracy rates, standard deviations, p-values, and uncertainty degrees for the human heart rhythms clustering

Investigated method	q_{total}	s	p-value	U_d
TASOM	0.8175	1.1933	$< 10^{-4}$	0.1687
Auto-SOM	0.6905	1.1952	$< 10^{-4}$	0.1270
SOAN	0.6595	1.1865	$< 10^{-4}$	0.2063
Batch SOM	0.7778	1.2164	$< 10^{-4}$	0.1905
ADSOM	0.7857	1.1911	$< 10^{-4}$	0.2063
BTASOM	0.7937	1.2014	$< 10^{-4}$	0.1429
A-GHSOM	0.7302	1.1851	$< 10^{-4}$	0.1190
t-SNE and LDA	0.8095	1.3248	$< 10^{-4}$	0.0079
t-SNE and NMF	0.8071	1.3529	$< 10^{-4}$	0.0079
NeRV and LDA	0.7952	1.1493	$< 10^{-4}$	0.1349
NeRV and NMF	0.7778	1.2349	$< 10^{-4}$	0.1587
Proposed adaptive SOM	0.9286	1.3334	–	0.0476

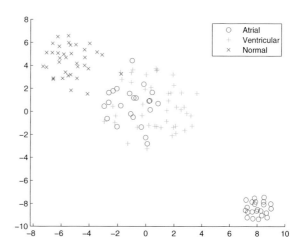

Fig. 9 The t-SNE visualization of the ECG dataset

Experimental Results The results of this part of our experiments are presented in Fig. 8 and in Table 4, which are constructed in the same way as in Sect. 5.8. Each of the examined methods was run 50 times.

The t-SNE and NeRV visualizations (single run of each of them) are depicted in Figs. 9 and 10, respectively.

Figure 11 illustrates the relationship between the total accuracy rate and the values of the parameter τ. The values of τ belong to the range from 0.0 to 0.08 (with the step 0.002).

Fig. 10 The NeRV
visualization of the ECG
dataset

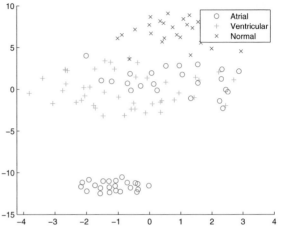

Fig. 11 The function of the
total accuracy rate versus the
parameter τ for the ECG
dataset

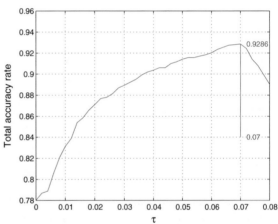

Figure 12 shows the function of the total accuracy rate versus the values of the
parameter σ_f. The values of σ_f belong to the range from 2.0 to 4.0M (with the step
0.1M).

In the last part of our empirical study, the t-SNE visualization technique clustered by means of the LDA-based clustering method and the NMF method (i.e., the
combinations t-SNE and LDA and t-SNE and NMF) produced lower values of the
uncertainty degree than our approach (0.0079 and 0.0079 versus 0.0476). However,
our approach provided higher total accuracy rate (0.9286 versus 0.8095 and 0.8071),
and the total accuracy rate is, naturally, the primary and the most important evaluation criterion in our experimental research. Therefore, taking this into consideration
together with the fact of outperforming all the other assessed methods, the overall
results of this part of the experiments may be recognized as satisfactory.

Fig. 12 The function of the total accuracy rate versus the parameter σ_f for the ECG dataset

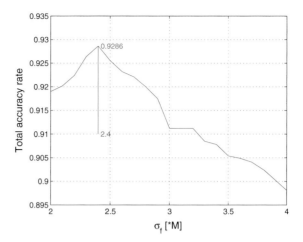

Defeating the NeRV visualization technique in our experimental study confirms the observations and conclusions of the work [23], where the authors compare NeRV and SOM. The comparison leads to a conclusion similar to our outcomes reporting the superiority of SOM with respect to NeRV. According to the results of [23], SOM produced the most trustworthy projection by far, compared to other discussed methods (NeRV, among others), for two real-world datasets analyzed in [23].

6 Conclusion and Future Study

In this paper, a novel adaptive SOM version was proposed. In the introduced approach, the neurons' neighborhood widths are determined using the information about the frequencies of occurrences of input patterns in the input space. The neighborhood widths are determined independently for each neuron in the SOM grid. In case of input patterns appearing frequently in the input space, the neighborhood of the corresponding BMU is wider than in case of the input patterns occurring rarely in the input space. Consequently, the patterns frequent in the input space will receive larger area for their prototypes in the SOM grid, in contrast to the patterns rare in the input space, which will get less place for their prototypes in the grid. In this way, the proposed method provides a proper visualization of the input data, especially, when there are significant differences in the frequencies of occurrences of input patterns, and consequently, our proposal can be regarded as superior over the traditional adaptive SOM technique.

The experimental research on real data in three fields (i.e., in the field of words analysis, in the field of piano music analysis, and in the field of human heart rhythms analysis) confirmed that the approach developed in this paper outperforms the 11 different visualization and clustering methods. The superiority was ascertained on the basis of the clustering performance of the all examined approaches. We investigated 7 reference adaptive SOM versions (clustered by means of the k-means clustering algorithm), 2 different visualization techniques (i.e., t-SNE and NeRV), and 2 clustering competitors (i.e., LDA-based clustering and NMF). In all investigated experimental cases, all the reference methods remained inferior with respect to the proposed solution, when taking into account the total accuracy rate. Only in 2 cases (t-SNE and LDA and t-SNE and NMF) our method returned higher values of the uncertainty degree.

In the future study, there are two possible directions of research. On one hand, the SOM neurons' neighborhood function could be adapted differently, i.e., their parameters might be updated on the basis of certain other (not only frequency-based) properties of the input data. That could essentially improve the quality of the resulting visualization. On the other hand, the information about the frequencies of occurrences of input patterns in the input space might be utilized in order to update and tune other (not only neurons' neighborhood width) SOM parameters, and affect other SOM aspects. In this way, the frequency-based properties of the input data would have a stronger impact on the form and shape of the generated SOM, and consequently, a higher accuracy of the final visualization may be obtained.

Acknowledgments This contribution is supported by the Foundation for Polish Science under International PhD Projects in Intelligent Computing. Project financed from The European Union within the Innovative Economy Operational Programme 2007–2013 and European Regional Development Fund.

References

1. Berglund, E., Sitte, J.: The parameterless self-organizing map algorithm. IEEE Trans. Neural Netw. **17**(2), 305–316 (2006)
2. Blei, D.M., Ng, A.Y., Jordan, M.I.: Latent Dirichlet allocation. J. Mach. Learn. Res. **3**, 993–1022 (2003)
3. Chen, S., Zhou, Z., Hu, D.: Diffusion and growing self-organizing map: a nitric oxide based neural model. In: Advances in Neural Networks—ISNN 2004, Lecture Notes in Computer Science, vol. 3173 (2004)
4. Chengalvarayan, R., Deng, L.: HMM-based speech recognition using state-dependent, discriminatively derived transforms on mel-warped DFT features. IEEE Trans. Speech Audio Process. **2**(3), 243–256 (1997)
5. DeSieno, D.: Adding a conscience to competitive learning. In: Proceedings of the Second IEEE International Conference on Neural Networks (ICNN-88). vol. 1, pp. 117–124. IEEE (July 1988)
6. Frank, A., Asuncion, A.: UCI machine learning repository (2010), http://archive.ics.uci.edu/ml
7. Goldberger, A.L., Amaral, L.A.N., Glass, L., Hausdorff, J.M., Ivanov, P.C., Mark, R.G., Mietus, J.E., Moody, G.B., Peng, C.K., Stanley, H.E.: PhysioBank, PhysioToolkit, and PhysioNet: components of a new research resource for complex physiologic signals. Circulation 101(23), e215–e220 (2000), http://circ.ahajournals.org/cgi/content/full/101/23/e215, circulation Electronic Pages

8. Haese, K., Goodhill, G.J.: Auto-SOM: recursive parameter estimation for guidance of self-organizing feature maps. Neural Comput. **13**(3), 595–619 (2001)
9. Halkidi, M., Batistakis, Y., Vazirgiannis, M.: On clustering validation techniques. J. Intell. Inf. Syst. **17**(2/3), 107–145 (2001)
10. Handl, J., Knowles, J., Kell, D.B.: Computational cluster validation in post-genomic data analysis. Bioinformatics **21**(15), 3201–3212 (2005)
11. Heskes, T.: Self-organizing maps, vector quantization, and mixture modeling. IEEE Trans. Neural Netw. **12**(6), 1299–1305 (2001)
12. van Hulle, M.M.: Faithful Representations and Topographic Maps: From Distortion- to Information-Based Self-Organization. Wiley, New York (2000)
13. Iglesias, R., Barro, S.: SOAN: self organizing with adaptive neighborhood neural network. In: Mira, J., Sánchez-Andrés, J. (eds.) Foundations and Tools for Neural Modeling. Lecture Notes in Computer Science, vol. 1606, pp. 591–600. Springer, Heidelberg (1999)
14. Ippoliti, D., Zhou, X.: A-GHSOM: an adaptive growing hierarchical self organizing map for network anomaly detection. J. Parallel Distrib. Comput. **72**(12), 1576–1590 (2012)
15. Kohonen, T.: Self-Organizing Maps. 3rd edn, Springer, Heidelberg (2001)
16. Kohonen, T.: Essentials of the self-organizing map. Neural Netw. **37**, 52–65 (2013)
17. Kohonen, T.: Self-organized formation of topologically correct feature maps. Biol. Cybern. **43**(1), 59–69 (1982)
18. Kohonen, T.: The self-organizing map. Proc. IEEE **28**, 1464–1480 (1990)
19. Lee, D.D., Seung, H.S.: Learning the parts of objects by non-negative matrix factorization. Nature **401**(6755), 788–791 (1999)
20. van der Maaten, L., Hinton, G.E.: Visualizing data using t-SNE. J. Mach. Learn. Res. **9**, 2579–2605 (2008)
21. Martín-Merino, M., Muñoz, A.: Visualizing asymmetric proximities with SOM and MDS models. Neurocomputing **63**, 171–192 (2005)
22. Mulier, F., Cherkassky, V.: Self-organization as an iterative Kernel smoothing process. Neural Comput. **7**(6), 1165–1177 (1995)
23. Nybo, K., Venna, J., Kaski, S.: The self-organizing map as a visual neighbor retrieval method. In: Proceedings of the 6th International Workshop on Self-Organizing Maps (WSOM 2007). pp. 1–8 (2007)
24. Olszewski, D.: An experimental study on asymmetric self-organizing map. In: Yin, H., Wang, W., Rayward-Smith, V. (eds.) Intelligent Data Engineering and Automated Learning—IDEAL 2011. Lecture Notes in Computer Science, vol. 6936, pp. 42–49 (2011)
25. Olszewski, D.: k-Means clustering of asymmetric data. In: Corchado, E., Snášel, V., Abraham, A., Woźniak, M., Grana, M., Cho, S.B. (eds.) Hybrid Artificial Intelligent Systems. Lecture Notes in Computer Science, vol. 7208, pp. 243–254 (2012)
26. Olszewski, D.: Fraud detection using self-organizing map visualizing the user profiles. Knowl. Based Syst. **70**, 324–334 (2014)
27. Olszewski, D., Kacprzyk, J., Zadrożny, S.: Time series visualization using asymmetric self-organizing map. In: Tomassini, M., Antonioni, A., Daolio, F., Buesser, P. (eds.) Adaptive and Natural Computing Algorithms. Lecture Notes in Computer Science, vol. 7824, pp. 40–49. Springer, Heidelberg (2013)
28. Olszewski, D., Šter, B.: Asymmetric clustering using the alpha-beta divergence. Pattern Recog. **47**(5), 2031–2041 (2014)
29. Paatero, P., Tapper, U.: Positive matrix factorization: A non-negative factor model with optimal utilization of error estimates of data values. Environmetrics **5**(2), 111–126 (1994)
30. Piastra, M.: Self-organizing adaptive map: Autonomous learning of curves and surfaces from point samples. Neural Netw. **41**, 96–112 (2013)
31. Rauber, A., Merkl, D., Dittenbach, M.: The growing hierarchical self-organizing map: exploratory analysis of high-dimensional data. IEEE Trans. Neural Netw. **13**(6), 1331–1341 (2002)
32. Ressom, H., Wang, D., Natarajan, P.: Adaptive double self-organizing maps for clustering gene expression profiles. Neural Netw. **16**(5–6), 633–640 (2003)

33. Segev, A., Kantola, J.: Identification of trends from patents using self-organizing maps. Expert Syst. Appl. **39**(18), 13235–13242 (2012)
34. Shah-Hosseini, H., Safabakhsh, R.: TASOM: A new time adaptive self-organizing map. IEEE Trans. Syst. Man Cybern. Part B Cybern. **33**(2), 271–282 (2003)
35. Shah-Hosseini, H.: Binary tree time adaptive self-organizing map. Neurocomputing **74**(11), 1823–1839 (2011)
36. Venna, J., Peltonen, J., Nybo, K., Aidos, H., Kaski, S.: Information retrieval perspective to nonlinear dimensionality reduction for data visualization. J. Mach. Learn. Res. **11**, 451–490 (2010)
37. Vesanto, J., Himberg, J., Alhoniemi, E., Parhankangas, J.: Self-organizing map in Matlab: the SOM Toolbox. In: Proceedings of the Matlab DSP Conference. pp. 35–40 (1999)
38. Vesanto, J., Himberg, J., Alhoniemi, E., Parhankangas, J.: SOM Toolbox for Matlab 5. Tech. Rep. Report A57, Helsinki University of Technology (2000)
39. Villmann, T., Claussen, J.C.: Magnification control in self-organizing maps and neural gas. Neural Comput. **18**(2), 446–469 (2006)

Support Vector Machines in Fuzzy Regression

Paulina Wieszczy and Przemysław Grzegorzewski

Abstract This paper presents methods of estimating fuzzy regression models based on support vector machines. Starting from the approaches known from the literature and dedicated to triangular fuzzy numbers and based on linear and quadratic loss, a new method applying loss function based on the Trutschnig distance is proposed. Furthermore, a generalization of those models for fuzzy numbers with trapezoidal membership function is given. Finally, the proposed models are illustrated and compared in the examples and some of their properties are discussed.

Keywords Fuzzy numbers · Fuzzy regression · Loss function · Support vector machines

1 Introduction

Regression analysis is one of the most often used statistical tools applied in engineering, biology, economics and other sciences. Its goal is to create a model describing a relationship between a response (dependent) variable and one or more explanatory (independent) variables. Many approaches tried to extend the classical regression into a fuzzy environment to cope with the problems delivered by imprecise data. In the paper by Tanaka et al. [27], probably the first on this topic, a regression problem with fuzzy dependent and crisp independent variable was formulated as a mathe-

P. Wieszczy · P. Grzegorzewski (✉)
Faculty of Mathematics and Information Science, Warsaw University of Technology,
Koszykowa 75, 00-662 Warsaw, Poland
e-mail: pgrzeg@ibspan.waw.pl

P. Wieszczy
Department of Gastroenterology, Hepatology and Clinical Oncology, Medical Center
for Postgraduate Education, Roentgen 5, 02-781 Warsaw, Poland
e-mail: p.wieszczy@gmail.com

P. Grzegorzewski
Systems Research Institute, Polish Academy of Sciences, Newelska 6,
01-447 Warsaw, Poland

© Springer International Publishing Switzerland 2016
G. De Tré et al. (eds.), *Challenging Problems and Solutions
in Intelligent Systems*, Studies in Computational Intelligence 634,
DOI 10.1007/978-3-319-30165-5_6

matical programming problem. On the other hand Diamond [5, 6] proposed a least squares approach for both fuzzy inputs and outputs. Some other approaches were suggested e.g. in [2, 13, 14, 21, 23]. For the review of fuzzy regression models we refer the reader to [4, 20, 24].

Support vector machines (SVMs), introduced by V.N. Vapnik, appeared very successful in many areas of statistical learning including classification, pattern recognition and function estimation problems (see, e.g. [17, 30, 31]). Because of their interesting properties SVMs they were also found to be useful in regression analysis or rank regression. Recently, Hong and Hwang [18, 19] applied SVMs for fuzzy regression. However, in their paper they considered triangular fuzzy numbers only. Thus, the first goal of our paper is to generalize their approach for trapezoidal fuzzy numbers. The second aim is to consider another fuzzy regression method utilizing SVMs with a loss function based on the Trutschnig distance [29] which is recently commonly used in fuzzy statistics.

The paper is organized as follows: In Sect. 2 we recall basic notation and terminology connected with fuzzy numbers. A short introduction to fuzzy regression and support vector machines is given in Sects. 3 and 4, respectively. Then, in Sect. 5 we consider SVMs in fuzzy linear regression with triangular fuzzy numbers. Starting from the ideas described in [18, 19] for the linear and quadratic loss we propose another approach utilizing the loss function based on the Trutschnig distance. Next, in Sect. 6 we generalize the methods shown in the previous section for the trapezoidal fuzzy numbers. Some examples illustrating our results and comparing the suggested models are given in Sect. 7.

2 Fuzzy Numbers

Let A denote a **fuzzy number**, i.e. such fuzzy subset A of the real line $\mathbb{R}$ with membership function $\mu_A : \mathbb{R} \to [0, 1]$ which is (see [8]):

- normal (i.e. there exist an element x_0 such that $\mu_A(x_0) = 1$),
- fuzzy convex (i.e. $\mu_A(\lambda x_1 + (1 - \lambda)x_2) \geq \mu_A(x_1) \wedge \mu_A(x_2)$, $\forall x_1, x_2 \in \mathbb{R}$, $\forall \lambda \in [0, 1]$),
- μ_A is upper semicontinuous,
- supp(A) is bounded, where supp$(A) = cl(\{x \in \mathbb{R} : \mu_A(x) > 0\})$, and cl is the closure operator.

For any fuzzy number A there exist four numbers $a_1, a_2, a_3, a_4 \in \mathbb{R}$ and two functions $l_A, r_A : \mathbb{R} \to [0, 1]$, where l_A is nondecreasing and r_A is nonincreasing, such that we can describe a membership function μ_A in a following manner

$$
\mu_A(x) = \begin{cases}
0 & \text{if } x < a_1 \\
l_A(x) & \text{if } a_1 \leq x < a_2 \\
1 & \text{if } a_2 \leq x \leq a_3 \\
r_A(x) & \text{if } a_3 < x \leq a_4 \\
0 & \text{if } a_4 < x.
\end{cases}
$$

Functions l_A and r_A are called the left side and the right side of a fuzzy number A, respectively. A space of all fuzzy numbers will be denoted by $\mathbb{F}(\mathbb{R})$.

In many situations connected with modeling and processing imprecise information we use fuzzy numbers with relatively simple sides. This is obvious since simple membership functions not only make calculations easier but also simplify computer applications and give more intuitive and natural interpretation. For this reason fuzzy numbers with linear sides, called **trapezoidal fuzzy numbers**, are most frequently used in practice. The sides of such trapezoidal fuzzy number are linear functions given as follows

$$l_A(x) = \frac{x - a_1}{a_2 - a_1},$$
$$r_A(x) = \frac{a_4 - x}{a_4 - a_3}.$$

Thus it is clear that any trapezoidal fuzzy number is completely defined by four real values $a_1 \leq a_2 \leq a_3 \leq a_4$. Hence a trapezoidal fuzzy number A based on these four values we often denote by $A = T(a_1, a_2, a_3, a_4)$. Let us denote a subfamily of the trapezoidal fuzzy numbers by $\mathbb{F}^T(\mathbb{R})$.

A particular useful subfamily of $\mathbb{F}^T(\mathbb{R})$ is given by the **triangular fuzzy numbers**, i.e. such trapezoidal fuzzy numbers for which $a_2 = a_3$. So the notation $A = T(a_1, a_2, a_3)$ would indicate that given A is a triangular fuzzy number. Note also that the family of all closed intervals in the real line is isomorphic with the subfamily of such fuzzy numbers for which $a_1 = a_2$ and $a_3 = a_4$, while each real number can be treated as a particular fuzzy number for such that $a_1 = a_2 = a_3 = a_4$.

Moreover, let $A_\alpha = \{x \in \mathbb{R} : \mu_A(x) \geq \alpha\}$, $\alpha \in (0, 1]$, and $A_0 = \text{supp}(A)$, denote an α-cut of a fuzzy number A. As it is known, every α-cut of a fuzzy number is a closed interval, i.e. $A_\alpha = [A_L(\alpha), A_U(\alpha)]$, where $A_L(\alpha) = \inf\{x \in \mathbb{R} : \mu_A(x) \geq \alpha\}$ and $A_U(\alpha) = \sup\{x \in \mathbb{R} : \mu_A(x) \geq \alpha\}$.

The core of a fuzzy number A is the set of all points that surely belong to A, i.e. $\text{core}(A) = \{x \in \mathbb{R} : \mu_A(x) = 1\} = A_{\alpha=1}$.

Having any two fuzzy numbers we often need a distance between them. A suitable distance on the family $\mathbb{F}(\mathbb{R})$, a distance that is both not too hard to calculate and which reflects the intuitive meaning of fuzzy sets. The most popular one is the distance $d_2 : \mathbb{F}(\mathbb{R}) \times \mathbb{F}(\mathbb{R}) \to [0, +\infty)$ which is actually the L_2-distance in $\mathbb{F}(\mathbb{R})$ defined for two arbitrary fuzzy numbers A and B as follows (see, e.g., [10])

$$d_2(A, B) = \sqrt{\int_0^1 (A_L(\alpha) - B_L(\alpha))^2 d\alpha + \int_0^1 (A_U(\alpha) - B_U(\alpha))^2 d\alpha}, \quad (1)$$

where $[A_L(\alpha), A_U(\alpha)]$ and $[B_L(\alpha), B_U(\alpha)]$ are the α-cuts of A and B, respectively. Distance (1) is a particular case of the weighted distance

$$d_{ZL}(A, B) = \sqrt{\int_0^1 (A_L(\alpha) - B_L(\alpha))^2 \lambda(\alpha) d\alpha + \int_0^1 (A_U(\alpha) - B_U(\alpha))^2 \lambda(\alpha) d\alpha}$$

where λ is an integrable and nonnegative weighting function. Some authors assume that a weighting function λ is increasing on $[0, 1]$ and such that $\lambda(0) = 0$. These properties mean that higher weight and thus greater importance is attributed to higher α-cuts. Such typical weighted function applied by many authors is simply $\lambda(\alpha) = \alpha$ (e.g. [32]). However, non-monotonic weighting function may be also of interest like bi-symmetrical weighting functions (see [15, 16]).

Another interesting distance was proposed by Bertoluzza et al. [3] and generalized by Trutschnig et al. [29] which can be expressed in terms of the squared Euclidean distance between the mids and spreads of the interval level sets of the two fuzzy numbers involved. For each fuzzy number $A \in \mathbb{F}(\mathbb{R})$ we define the following two functions: its mid

$$\text{mid}(A(\alpha)) = \frac{1}{2}(A_L(\alpha) + A_U(\alpha)) \tag{2}$$

and spread

$$\text{spr}(A(\alpha)) = \frac{1}{2}(A_U(\alpha) - A_L(\alpha)), \tag{3}$$

which attribute to each α-cut A_α the center and the half of the length of that α-cut, respectively. Then for two arbitrary fuzzy numbers A and B we get the distance $\delta_\theta : \mathbb{F}(\mathbb{R}) \times \mathbb{F}(\mathbb{R}) \to [0, +\infty)$ as follows

$$t_\theta(A, B) = \left(\int_0^1 \left([\text{mid}(A_\alpha) - \text{mid}(B_\alpha)]^2 + \theta[\text{spr}(A_\alpha) - \text{spr}(B_\alpha)]^2 \right) d\alpha \right)^{1/2}, \tag{4}$$

where $\theta \in (0, 1]$ is a parameter indicating the relative importance of the spreads against the mids.

One can easily check that for $\theta = 1$ the Trutschnig distance (4) is equivalent to the most widespread distance (1), i.e. $t_1^2(A, B) = \frac{1}{2}d_2^2(A, B)$ for any $A, B \in \mathbb{F}(\mathbb{R})$. It means that (1) attaches the same importance to mids and spreads. However, it seems that the distance between mids is often more important than that of the spreads because especially mids determine the position of the set. Hence distances $A \in \mathbb{F}(\mathbb{R})$ with $\theta < 1$ should be rather of interest. As the Euclidean metric is dominating in fuzzy number approximation (see, e.g., [11, 12]), the Trutschnig distance is widespread in fuzzy statistics.

Finally, let us mention that another parametrization of the trapezoidal fuzzy numbers is sometimes useful to simplify some operations. Having $A = T(a_1, a_2, a_3, a_4)$ such that $a_1 \leq a_2 < a_3 \leq a_4$ let us denote the borders of the core (i.e. a_2 and a_3) by m and n, respectively, and moreover, let $a = m - a_1$ and $b = a_4 - n$ stand for the spread of both arms, i.e. $a \geq 0$ and $b \geq 0$ are the projections of the left and right side of the fuzzy number A, respectively, onto the real line. Using this notation one may write down the membership of A as

$$\mu_A(x) = \begin{cases} 0 & \text{if} \quad x < m - a, \\ \frac{x-(m-a)}{a} & \text{if} \quad m - a \le x < m, \\ 1 & \text{if} \quad m \le x \le n, \\ \frac{n+b-x}{b} & \text{if} \quad n < x \le n + b, \\ 0 & \text{if} \quad n + b < x, \end{cases}$$

and denote such trapezoidal fuzzy number by $A = \tilde{T}(m, n, a, b)$. It is clear that for $m = n$ we get a triangular fuzzy number which will be denoted as $A = \tilde{T}(m, a, b)$. Of course, if both $m = n$ and $a = b = 0$ then our fuzzy number A reduces to the real number m. Further on, when discussing on fuzzy numbers, we will generally use this very notation.

3 Fuzzy Regression Analysis

The key idea in regression analysis is to find a relationship between a *dependent* (*response*) variable y and some *independent* (*explanatory, predictor*) variables $x_1, \dots, x_p$. Knowing such relationship one can predict a value of the response corresponding to given values of the explanatory variables. In the most often used linear regression model we consider the following relationship

$$y_i = \beta_0 + \beta_1 x_{i1} + \cdots + \beta_p x_{ip} + \varepsilon_i \qquad \text{for } i = 1, \dots, N,$$

where $(y_i)_{i=1,\dots,N}$ denote observed values of the response obtained for given values $(x_{ij})_{i,j=1,\dots,N}$ of predictors and where $(\varepsilon_i)_{i=1,\dots,N}$ represent random errors which are assumed to be independent and identically distributed random variables with mean $\mathbb{E}(\varepsilon_i) = 0$ and variance $\text{Var}(\varepsilon_i) = \sigma^2$.

A method that gives us usually good estimates of the regression coefficients $\beta = (\beta_0, \dots, \beta_p)$ is the *method of least squares* which minimizes the sum of the squared errors, i.e.

$$S(\beta) = \sum_{i=1}^{N} (y_i - \beta_0 - \beta_1 x_{i1} + \cdots - \beta_p x_{ip})^2.$$

This way we get the best linear unbiased estimator (BLUE) of β, i.e. the estimator having the smallest variance among all linear unbiased estimators of β.

In practice one often meets situations where observations are not precise or they are rather subjective perceptions than strict measurements. Thousands of papers proved that fuzzy theory might be helpful in modeling such data. In particular, we can distinguish the following three cases met in regression analysis:

- CICO (i.e. *Crisp-Inputs, Crisp-Output*), where both predictors (inputs) and responses (outputs) are real numbers,
- CIFO (i.e. *Crisp-Inputs, Fuzzy-Output*), where predictors (inputs) are real numbers but responses (outputs) are modeled by fuzzy numbers
- FIFO (i.e. *Fuzzy-Inputs, Fuzzy-Output*), where both predictors (inputs) and responses (outputs) are modeled by fuzzy numbers.

Two main approaches to fuzzy regression can be found in the literature. The first one, called sometimes the *possibilistic regression*, proposed by Tanaka et al. [26], is based on the extension principle and can be treated somehow as a minimization of the response spread. The second approach, developed by Diamond [5] and his followers, generalizes the classical least square method into the fuzzy context.

In fact, possibilistic regression reduces to the particular linear program (see [7]). The simplest version of the possibilistic regression is *interval regression* expressed by

$$Y_i = A_1 x_{i1} + \cdots + A_p x_{ip}, \qquad i = 1, \ldots, N,$$

with the interval response Y_i, real predictors x_{ip} and interval coefficients A_i. Adopting the following notation of $A_j = (a_{cj}, a_{wj})$, where a_{cj} is the center of the interval and $a_{wj} \geq 0$ its radius, i.e. the half of its width (spread), we may consider A_j as

$$A_j = \{a \; : \; a_{cj} - a_{wj} \leq a \leq a_{cj} + a_{wj}\}.$$

Then, using well-known operations of the interval arithmetic we get

$$Y_i = (a_{c1} x_{i1} + \cdots + a_{cp} x_{ip}, \; a_{w1}|x_{i1}| + \cdots + a_{wp}|x_{ip}|).$$

Additionally, we assume that given output y_i should be included in the estimation interval Y_i, i.e.

$$\sum_{j=1}^{p}(a_{cj} x_{ij} - a_{wj}|x_{ij}|) \leq y_i \leq \sum_{j=1}^{p}(a_{cj} x_{ij} + a_{wj}|x_{ij}|), \tag{5}$$

for $i = 1, \ldots, N$ but, on the other hand, the length of Y_i should be as small as possible. Therefore, to estimate intervals A_j we have to minimize the following expression

$$S(a_{w1}, \ldots, a_{wp}) = \sum_{i=1}^{N}\left(a_{w1}|x_{i1}| + \cdots + a_{wp}|x_{ip}|\right)$$

for all a_{cj}, a_{wj} with respect to constraint (5) and $a_{wj} \geq 0$ for $j = 1, \ldots, p$.

If the inputs are real numbers but the outputs are intervals we may, in fact, consider two regression models: **possibility regression model**

$$\overline{Y}_i = \overline{A}_1 x_{i1} + \cdots + \overline{A}_p x_{ip}$$

and **necessity regression model**

$$\underline{Y}_i = \underline{A}_1 x_{i1} + \cdots + \underline{A}_p x_{ip}$$

for $i = 1, \ldots, N$. In the possibility regression model the problem is to find such intervals $\overline{A}_i$ which satisfy

$$Y_i \subset \overline{Y}_i \qquad \text{for } i = 1, \ldots, N,$$

and minimize the sum of the widths, i.e.

$$\sum_{i=1}^{N} \left(\overline{a}_{w1} |x_{i1}| + \cdots + \overline{a}_{wp} |x_{ip}| \right) \longrightarrow \min.$$

Similarly, in the necessity regression model the problem is to find such intervals $\overline{A}_i$ which satisfy

$$\underline{Y}_i \subset Y_i \qquad \text{for } i = 1, \ldots, N,$$

and maximize the sum of the widths, i.e.

$$\sum_{i=1}^{N} \left(\underline{a}_{w1} |x_{i1}| + \cdots + \underline{a}_{wp} |x_{ip}| \right) \longrightarrow \max.$$

If the given data $(Y_i^0, x_{i1}^0, \ldots, x_{ip}^0)$ for $i = 1, \ldots, N$, satisfy

$$Y_i^0 = A_1^0 x_{i1} + \cdots + A_p^0 x_{ip},$$

then the intervals obtained from the possibility model are the same as obtained under the necessity model, i.e. $\overline{A}_j = \underline{A}_j = A_j^0$ for $j = 1, \ldots, p$.

For more general case with crisp inputs and fuzzy outputs we may consider the following model (see [28])

$$\hat{Y}_i = A_1 x_{i1} + \cdots + A_p x_{ip} \qquad \text{dla } i = 1, \ldots, N,$$

where, for simplicity, we assume that $A_j = T(a_{j1}, a_{j2}, a_{j3})$ are symmetrical triangular fuzzy numbers. Then, to estimate A_j we will minimize

$$S(a_{11}, a_{12}, \ldots, a_{p1}, a_{p2}) = \sum_{i=1}^{N} \left((a_{12} - a_{11}) |x_{i1}| + \cdots + (a_{p2} - a_{p1}) |x_{ip}| \right)$$

with the following constraints: $a_{j2} - a_{j1} \geq 0$ for $j = 1, \ldots, p$ and $(Y_i)_\alpha \subset (\hat{Y}_i)_\alpha$ for $i = 1, \ldots, N$ and for each $\alpha \in (0, 1]$.

Another approach, suggested by Diamond [5], consists in the minimization of a loss function which is usually the square of the distance $\delta_{2,1/2}$ between the observed and estimated value of the response. Here, we may consider the following model

$$\hat{Y}_i = a_0 + a_1 X_{i1} + \cdots + a_p X_{ip} \qquad \text{for } i = 1, \ldots, N,$$

where $\hat{Y}_i$ and $X_{i1}, \ldots, X_{ip}$ are fuzzy numbers while $a_0, a_1, \ldots, a_p$ are reals. In order to estimate $a_0, a_1, \ldots, a_p$ we have to minimize

$$S(a_0, \ldots, a_p) = \sum_{i=1}^{N} \left(\frac{1}{2} \int_0^1 [(Y_i)_L(\alpha) - (\hat{Y}_i)_L(\alpha)]^2 d\alpha + \frac{1}{2} \int_0^1 [(Y_i)_U(\alpha) - (\hat{Y}_i)_U(\alpha)]^2 d\alpha \right)$$

with respect to all $a_0, a_1, \ldots, a_p$ (see [2]).

4 Support Vector Machines

Support vector machines are widely used in different areas of statistical learning. To explain their application in regression analysis let us consider the simple linear regression.

The main idea of the SVM goes back to the *ε-intensive loss function* leading in practise to the best separating hyperplane for the binary data (see [25]). Let us consider

$$|y - \hat{y}|_\varepsilon = \begin{cases} 0 & \text{if } |y - \hat{y}| < \varepsilon, \\ |y - \hat{y}| - \varepsilon & \text{if } |y - \hat{y}| \geq \varepsilon, \end{cases}$$

where $\hat{y}$ stands for an estimator of y. We may list some ε-intensive loss functions like

1. linear ε-intensive loss function

$$L(y, \hat{y}) = |y - \hat{y}|_\varepsilon, \tag{6}$$

2. quadratic ε-intensive loss function

$$L(y, \hat{y}) = |y - \hat{y}|_\varepsilon^2, \tag{7}$$

3. the Huber loss function

$$L(|y, \hat{y}|) = \begin{cases} \frac{1}{2}|y - \hat{y}|^2 & \text{if } |y - \hat{y}| \leq \varepsilon, \\ \varepsilon|y - \hat{y}| - \frac{\varepsilon^2}{2} & \text{if } |y - \hat{y}| > \varepsilon. \end{cases}$$

Note that for $\varepsilon = 0$ the ε-intensive loss functions have their counterparts used in the classical regression analysis.

Suppose $\{x_i, y_i\}_{i=1,\dots,N} \in \mathbb{R}^p \times \mathbb{R}$ are given training data. We wish to find such vector $w \in \mathbb{R}^p$ and parameter b that minimize the norm $||w||^2 := \sum_{j=1}^{p} w_j^2$ and empirical risk

$$R_{\mathrm{emp}}(w, b) = \frac{1}{N} \sum_{i=1}^{N} |y_i - \langle w, x_i \rangle - b|_{\varepsilon_i}^k,$$

where $\langle w, x_i \rangle = \sum_{j=1}^{p} w_j x_{ij}$. It is easily seen that for $k = 1$ we get the risk corresponding to the linear ε-intensive loss function, while for $k = 2$ to the quadratic ε-intensive loss function.

Let d denote a hyperplane given by the equation $\langle w, x \rangle + b = 0$ (see Fig. 1). Then each point x in the area restricted by hyperplanes $d - \varepsilon$ and $d + \varepsilon$ satisfies $|y - \langle w, x \rangle - b|_{\varepsilon}^k = 0$.

Thus our problem reduces to the minimization of the expression

$$S(w, \xi, \xi^*) = \frac{1}{2}||w||^2 + \frac{C}{k} \sum_{i=1}^{N} ((\xi_i)^k + (\xi_i^*)^k) \tag{8}$$

subject to

$$\begin{cases} \langle w, x_i \rangle + b - y_i \leq \varepsilon + \xi_i & \text{if } i = 1, \dots, N, \\ \langle w, x_i \rangle + b - y_i \geq -\varepsilon - \xi_i^* & \text{if } i = 1, \dots, N, \end{cases} \tag{9}$$

where $\xi_i, \xi_i^* \geq 0$ for $i = 1, \dots, N$, and C is a positive constant called soft margin parameter (see [9]).

Fig. 1 Regression with ε-intensive tube

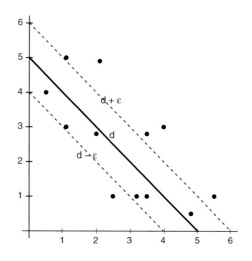

5 SVM in Fuzzy Linear Regression with Triangular Fuzzy Numbers

5.1 Fuzzy Regression with Triangular Fuzzy Numbers

In this section we present the SVM based fuzzy linear regression for the case of FIFO data and triangular fuzzy numbers. Let $\{X_i, Y_i\}_{i=1}^N$ denote a training data set, where $X_i = \widetilde{T}((m_{X_{i1}}, a_{X_{i1}}, b_{X_{i1}}), \dots, \widetilde{T}(m_{X_{ip}}, a_{X_{ip}}, b_{X_{ip}}))$ and $Y_i = \widetilde{T}(m_{Y_i}, a_{Y_i}, b_{Y_i})$. Moreover, let $\boldsymbol{m_{X_i}} = (m_{X_{i1}}, \dots, m_{X_{ip}})$, $\boldsymbol{a_{X_i}} = (a_{X_{i1}}, \dots, a_{X_{ip}})$, $\boldsymbol{b_{X_i}} = (b_{X_{i1}}, \dots, b_{X_{ip}})$, $B = \widetilde{T}(m_B, a_B, b_B)$ and $\boldsymbol{w} = (w_1, \dots, w_p)$. We consider the following model

$$Y(X) = \langle \boldsymbol{w}, X \rangle + B = (\langle \boldsymbol{w}, \boldsymbol{m_X} \rangle + m_B, \langle |\boldsymbol{w}|, \boldsymbol{a_X} \rangle + a_B, \langle |\boldsymbol{w}|, \boldsymbol{b_X} \rangle + b_B),$$

where $|\boldsymbol{w}| = (|w_1|, \dots, |w_p|)$. Our aim is to minimize the distance between Y and $Y(X)$. Using d_2 norm defined by (1) we get

$$d_2^2(Y, Y(X)) = \frac{1}{3}\left((m_Y - a_Y) - (m_{Y(X)} - a_{Y(X)})\right)^2 + \frac{1}{3}\left(m_Y - m_{Y(X)}\right)^2$$
$$+ \frac{1}{3}\left((m_Y + b_Y) - (m_{Y(X)} + b_{Y(X)})\right)^2.$$

Note that minimization of $d_2(Y, Y(X))$ is equivalent to minimization of each of the addends separately. Hence, the actual difference between SVM method for the crisp and fuzzy linear regression comes down to the constraints.

Below we consider three situations corresponding to the linear, quadratic loss function and the loss function based on the Trutschnig distance.

5.2 Linear Loss Function

Hong and Hwang [18] by modifying the SVM method for the linear regression and using linear loss function obtained the optimization problem: minimize

$$S(w, \xi_1, \xi_1^*, \xi_2, \xi_2^*, \xi_3, \xi_3^*) = \frac{1}{2}\|w\|^2 + C\sum_{k=1}^{3}\sum_{i=1}^{N}(\xi_{ki} + \xi_{ki}^*), \qquad (10)$$

subject to

$$
\left\{
\begin{aligned}
& m_{Y_i} - \langle \boldsymbol{w}, \boldsymbol{m}_{X_i} \rangle - m_B \le \varepsilon + \xi_{1i}, \\
& \langle \boldsymbol{w}, \boldsymbol{m}_{X_i} \rangle + m_B - m_{Y_i} \le \varepsilon + \xi_{1i}^*, \\
& (m_{Y_i} - a_{Y_i}) - (\langle \boldsymbol{w}, \boldsymbol{m}_{X_i} \rangle + m_B - \langle |\boldsymbol{w}|, \boldsymbol{a}_{X_i} \rangle - a_B) \le \varepsilon + \xi_{2i}, \\
& (\langle \boldsymbol{w}, \boldsymbol{m}_{X_i} \rangle + m_B - \langle |\boldsymbol{w}|, \boldsymbol{a}_{X_i} \rangle - a_B) - (m_{Y_i} - a_{Y_i}) \le \varepsilon + \xi_{2i}^*, \\
& (m_{Y_i} + b_{Y_i}) - (\langle \boldsymbol{w}, \boldsymbol{m}_{X_i} \rangle + m_B + \langle |\boldsymbol{w}|, \boldsymbol{b}_{X_i} \rangle + b_B) \le \varepsilon + \xi_{3i}, \\
& (\langle \boldsymbol{w}, \boldsymbol{m}_{X_i} \rangle + m_B + \langle |\boldsymbol{w}|, \boldsymbol{b}_{X_i} \rangle + b_B) - (m_{Y_i} + b_{Y_i}) \le \varepsilon + \xi_{3i}^*, \\
& \xi_{ki}, \ \xi_{ki}^* \ge 0,
\end{aligned}
\right.
\tag{11}
$$

where C is a soft margin parameter.

The Lagrange function for (10) minimization subject to (11) is as follows

$$
L(\boldsymbol{w}, m_B, a_B, b_B, \xi, \xi^*) = \frac{1}{2}||w||^2 + C \sum_{k=1}^{3} \sum_{i=1}^{N} (\xi_{ki} + \xi_{ki}^*)
$$

$$
- \sum_{i=1}^{N} \gamma_{1i}\left(\varepsilon + \xi_{1i} - m_{Y_i} + \langle \boldsymbol{w}, \boldsymbol{m}_{X_i} \rangle + m_B\right) - \sum_{i=1}^{N} \gamma_{1i}^*\left(\varepsilon + \xi_{1i}^* - \langle \boldsymbol{w}, \boldsymbol{m}_{X_i} \rangle - m_B + m_{Y_i}\right)
$$

$$
- \sum_{i=1}^{N} \gamma_{2i}\left(\varepsilon + \xi_{2i} - (m_{Y_i} - a_{Y_i}) + (\langle \boldsymbol{w}, \boldsymbol{m}_{X_i} \rangle + m_B - \langle |\boldsymbol{w}|, \boldsymbol{a}_{X_i} \rangle - a_B)\right)
$$

$$
- \sum_{i=1}^{N} \gamma_{2i}^*\left(\varepsilon + \xi_{2i}^* - (\langle \boldsymbol{w}, \boldsymbol{m}_{X_i} \rangle + m_B - \langle |\boldsymbol{w}|, \boldsymbol{a}_{X_i} \rangle - a_B) + (m_{Y_i} - a_{Y_i})\right)
$$

$$
- \sum_{i=1}^{N} \gamma_{3i}\left(\varepsilon + \xi_{3i} - (m_{Y_i} + b_{Y_i}) + (\langle \boldsymbol{w}, \boldsymbol{m}_{X_i} \rangle + m_B + \langle |\boldsymbol{w}|, \boldsymbol{b}_{X_i} \rangle + b_B)\right)
$$

$$
- \sum_{i=1}^{N} \gamma_{3i}^*\left(\varepsilon + \xi_{3i}^* - (\langle \boldsymbol{w}, \boldsymbol{m}_{X_i} \rangle + m_B + \langle |\boldsymbol{w}|, \boldsymbol{b}_{X_i} \rangle + b_B) + (m_{Y_i} + b_{Y_i})\right)
$$

$$
- \sum_{k=1}^{3} \sum_{i=1}^{N} \left(\eta_{ki}\xi_{ki} + \eta_{ki}^*\xi_{ki}^*\right).
$$

By differentiating the above function we get

$$
\frac{\partial L}{\partial \boldsymbol{w}} = \boldsymbol{w} - \sum_{i=1}^{N} \Big((\gamma_{1i} - \gamma_{1i}^*)\boldsymbol{m}_{X_i} + (\gamma_{2i} - \gamma_{2i}^*)(\boldsymbol{m}_{X_i} - \mathrm{sgn}(\boldsymbol{w}) * \boldsymbol{a}_{X_i})
$$

$$
+ (\gamma_{3i} - \gamma_{3i}^*)(\boldsymbol{m}_{X_i} + \mathrm{sgn}(\boldsymbol{w}) * \boldsymbol{b}_{X_i})\Big),
$$

$$
\frac{\partial L}{\partial m_B} = - \sum_{i=1}^{N} \left(\gamma_{1i} - \gamma_{1i}^* + \gamma_{2i} - \gamma_{2i}^*\right),
$$

$$
\frac{\partial L}{\partial a_B} = - \sum_{i=1}^{N} \left(\gamma_{2i} - \gamma_{2i}^*\right),
$$

$$\frac{\partial L}{\partial b_B} = -\sum_{i=1}^{N}\left(\gamma_{3i} - \gamma_{3i}^*\right),$$

$$\frac{\partial L}{\partial \xi_{ki}} = C - \gamma_{ki} - \eta_{ki},$$

$$\frac{\partial L}{\partial \xi_{ki}^*} = C - \gamma_{ki}^* - \eta_{ki}^*,$$

where $(u_1, \ldots, u_p)' * (v_1, \ldots, v_p)' = (u_1 v_1, \ldots, u_p v_p)'$. After comparing the above functions with zero we get

$$\boldsymbol{w} = \sum_{i=1}^{N}\left((\gamma_{1i} - \gamma_{1i}^*)\boldsymbol{m}_{X_i} + (\gamma_{2i} - \gamma_{2i}^*)(\boldsymbol{m}_{X_i} - \operatorname{sgn}(\boldsymbol{w}) * \boldsymbol{a}_{X_i})\right. \tag{12}$$

$$\left. + (\gamma_{3i} - \gamma_{3i}^*)(\boldsymbol{m}_{X_i} + \operatorname{sgn}(\boldsymbol{w}) * \boldsymbol{b}_{X_i})\right),$$

$$\sum_{i=1}^{N}\left(\gamma_{1i} - \gamma_{1i}^*\right) = 0, \quad \sum_{i=1}^{N}\left(\gamma_{2i} - \gamma_{2i}^*\right) = 0, \quad \sum_{i=1}^{N}\left(\gamma_{3i} - \gamma_{3i}^*\right) = 0$$

and $\gamma_{ki}, \gamma_{ki}^* \leq C$ dla $\xi_{ki} = 0,\ \xi_{ki}^* = 0$.

Hence, we can construct the problem dual to (10) minimization subject to (11) by putting the above values into the Lagrange function of the previous problem:

$$W(\gamma, \gamma^*) = \frac{1}{2}\Big(\sum_{i=1}^{N}\sum_{j=1}^{N}(\gamma_{1i} - \gamma_{1i}^*)(\gamma_{1j} - \gamma_{1j}^*)\langle \boldsymbol{m}_{X_i}, \boldsymbol{m}_{X_j}\rangle$$

$$+2\sum_{i=1}^{N}\sum_{j=1}^{N}(\gamma_{1i} - \gamma_{1i}^*)(\gamma_{2j} - \gamma_{2j}^*)\langle \boldsymbol{m}_{X_i}, \boldsymbol{m}_{X_j} - \operatorname{sgn}(\boldsymbol{w}) * \boldsymbol{a}_{X_j}\rangle$$

$$+2\sum_{i=1}^{N}\sum_{j=1}^{N}(\gamma_{1i} - \gamma_{1i}^*)(\gamma_{3j} - \gamma_{3j}^*)\langle \boldsymbol{m}_{X_i}, \boldsymbol{m}_{X_j} + \operatorname{sgn}(\boldsymbol{w}) * \boldsymbol{b}_{X_j}\rangle$$

$$+\sum_{i=1}^{N}\sum_{j=1}^{N}(\gamma_{2i} - \gamma_{2i}^*)(\gamma_{2j} - \gamma_{2j}^*)\langle \boldsymbol{m}_{X_i} - \operatorname{sgn}(\boldsymbol{w}) * \boldsymbol{a}_{X_i}, \boldsymbol{m}_{X_j} - \operatorname{sgn}(\boldsymbol{w}) * \boldsymbol{a}_{X_j}\rangle$$

$$+2\sum_{i=1}^{N}\sum_{j=1}^{N}(\gamma_{2i} - \gamma_{2i}^*)(\gamma_{3j} - \gamma_{3j}^*)\langle \boldsymbol{m}_{X_i} - \operatorname{sgn}(\boldsymbol{w}) * \boldsymbol{a}_{X_i}, \boldsymbol{m}_{X_j} + \operatorname{sgn}(\boldsymbol{w}) * \boldsymbol{b}_{X_j}\rangle$$

$$+\sum_{i=1}^{N}\sum_{j=1}^{N}(\gamma_{3i} - \gamma_{3i}^*)(\gamma_{3j} - \gamma_{3j}^*)\langle \boldsymbol{m}_{X_i} + \operatorname{sgn}(\boldsymbol{w}) * \boldsymbol{b}_{X_i}, \boldsymbol{m}_{X_j} + \operatorname{sgn}(\boldsymbol{w}) * \boldsymbol{b}_{X_j}\rangle\Big)$$

$$+\sum_{i=1}^{N}(\gamma_{1i} - \gamma_{1i}^*)m_{Y_i}$$

$$-\sum_{i=1}^{N}(\gamma_{1i}-\gamma_{1i}^{*})\Big(\sum_{j=1}^{N}\big((\gamma_{1j}-\gamma_{1j}^{*})\langle m_{X_i},m_{X_j}\rangle$$
$$+(\gamma_{2j}-\gamma_{2j}^{*})\langle m_{X_i},m_{X_j}-\operatorname{sgn}(w)*a_{X_j}\rangle$$
$$+(\gamma_{3j}-\gamma_{3j}^{*})\langle m_{X_i},m_{X_j}+\operatorname{sgn}(w)*b_{X_j}\rangle\big)\Big)$$
$$+\sum_{i=1}^{N}(\gamma_{2i}-\gamma_{2i}^{*})(m_{Y_i}-a_{Y_i})$$
$$-\sum_{i=1}^{N}(\gamma_{2i}-\gamma_{2i}^{*})\Big(\sum_{j=1}^{N}\big((\gamma_{1j}-\gamma_{1j}^{*})\langle m_{X_i}-\operatorname{sgn}(w)*a_{X_i},m_{X_j}\rangle$$
$$+(\gamma_{2j}-\gamma_{2j}^{*})\langle m_{X_i}-\operatorname{sgn}(w)*a_{X_i},m_{X_j}-\operatorname{sgn}(w)*a_{X_j}\rangle$$
$$+(\gamma_{3j}-\gamma_{3j}^{*})\langle m_{X_i}-\operatorname{sgn}(w)*a_{X_i},m_{X_j}+\operatorname{sgn}(w)*b_{X_j}\rangle\big)\Big)$$
$$-\sum_{i=1}^{N}(\gamma_{3i}-\gamma_{3i}^{*})\Big(\sum_{j=1}^{N}\big((\gamma_{1j}-\gamma_{1j}^{*})\langle m_{X_i}+\operatorname{sgn}(w)*b_{X_i},m_{X_j}\rangle$$
$$+(\gamma_{2j}-\gamma_{2j}^{*})\langle m_{X_i}+\operatorname{sgn}(w)*b_{X_i},m_{X_j}-\operatorname{sgn}(w)*a_{X_j}\rangle$$
$$+(\gamma_{3j}-\gamma_{3j}^{*})\langle m_{X_i}+\operatorname{sgn}(w)*b_{X_i},m_{X_j}+\operatorname{sgn}(w)*b_{X_j}\rangle\big)\Big)$$
$$-\varepsilon\sum_{k=1}^{3}\sum_{i=1}^{N}(\gamma_{ki}+\gamma_{ki}^{*})$$
$$=-\frac{1}{2}||w||^2-\varepsilon\sum_{k=1}^{3}\sum_{i=1}^{N}(\gamma_{ki}+\gamma_{ki}^{*})+\sum_{i=1}^{N}(\gamma_{1i}-\gamma_{1i}^{*})m_{Y_i}$$
$$+\sum_{i=1}^{N}(\gamma_{2i}-\gamma_{2i}^{*})(m_{Y_i}-a_{Y_i})+\sum_{i=1}^{N}(\gamma_{3i}-\gamma_{3i}^{*})(m_{Y_i}+b_{Y_i}).$$

Therefore, the Lagrange multipliers $\gamma_{ki},\gamma_{ki}^{*}$ can be obtained by maximizing the following equation

$$W(w,\gamma_1,\gamma_1^*,\gamma_2,\gamma_2^*,\gamma_3,\gamma_3^*)=-\frac{1}{2}||w||^2-\varepsilon\sum_{k=1}^{3}\sum_{i=1}^{N}(\gamma_{ki}+\gamma_{ki}^{*})$$
$$+\sum_{i=1}^{N}(\gamma_{1i}-\gamma_{1i}^{*})m_{Y_i}+\sum_{i=1}^{N}(\gamma_{2i}-\gamma_{2i}^{*})(m_{Y_i}-a_{Y_i})+\sum_{i=1}^{N}(\gamma_{3i}-\gamma_{3i}^{*})(m_{Y_i}+b_{Y_i})$$

subject to

$$\begin{cases}\sum_{i=1}^{N}(\gamma_{ki}-\gamma_{ki}^{*})=0,\\ \gamma_{ki},\gamma_{ki}^{*}\in[0,C],\end{cases}\tag{13}$$

for $k=1,2,3,\ i=1,\ldots,N$. The Lagrangian function for the above problem looks as follows

$$R(\gamma_1, \gamma_1^*, \gamma_2, \gamma_2^*, \gamma_3, \gamma_3^*) = -\frac{1}{2}||w||^2 - \varepsilon \sum_{k=1}^{3} \sum_{i=1}^{N} (\gamma_{ki} + \gamma_{ki}^*) + \sum_{i=1}^{N} (\gamma_{1i} - \gamma_{1i}^*) m_{Y_i}$$

$$+ \sum_{i=1}^{N} (\gamma_{2i} - \gamma_{2i}^*)(m_{Y_i} - a_{Y_i}) + \sum_{i=1}^{N} (\gamma_{3i} - \gamma_{3i}^*)(m_{Y_i} + b_{Y_i}) + \sum_{k=1}^{3} \lambda_k \left(\sum_{i=1}^{N} (\gamma_{ki} - \gamma_{ki}^*) \right).$$

By differentiating the above function with respect to $\gamma_1, \gamma_1^*, \gamma_2, \gamma_2^*, \gamma_3, \gamma_3^*$ and comparing the derivatives with zero we obtain linear equations as follows:

$$\begin{pmatrix} S_{11} & -S_{11} & S_{12} & -S_{12} & S_{13} & -S_{13} & -1 & 0 & 0 \\ S_{12}' & -S_{12}' & S_{22} & -S_{22} & S_{23} & -S_{23} & 0 & -1 & 0 \\ S_{13}' & -S_{13}' & S_{23}' & -S_{23}' & S_{33} & -S_{33} & 0 & 0 & -1 \\ 1' & -1' & 0' & 0' & 0' & 0' & 0 & 0 & 0 \\ 0' & 0' & 1' & -1' & 0' & 0' & 0 & 0 & 0 \\ 0' & 0' & 0' & 0' & 1' & -1' & 0 & 0 & 0 \end{pmatrix} \begin{pmatrix} \gamma_1 \\ \gamma_1^* \\ \gamma_2 \\ \gamma_2^* \\ \gamma_3 \\ \gamma_3^* \\ \lambda_1 \\ \lambda_2 \\ \lambda_3 \end{pmatrix} = \begin{pmatrix} m_Y \\ m_Y - a_Y \\ m_Y + b_Y \\ 0 \\ 0 \\ 0 \end{pmatrix},$$

with

$$S_{11} = m_X m_X' \left(\frac{1}{2}J + \frac{1}{2}I \right),$$

$$S_{12} = m_X (m_X - \mathrm{sgn}(w) * a_X)',$$

$$S_{13} = m_X (m_X + \mathrm{sgn}(w) * b_X)',$$

$$S_{22} = (m_X - \mathrm{sgn}(w) * a_X)(m_X - \mathrm{sgn}(w) * a_X)'(\frac{1}{2}J + \frac{1}{2}I),$$

$$S_{23} = (m_X - \mathrm{sgn}(w) * a_X)(m_X + \mathrm{sgn}(w) * b_X)',$$

$$S_{33} = (m_X + \mathrm{sgn}(w) * b_X)(m_X + \mathrm{sgn}(w) * b_X)'(\frac{1}{2}J + \frac{1}{2}I),$$

where $\mathbf{1} = (1, \ldots, 1)', \mathbf{0} = (0, \ldots, 0)'$ are vectors of length N, I is the identity matrix of size $N \times N$, J is the matrix of ones of size $N \times N$, i.e. $J = [1]_{N \times N}$.

By (12) we get

$$w = \sum_{i=1}^{N} \left((\gamma_{1i} - \gamma_{1i}^*) m_{X_i} + (\gamma_{2i} - \gamma_{2i}^*)(m_{X_i} - \mathrm{sgn}(w) * a_{X_i}) \right.$$

$$\left. + (\gamma_{3i} - \gamma_{3i}^*)(m_{X_i} + \mathrm{sgn}(w) * b_{X_i}) \right),$$

where only $B = (m_B, a_B, b_B)$ is unknown. By the Karush–Kuhn–Tucker theorem we get

$$\begin{cases} \gamma_{1i}(\varepsilon + \xi_{1i} - m_{Y_i} + \langle w, m_{X_i} \rangle + m_B) = 0, \\ \gamma_{1i}^*(\varepsilon + \xi_{1i}^* + m_{Y_i} - \langle w, m_{X_i} \rangle - m_B) = 0, \\ \qquad\qquad\qquad (C - \gamma_{1i})\xi_{1i} = 0, \\ \qquad\qquad\qquad (C - \gamma_{1i}^*)\xi_{1i}^* = 0, \end{cases}$$

and

$$\begin{cases} m_B = m_{Y_i} - \langle w, m_{X_i} \rangle - \varepsilon \ \text{dla} \ \gamma_{1i} \in (0, C), \\ m_B = m_{Y_i} - \langle w, m_{X_i} \rangle + \varepsilon \ \text{dla} \ \gamma_{1i}^* \in (0, C). \end{cases}$$

To obtain a_B, b_B we solve problems

$$\min_{a_B \geq 0} \sum_{i=1}^{N} \left(|m_{Y_i} - a_{Y_i} - (\langle w, m_{X_i} \rangle + m_B - \langle |w|, a_{X_i} \rangle - a_B)|_\varepsilon \right),$$

$$\min_{b_B \geq 0} \sum_{i=1}^{N} \left(|m_{Y_i} + b_{Y_i} - (\langle w, m_{X_i} \rangle + m_B + \langle |w|, b_{X_i} \rangle + b_B)|_\varepsilon \right).$$

5.3 Quadratic Loss Function

In the case of quadratic loss function all inequality constraints simplify to equality constraints (since after squaring residuals its sign is no longer significant). Hence the problem now is to minimize the following function

$$S(w, \xi_1, \xi_2, \xi_3) = \frac{1}{2}||w||^2 + \frac{C}{2} \sum_{k=1}^{3} \sum_{i=1}^{N} \xi_{ki}^2 \qquad (14)$$

subject to

$$\begin{cases} m_{Y_i} - \langle w, m_X \rangle - m_B = \varepsilon + \xi_{1i}, \\ (m_{Y_i} - a_{Y_i}) - (\langle w, m_{X_i} \rangle + m_B - \langle |w|, a_{X_i} \rangle - a_B) = \varepsilon + \xi_{2i}, \\ (m_{Y_i} + b_{Y_i}) - (\langle w, m_{X_i} \rangle + m_B + \langle |w|, b_{X_i} \rangle + b_B) = \varepsilon + \xi_{3i}, \end{cases} \qquad (15)$$

where C is positive constant. The Lagrangian function now looks as follows

$$L(w, \xi_1, \xi_2, \xi_3) = \frac{1}{2}||w||^2 + \frac{C}{2} \sum_{k=1}^{3} \sum_{i=1}^{N} \xi_{ki}^2$$
$$- \sum_{i=1}^{N} \gamma_{1i}\left(\varepsilon + \xi_{1i} - m_{Y_i} + \langle w, m_X \rangle + m_B\right)$$

$$-\sum_{i=1}^{N} \gamma_{2i}\left(\varepsilon + \xi_{2i} - (m_{Y_i} - a_{Y_i}) + (\langle w, m_{X_i}\rangle + m_B - \langle |w|, a_{X_i}\rangle - a_B)\right)$$

$$-\sum_{i=1}^{N} \gamma_{3i}\left(\varepsilon + \xi_{3i} - (m_{Y_i} + b_{Y_i}) + (\langle w, m_{X_i}\rangle + m_B + \langle |w|, b_{X_i}\rangle + b_B)\right),$$

where $\gamma_{ki} \in \mathbb{R}$ for $k = 1, 2, 3$, $i = 1, \ldots, N$. After calculating derivatives and comparing them with zero we get

$$\begin{pmatrix} 0 & 0 & 0 & \mathbf{1}' & \mathbf{1}' & \mathbf{1}' \\ 0 & 0 & 0 & \mathbf{0}' & \mathbf{1}' & \mathbf{0}' \\ 0 & 0 & 0 & \mathbf{0}' & \mathbf{0}' & \mathbf{1}' \\ \mathbf{1} & 0 & 0 & S_{11} & S_{12} & S_{13} \\ \mathbf{1} & -\mathbf{1} & 0 & S'_{12} & S_{22} & S_{23} \\ \mathbf{1} & 0 & \mathbf{1} & S'_{13} & S'_{23} & S_{33} \end{pmatrix} \begin{pmatrix} m_B + \varepsilon \\ a_B - \varepsilon \\ b_B + \varepsilon \\ \gamma_1 \\ \gamma_2 \\ \gamma_3 \end{pmatrix} = \begin{pmatrix} 0 \\ 0 \\ 0 \\ m_Y \\ m_Y - a_Y \\ m_Y + b_Y \end{pmatrix},$$

with

$$S_{11} = m_X m'_X + I/C,$$
$$S_{12} = m_X(m_X - \mathrm{sgn}(w) * a_X)',$$
$$S_{13} = m_X(m_X + \mathrm{sgn}(w) * b_X)',$$
$$S_{22} = (m_X - \mathrm{sgn}(w) * a_X)(m_X - \mathrm{sgn}(w) * a_X)',$$
$$S_{23} = (m_X - \mathrm{sgn}(w) * a_X)(m_X + \mathrm{sgn}(w) * b_X)',$$
$$S_{33} = (m_X + \mathrm{sgn}(w) * b_X)(m_X + \mathrm{sgn}(w) * b_X)',$$

where $\mathbf{1} = (1, \ldots, 1)'$, $\mathbf{0} = (0, \ldots, 0)'$ are vectors of length N, I is the identity matrix of size $N \times N$. Hence

$$w = \sum_{i=1}^{N} \left(\gamma_{1i} m_{X_i} + \gamma_{2i}(m_{X_i} - \mathrm{sgn}(w) * a_{X_i}) + \gamma_{3i}(m_{X_i} + \mathrm{sgn}(w) * b_{X_i})\right)$$

and for a new observation X we get

$$Y(X) = (\langle w, m_X\rangle + m_B, \langle |w|, a_X\rangle + a_B, \langle |w|, b_X\rangle + b_B).$$

5.4 A Loss Function Based on the Trutschnig Distance

In this section we propose a fuzzy linear regression with the loss function based on the Trutschnig distance (4). This approach allows to attache more importance to the distance between mids of the observed and estimated fuzzy number than to the distance between their spreads.

Let $A = \widetilde{T}(m_A, a_A, b_A)$ and $B = \widetilde{T}(m_B, a_B, b_B)$ be triangular fuzzy numbers. By (4) we get

$$t_\theta^2(A, B) = \int_0^1 \left([\mathrm{mid}(A_\alpha) - \mathrm{mid}(B_\alpha)]^2 + \theta[\mathrm{spr}(A_\alpha) - \mathrm{spr}(B_\alpha)]^2\right)d\alpha.$$

Let us introduce the following notation

$$I_1 := \int_0^1 [\mathrm{mid}(A_\alpha) - \mathrm{mid}(B_\alpha)]^2 d\alpha,$$

$$I_2 := \int_0^1 [\mathrm{spr}(A_\alpha) - \mathrm{spr}(B_\alpha)]^2 d\alpha.$$

By the definition of triangular fuzzy numbers we get

$$\begin{aligned}
I_1 &= \frac{1}{4}\big((2m_A - a_A + b_A - 2m_B + a_B - b_B)^2 \\
&\quad + (2m_A - a_A + b_A - 2m_B + a_B - b_B)(a_A - b_A - a_B + b_B) \\
&\quad + \frac{1}{3}(a_A - b_A - a_B + b_B)^2\big) \\
&= \frac{1}{4}\left(\left(2m_A - \frac{1}{2}a_A + \frac{1}{2}b_A - 2m_B + \frac{1}{2}a_B - \frac{1}{2}b_B\right)^2 + \frac{1}{12}(a_A - b_A - a_B + b_B)^2\right), \\
I_2 &= \frac{1}{12}(a_B + b_B - a_A - b_A)^2.
\end{aligned}$$

Hence, the problem of minimizing the distance $t_\theta^2(A, B)$ for triangular fuzzy numbers is equivalent to the minimization of

$$\begin{aligned}
E_1 &= \left(2m_A - \frac{1}{2}a_A + \frac{1}{2}b_A - 2m_B + \frac{1}{2}a_B - \frac{1}{2}b_B\right)^2, \\
E_2 &= (a_A - b_A - a_B + b_B)^2, \\
E_3 &= (a_B + b_B - a_A - b_A)^2.
\end{aligned}$$

Similarly as for the quadratic loss function to estimate a vector $\boldsymbol{w}$ and fuzzy number $B = \widetilde{T}(m_B, a_B, b_B)$ we have to minimize

$$S(\boldsymbol{w}, \boldsymbol{\xi_1}, \boldsymbol{\xi_2}, \boldsymbol{\xi_3}) = \frac{1}{2}||\boldsymbol{w}||^2 + \frac{C}{2}\sum_{i=1}^N (\xi_{1i}^2 + \xi_{2i}^2 + \theta\xi_{3i}^2) \tag{16}$$

subject to

$$
\left\{
\begin{array}{l}
2m_{Y_i} - \frac{1}{2}a_{Y_i} + \frac{1}{2}b_{Y_i} - 2(\langle w, m_{X_i}\rangle + m_B) \\
\quad + \frac{1}{2}(\langle |w|, a_{X_i}\rangle + a_B) - \frac{1}{2}(\langle |w|, b_{X_i}\rangle + b_B) = \varepsilon + \xi_{1i}, \\
a_{Y_i} - b_{Y_i} - \langle |w|, a_{X_i}\rangle - a_B + \langle |w|, b_{X_i}\rangle + b_B = \varepsilon + \xi_{2i}, \\
\langle |w|, a_{X_i}\rangle + a_B + \langle |w|, b_{X_i}\rangle + b_B - a_{Y_i} - b_{Y_i} = \varepsilon + \xi_{3i},
\end{array}
\right.
\tag{17}
$$

where C is a positive constant. We obtain the following Lagrangian function for the above problem

$$
\begin{aligned}
L(w, \xi_1, \xi_2, \xi_3) = {} & \frac{1}{2}\|w\|^2 + \frac{C}{2}\sum_{i=1}^{N}(\xi_{1i}^2 + \xi_{2i}^2 + \theta\xi_{3i}^2) \\
& - \sum_{i=1}^{N}\gamma_{1i}\Big(\varepsilon + \xi_{1i} - 2m_{Y_i} + \frac{1}{2}a_{Y_i} - \frac{1}{2}b_{Y_i} + 2\left(\langle w, m_{X_i}\rangle + m_B\right) \\
& \qquad - \frac{1}{2}\left(\langle |w|, a_{X_i}\rangle + a_B\right) + \frac{1}{2}\left(\langle |w|, b_{X_i}\rangle + \frac{1}{2}b_B\right)\Big) \\
& - \sum_{i=1}^{N}\gamma_{2i}\Big(\varepsilon + \xi_{2i} - a_{Y_i} + b_{Y_i} + \langle |w|, a_{X_i}\rangle + a_B - \langle |w|, b_{X_i}\rangle - b_B\Big) \\
& - \sum_{i=1}^{N}\gamma_{3i}\Big(\varepsilon + \xi_{3i} - \langle |w|, a_{X_i}\rangle - a_B - \langle |w|, b_{X_i}\rangle - b_B + a_{Y_i} + b_{Y_i}\Big).
\end{aligned}
$$

After calculating the desired derivatives and comparing them with zero we get

$$
\begin{pmatrix}
0 & 0 & 0 & 1' & 0' & 0' \\
0 & 0 & 0 & 0' & 1' & 0' \\
0 & 0 & 0 & 0' & 0' & 1' \\
1 & 0 & 0 & S_{11} & S_{12} & S_{13} \\
0 & 1 & 0 & S_{21} & S_{22} & S_{23} \\
0 & 0 & 1 & S_{31} & S_{32} & S_{33}
\end{pmatrix}
\begin{pmatrix}
m_B + \frac{3}{4}\varepsilon \\
a_B \\
b_B - \varepsilon \\
\gamma_1 \\
\gamma_2 \\
\gamma_3
\end{pmatrix}
=
\begin{pmatrix}
0 \\
0 \\
0 \\
m_Y \\
a_Y \\
b_Y
\end{pmatrix},
$$

with

$$
\begin{aligned}
S_{11} &= (2m_X - \frac{1}{2}sgn(w) * a_X + \frac{1}{2}sgn(w) * b_X)m_X' + I/(2C), \\
S_{12} &= (sgn(w) * a_X - sgn(w) * b_X)m_X' + I/(4C), \\
S_{13} &= -(sgn(w) * a_X + sgn(w) * b_X)m_X', \\
S_{21} &= (2m_X - \frac{1}{2}sgn(w) * a_X + \frac{1}{2}sgn(w) * b_X)(sgn(w) * a_X)', \\
S_{22} &= (a_X - b_X)a_X' + I/(2C), \\
S_{23} &= -(a_X + b_X)a_X' - I/(2\theta C), \\
S_{31} &= (2m_X - \frac{1}{2}sgn(w) * a_X + \frac{1}{2}sgn(w) * b_X)(sgn(w) * b_X)',
\end{aligned}
$$

$$S_{32} = (a_X - b_X)b_X' - I/(2C),$$
$$S_{33} = -(a_X + b_X)b_X' - I/(2\theta C),$$

where $\mathbf{1} = (1, \ldots, 1)'$, $\mathbf{0} = (0, \ldots, 0)'$ are vectors of length N, I is the identity matrix of size $N \times N$. Therefore,

$$
\begin{aligned}
w = &\sum_{i=1}^{N} \gamma_{1i} \left(2m_{X_i} - \frac{1}{2} \operatorname{sgn}(w) * a_{X_i} + \frac{1}{2} \operatorname{sgn}(w) * b_{X_i} \right) \\
&+ \sum_{i=1}^{N} \gamma_{2i} \left(\operatorname{sgn}(w) * a_{X_i} - \operatorname{sgn}(w) * b_{X_i} \right) \\
&- \sum_{i=1}^{N} \gamma_{3i} \left(\operatorname{sgn}(w) * a_{X_i} + \operatorname{sgn}(w) * b_{X_i} \right)
\end{aligned}
$$

and for a new observation X we get

$$Y(X) = (\langle w, m_X \rangle + m_B, \langle |w|, a_X \rangle + a_B, \langle |w|, b_X \rangle + b_B).$$

6 SVM in Fuzzy Linear Regression with Trapezoidal Fuzzy Numbers

6.1 Fuzzy Regression with Trapezoidal Fuzzy Numbers

In this section we generalize the results presented in Sect. 5 for the SVM fuzzy linear regression with trapezoidal fuzzy numbers. Let $\{X_i, Y_i\}_{i=1}^{N}$ denote a training data set, where $X_i = (\widetilde{T}(m_{X_{i1}}, n_{X_{i1}}, a_{X_{i1}}, b_{X_{i1}}), \ldots, \widetilde{T}(m_{X_{ip}}, n_{X_{ip}}, a_{X_{ip}}, b_{X_{ip}}))$ and $Y_i = \widetilde{T}(m_{Y_i}, n_{X_{i1}}, a_{Y_i}, b_{Y_i})$. Moreover, let $\boldsymbol{m}_{X_i} = (m_{X_{i1}}, \ldots, m_{X_{ip}})$, $\boldsymbol{n}_{X_i} = (n_{X_{i1}}, \ldots, n_{X_{ip}})$, $\boldsymbol{a}_{X_i} = (a_{X_{i1}}, \ldots, a_{X_{ip}})$, $\boldsymbol{b}_{X_i} = (b_{X_{i1}}, \ldots, b_{X_{ip}})$, $B = \widetilde{T}(m_B, n_B, a_B, b_B)$ and $w = (w_1, \ldots, w_p)$. We consider the following model

$$Y(X) = \langle w, X \rangle + B = (\langle w, m_X \rangle + m_B, \langle w, n_X \rangle + n_B, \langle |w|, a_X \rangle + a_B, \langle |w|, b_X \rangle + b_B),$$

where $|w| = (|w_1|, \ldots, |w_p|)$. Our goal is to minimize the distance between Y and $Y(X)$. Using d_2 norm definition (1) we get

$$
\begin{aligned}
d_2^2(Y, Y(X)) = &\frac{1}{4} \big((m_Y - a_Y) - (m_{Y(X)} - a_{Y(X)}) \big)^2 + \frac{1}{4} \big(m_Y - m_{Y(X)} \big)^2 \\
&+ \frac{1}{4} \big(n_Y - n_{Y(X)} \big)^2 + \frac{1}{4} \big((n_Y + b_Y) - (n_{Y(X)} + b_{Y(X)}) \big)^2.
\end{aligned}
$$

Hence the minimization of the distance $d_2(Y, Y(X))$ is equivalent to the minimization of each of the addends separately. In comparison with the triangular fuzzy numbers case here we have a one constraint more.

6.2 Linear Loss Function

By modifying the SVM method for the linear regression and linear loss function assumptions we obtain the following optimization problem: minimize

$$S(w, \xi_1, \xi_1^*, \xi_2, \xi_2^*, \xi_3, \xi_3^*, \xi_4, \xi_4^*) = \frac{1}{2}||w||^2 + C\sum_{k=1}^{4}\sum_{i=1}^{N}(\xi_{ki} + \xi_{ki}^*), \qquad (18)$$

subject to

$$\begin{cases} m_{Y_i} - \langle w, m_{X_i}\rangle - m_B \leq \varepsilon + \xi_{1i}, \\ \langle w, m_{X_i}\rangle + m_B - m_{Y_i} \leq \varepsilon + \xi_{1i}^*, \\ n_{Y_i} - \langle w, n_{X_i}\rangle - n_B \leq \varepsilon + \xi_{2i}, \\ \langle w, n_{X_i}\rangle + n_B - n_{Y_i} \leq \varepsilon + \xi_{2i}^*, \\ (m_{Y_i} - a_{Y_i}) - (\langle w, m_{X_i}\rangle + m_B - \langle |w|, a_{X_i}\rangle - a_B) \leq \varepsilon + \xi_{3i}, \\ (\langle w, m_{X_i}\rangle + m_B - \langle |w|, a_{X_i}\rangle - a_B) - (m_{Y_i} - a_{Y_i}) \leq \varepsilon + \xi_{3i}^*, \\ (n_{Y_i} + b_{Y_i}) - (\langle w, n_{X_i}\rangle + n_B + \langle |w|, b_{X_i}\rangle + b_B) \leq \varepsilon + \xi_{4i}, \\ (\langle w, n_{X_i}\rangle + n_B + \langle |w|, b_{X_i}\rangle + b_B) - (n_{Y_i} + b_{Y_i}) \leq \varepsilon + \xi_{4i}^*, \\ \xi_{ki}, \ \xi_{ki}^* \geq 0, \end{cases} \qquad (19)$$

where C is a positive constant. The Lagrangian function $L := L(w, m_B, n_B, a_B, b_B, \xi, \xi^*)$ for minimizing (18) subject to (19) is

$$L = \frac{1}{2}||w||^2 + C\sum_{k=1}^{4}\sum_{i=1}^{N}(\xi_{ki} + \xi_{ki}^*)$$

$$- \sum_{i=1}^{N}\gamma_{1i}(\varepsilon + \xi_{1i} - m_{Y_i} + \langle w, m_{X_i}\rangle + m_B) - \sum_{i=1}^{N}\gamma_{1i}^*(\varepsilon + \xi_{1i}^* - \langle w, m_{X_i}\rangle - m_B + m_{Y_i})$$

$$- \sum_{i=1}^{N}\gamma_{2i}(\varepsilon + \xi_{2i} - n_{Y_i} + \langle w, n_{X_i}\rangle + n_B) - \sum_{i=1}^{N}\gamma_{2i}^*(\varepsilon + \xi_{2i}^* - \langle w, n_{X_i}\rangle - n_B + n_{Y_i})$$

$$- \sum_{i=1}^{N}\gamma_{3i}(\varepsilon + \xi_{3i} - (m_{Y_i} - a_{Y_i}) + (\langle w, m_{X_i}\rangle + m_B - \langle |w|, a_{X_i}\rangle - a_B))$$

$$- \sum_{i=1}^{N}\gamma_{3i}^*(\varepsilon + \xi_{3i}^* - (\langle w, m_{X_i}\rangle + m_B - \langle |w|, a_{X_i}\rangle - a_B) + (m_{Y_i} - a_{Y_i}))$$

$$-\sum_{i=1}^{N}\gamma_{4i}\left(\varepsilon+\xi_{4i}-(n_{Y_i}+b_{Y_i})+\left(\langle \boldsymbol{w}, \boldsymbol{n}_{X_i}\rangle+n_B+\langle |\boldsymbol{w}|, \boldsymbol{b}_{X_i}\rangle+b_B\right)\right)$$

$$-\sum_{i=1}^{N}\gamma_{4i}^{*}\left(\varepsilon+\xi_{4i}^{*}-\left(\langle \boldsymbol{w}, \boldsymbol{n}_{X_i}\rangle+n_B+\langle |\boldsymbol{w}|, \boldsymbol{b}_{X_i}\rangle+b_B\right)+(n_{Y_i}+b_{Y_i})\right)$$

$$-\sum_{k=1}^{4}\sum_{i=1}^{N}\left(\eta_{ki}\xi_{ki}+\eta_{ki}^{*}\xi_{ki}^{*}\right).$$

After differentiating the above function we get

$$\frac{\partial L}{\partial \boldsymbol{w}} = \boldsymbol{w} - \sum_{i=1}^{N}\left((\gamma_{1i}-\gamma_{1i}^{*})\boldsymbol{m}_{X_i}+(\gamma_{2i}-\gamma_{2i}^{*})\boldsymbol{n}_{X_i}+(\gamma_{3i}-\gamma_{3i}^{*})(\boldsymbol{m}_{X_i}-\mathrm{sgn}(\boldsymbol{w})*\boldsymbol{a}_{X_i})\right.$$

$$\left.+(\gamma_{4i}-\gamma_{4i}^{*})(\boldsymbol{n}_{X_i}+\mathrm{sgn}(\boldsymbol{w})*\boldsymbol{b}_{X_i})\right),$$

$$\frac{\partial L}{\partial m_B} = -\sum_{i=1}^{N}\left(\gamma_{1i}-\gamma_{1i}^{*}+\gamma_{3i}-\gamma_{3i}^{*}\right),$$

$$\frac{\partial L}{\partial n_B} = -\sum_{i=1}^{N}\left(\gamma_{2i}-\gamma_{2i}^{*}+\gamma_{4i}-\gamma_{4i}^{*}\right),$$

$$\frac{\partial L}{\partial a_B} = -\sum_{i=1}^{N}\left(\gamma_{3i}-\gamma_{3i}^{*}\right),$$

$$\frac{\partial L}{\partial b_B} = -\sum_{i=1}^{N}\left(\gamma_{4i}-\gamma_{4i}^{*}\right),$$

$$\frac{\partial L}{\partial \xi_{ki}} = C - \gamma_{ki} - \eta_{ki},$$

$$\frac{\partial L}{\partial \xi_{ki}^{*}} = C - \gamma_{ki}^{*} - \eta_{ki}^{*}$$

and comparing the above functions with zero we get

$$\boldsymbol{w} = \sum_{i=1}^{N}\left((\gamma_{1i}-\gamma_{1i}^{*})\boldsymbol{m}_{X_i}+(\gamma_{2i}-\gamma_{2i}^{*})\boldsymbol{n}_{X_i}+(\gamma_{3i}-\gamma_{3i}^{*})(\boldsymbol{m}_{X_i}-\mathrm{sgn}(\boldsymbol{w})*\boldsymbol{a}_{X_i})\right.$$

$$\left.+(\gamma_{4i}-\gamma_{4i}^{*})(\boldsymbol{n}_{X_i}+\mathrm{sgn}(\boldsymbol{w})*\boldsymbol{b}_{X_i})\right), \tag{20}$$

$$\sum_{i=1}^{N}\left(\gamma_{1i}-\gamma_{1i}^{*}\right)=0, \quad \sum_{i=1}^{N}\left(\gamma_{2i}-\gamma_{2i}^{*}\right)=0,$$

$$\sum_{i=1}^{N}\left(\gamma_{3i}-\gamma_{3i}^{*}\right)=0 \quad \sum_{i=1}^{N}\left(\gamma_{4i}-\gamma_{4i}^{*}\right)=0$$

where $\gamma_{ki}, \gamma_{ki}^{*} \leq C$ dla $\xi_{ki}=0$, $\xi_{ki}^{*}=0$. Hence, we can construct a dual problem by putting the above values into the Lagrange function considered above

$$
\begin{aligned}
W(\gamma, \gamma^*) = \frac{1}{2}\Big(& \sum_{i=1}^{N}\sum_{j=1}^{N}(\gamma_{1i} - \gamma_{1i}^*)(\gamma_{1j} - \gamma_{1j}^*)\langle m_{X_i}, m_{X_j}\rangle \\
& + 2\sum_{i=1}^{N}\sum_{j=1}^{N}(\gamma_{1i} - \gamma_{1i}^*)(\gamma_{2j} - \gamma_{2j}^*)\langle m_{X_i}, n_{X_j}\rangle \\
& + 2\sum_{i=1}^{N}\sum_{j=1}^{N}(\gamma_{1i} - \gamma_{1i}^*)(\gamma_{3j} - \gamma_{3j}^*)\langle m_{X_i}, m_{X_j} - \mathrm{sgn}(w)*a_{X_j}\rangle \\
& + 2\sum_{i=1}^{N}\sum_{j=1}^{N}(\gamma_{1i} - \gamma_{1i}^*)(\gamma_{4j} - \gamma_{4j}^*)\langle m_{X_i}, n_{X_j} + \mathrm{sgn}(w)*b_{X_j}\rangle \\
& + \sum_{i=1}^{N}\sum_{j=1}^{N}(\gamma_{2i} - \gamma_{2i}^*)(\gamma_{2j} - \gamma_{2j}^*)\langle n_{X_i}, n_{X_j}\rangle \\
& + 2\sum_{i=1}^{N}\sum_{j=1}^{N}(\gamma_{2i} - \gamma_{2i}^*)(\gamma_{3j} - \gamma_{3j}^*)\langle n_{X_i}, m_{X_j} - \mathrm{sgn}(w)*a_{X_j}\rangle \\
& + 2\sum_{i=1}^{N}\sum_{j=1}^{N}(\gamma_{2i} - \gamma_{2i}^*)(\gamma_{4j} - \gamma_{4j}^*)\langle n_{X_i}, n_{X_j} + \mathrm{sgn}(w)*b_{X_j}\rangle \\
& + \sum_{i=1}^{N}\sum_{j=1}^{N}(\gamma_{3i} - \gamma_{3i}^*)(\gamma_{3j} - \gamma_{3j}^*)\langle m_{X_j} - \mathrm{sgn}(w)*b_{X_i}, m_{X_j} - \mathrm{sgn}(w)*b_{X_j}\rangle \\
& + 2\sum_{i=1}^{N}\sum_{j=1}^{N}(\gamma_{3i} - \gamma_{3i}^*)(\gamma_{4j} - \gamma_{4j}^*)\langle m_{X_j} - \mathrm{sgn}(w)*b_{X_i}, n_{X_j} + \mathrm{sgn}(w)*b_{X_j}\rangle \\
& + \sum_{i=1}^{N}\sum_{j=1}^{N}(\gamma_{4i} - \gamma_{4i}^*)(\gamma_{4j} - \gamma_{4j}^*)\langle n_{X_j} + \mathrm{sgn}(w)*b_{X_i}, n_{X_j} + \mathrm{sgn}(w)*b_{X_j}\rangle \Big) \\
& + \sum_{i=1}^{N}(\gamma_{1i} - \gamma_{1i}^*)m_{Y_i} \\
& - \sum_{i=1}^{N}(\gamma_{1i} - \gamma_{1i}^*)\Big(\sum_{j=1}^{N}\big((\gamma_{1j} - \gamma_{1j}^*)\langle m_{X_i}, m_{X_j}\rangle + (\gamma_{2j} - \gamma_{2j}^*)\langle m_{X_i}, n_{X_j}\rangle \\
& + (\gamma_{3j} - \gamma_{3j}^*)\langle m_{X_i}, m_{X_j} - \mathrm{sgn}(w)*a_{X_j}\rangle \\
& + (\gamma_{4j} - \gamma_{4j}^*)\langle m_{X_i}, n_{X_j} + \mathrm{sgn}(w)*b_{X_j}\rangle\big)\Big) \\
& + \sum_{i=1}^{N}(\gamma_{2i} - \gamma_{2i}^*)n_{Y_i} - \sum_{i=1}^{N}(\gamma_{2i} - \gamma_{2i}^*)\Big(\sum_{j=1}^{N}\big((\gamma_{1j} - \gamma_{1j}^*)\langle n_{X_i}, m_{X_j}\rangle \\
& + (\gamma_{2j} - \gamma_{2j}^*)\langle n_{X_i}, n_{X_j}\rangle + (\gamma_{3j} - \gamma_{3j}^*)\langle n_{X_i}, m_{X_j} - \mathrm{sgn}(w)*a_{X_j}\rangle \\
& + (\gamma_{4j} - \gamma_{4j}^*)\langle n_{X_i}, n_{X_j} + \mathrm{sgn}(w)*b_{X_j}\rangle\big)\Big) + \sum_{i=1}^{N}(\gamma_{3i} - \gamma_{3i}^*)(m_{Y_i} - a_{Y_i})
\end{aligned}
$$

$$-\sum_{i=1}^{N}(\gamma_{3i}-\gamma_{3i}^{*})\Big(\sum_{j=1}^{N}\big((\gamma_{1j}-\gamma_{1j}^{*})\langle \boldsymbol{m}_{X_i}-\mathrm{sgn}(\boldsymbol{w})*\boldsymbol{a}_{X_i},\boldsymbol{m}_{X_j}\rangle$$

$$+(\gamma_{2j}-\gamma_{2j}^{*})\langle \boldsymbol{m}_{X_i}-\mathrm{sgn}(\boldsymbol{w})*\boldsymbol{a}_{X_i},\boldsymbol{n}_{X_j}\rangle$$

$$+(\gamma_{3j}-\gamma_{3j}^{*})\langle \boldsymbol{m}_{X_i}-\mathrm{sgn}(\boldsymbol{w})*\boldsymbol{a}_{X_i},\boldsymbol{m}_{X_j}-\mathrm{sgn}(\boldsymbol{w})*\boldsymbol{a}_{X_j}\rangle$$

$$+(\gamma_{4j}-\gamma_{4j}^{*})\langle \boldsymbol{m}_{X_i}-\mathrm{sgn}(\boldsymbol{w})*\boldsymbol{a}_{X_i},\boldsymbol{n}_{X_j}+\mathrm{sgn}(\boldsymbol{w})*\boldsymbol{b}_{X_j}\rangle\big)\Big)$$

$$+\sum_{i=1}^{N}(\gamma_{4i}-\gamma_{4i}^{*})(\boldsymbol{n}_{Y_i}+\boldsymbol{b}_{Y_i})$$

$$-\sum_{i=1}^{N}(\gamma_{4i}-\gamma_{4i}^{*})\Big(\sum_{j=1}^{N}\big((\gamma_{1j}-\gamma_{1j}^{*})\langle \boldsymbol{n}_{X_i}+\mathrm{sgn}(\boldsymbol{w})*\boldsymbol{b}_{X_i},\boldsymbol{m}_{X_j}\rangle$$

$$+(\gamma_{2j}-\gamma_{2j}^{*})\langle \boldsymbol{n}_{X_i}+\mathrm{sgn}(\boldsymbol{w})*\boldsymbol{b}_{X_i},\boldsymbol{n}_{X_j}\rangle$$

$$+(\gamma_{3j}-\gamma_{3j}^{*})\langle \boldsymbol{n}_{X_i}+\mathrm{sgn}(\boldsymbol{w})*\boldsymbol{b}_{X_i},\boldsymbol{m}_{X_j}-\mathrm{sgn}(\boldsymbol{w})*\boldsymbol{a}_{X_j}\rangle$$

$$+(\gamma_{4j}-\gamma_{4j}^{*})\langle \boldsymbol{n}_{X_i}+\mathrm{sgn}(\boldsymbol{w})*\boldsymbol{b}_{X_i},\boldsymbol{n}_{X_j}+\mathrm{sgn}(\boldsymbol{w})*\boldsymbol{b}_{X_j}\rangle\big)\Big)$$

$$-\varepsilon\sum_{k=1}^{4}\sum_{i=1}^{N}(\gamma_{ki}+\gamma_{ki}^{*})$$

$$=-\frac{1}{2}\|\boldsymbol{w}\|^{2}-\varepsilon\sum_{k=1}^{4}\sum_{i=1}^{N}(\gamma_{ki}+\gamma_{ki}^{*})+\sum_{i=1}^{N}(\gamma_{1i}-\gamma_{1i}^{*})\boldsymbol{m}_{Y_i}+\sum_{i=1}^{N}(\gamma_{2i}-\gamma_{2i}^{*})\boldsymbol{n}_{Y_i}$$

$$+\sum_{i=1}^{N}(\gamma_{3i}-\gamma_{3i}^{*})(\boldsymbol{m}_{Y_i}-\boldsymbol{a}_{Y_i})+\sum_{i=1}^{N}(\gamma_{4i}-\gamma_{4i}^{*})(\boldsymbol{n}_{Y_i}+\boldsymbol{b}_{Y_i}).$$

Therefore, the Lagrange multipliers $\gamma_{ki},\gamma_{ki}^{*}$ can be obtained by maximizing the following function

$$W(\boldsymbol{w},\gamma_{1},\gamma_{1}^{*},\gamma_{2},\gamma_{2}^{*},\gamma_{3},\gamma_{3}^{*},\gamma_{4},\gamma_{4}^{*})=-\frac{1}{2}\|\boldsymbol{w}\|^{2}-\varepsilon\sum_{k=1}^{4}\sum_{i=1}^{N}(\gamma_{ki}+\gamma_{ki}^{*})$$

$$+\sum_{i=1}^{N}(\gamma_{1i}-\gamma_{1i}^{*})m_{Y_i}+\sum_{i=1}^{N}(\gamma_{2i}-\gamma_{2i}^{*})n_{Y_i}$$

$$+\sum_{i=1}^{N}(\gamma_{3i}-\gamma_{3i}^{*})(m_{Y_i}-a_{Y_i})+\sum_{i=1}^{N}(\gamma_{4i}-\gamma_{4i}^{*})(n_{Y_i}+b_{Y_i})$$

subject to

$$\sum_{i=1}^{N} (\gamma_{ki} - \gamma_{ki}^*) = 0,$$

$$\gamma_{ki}, \gamma_{ki}^* \in [0, C], \quad k = 1, \ldots, 4 \ i = 1, \ldots, N.$$

The Lagrangian function for the above problem looks as follows

$$R(\gamma_1, \gamma_1^*, \gamma_2, \gamma_2^*, \gamma_3, \gamma_3^*, \gamma_4, \gamma_4^*) = -\frac{1}{2}\|w\|^2 - \varepsilon \sum_{k=1}^{4} \sum_{i=1}^{N} (\gamma_{ki} + \gamma_{ki}^*)$$

$$+ \sum_{i=1}^{N} (\gamma_{1i} - \gamma_{1i}^*)m_{Y_i} + \sum_{i=1}^{N} (\gamma_{2i} - \gamma_{2i}^*)n_{Y_i}$$

$$+ \sum_{i=1}^{N} (\gamma_{3i} - \gamma_{3i}^*)(m_{Y_i} - a_{Y_i}) + \sum_{i=1}^{N} (\gamma_{4i} - \gamma_{4i}^*)(n_{Y_i} + b_{Y_i})$$

$$+ \sum_{k=1}^{4} \lambda_k \left(\sum_{i=1}^{N} (\gamma_{ki} - \gamma_{ki}^*) \right). \tag{21}$$

By differentiating (21) with respect to $\gamma_1, \gamma_1^*, \gamma_2, \gamma_2^*, \gamma_3, \gamma_3^*, \gamma_4, \gamma_4^*$ and comparing the above functions with zero we obtain the following linear equations

$$\begin{pmatrix} S_{11} & -S_{11} & S_{12} & -S_{12} & S_{13} & -S_{13} & S_{14} & -S_{14} & -1 & 0 & 0 & 0 \\ S_{12}' & -S_{12}' & S_{22} & -S_{22} & S_{23} & -S_{23} & S_{24} & -S_{24} & 0 & -1 & 0 & 0 \\ S_{13}' & -S_{13}' & S_{23}' & -S_{23}' & S_{33} & -S_{33} & S_{34} & -S_{34} & 0 & 0 & -1 & 0 \\ S_{14}' & -S_{14}' & S_{24}' & -S_{24}' & S_{34}' & -S_{34}' & S_{44} & -S_{44} & 0 & 0 & 0 & -1 \\ 1' & -1' & 0' & 0' & 0' & 0' & 0' & 0' & 0 & 0 & 0 & 0 \\ 0' & 0' & 1' & -1' & 0' & 0' & 0' & 0' & 0 & 0 & 0 & 0 \\ 0' & 0' & 0' & 0' & 1' & -1' & 0' & 0' & 0 & 0 & 0 & 0 \\ 0' & 0' & 0' & 0' & 0' & 0' & 1' & -1' & 0 & 0 & 0 & 0 \end{pmatrix} \begin{pmatrix} \gamma_1 \\ \gamma_1^* \\ \gamma_2 \\ \gamma_2^* \\ \gamma_3 \\ \gamma_3^* \\ \gamma_4 \\ \gamma_4^* \\ \lambda_1 \\ \lambda_2 \\ \lambda_3 \\ \lambda_4 \end{pmatrix} = \begin{pmatrix} m_Y - \varepsilon \\ n_Y - \varepsilon \\ m_Y - a_Y - \varepsilon \\ n_Y + b_Y - \varepsilon \\ 0 \\ 0 \\ 0 \\ 0 \end{pmatrix},$$

with

$$S_{11} = m_X m_X' \left(\frac{1}{2}J + \frac{1}{2}I \right),$$

$$S_{12} = m_X n_X',$$

$$S_{13} = m_X (m_X - \text{sgn}(w) * a_X)',$$

$$S_{14} = m_X (n_X + \text{sgn}(w) * b_X)',$$

$$S_{22} = n_X m_X' \left(\frac{1}{2}J + \frac{1}{2}I \right),$$

$$S_{23} = n_X (m_X - \text{sgn}(w) * a_X)',$$

$$S_{24} = n_X (n_X + \text{sgn}(w) * b_X)',$$

$$S_{33} = (\boldsymbol{m}_X - \mathrm{sgn}(\boldsymbol{w}) * \boldsymbol{a}_X)(\boldsymbol{m}_X - \mathrm{sgn}(\boldsymbol{w}) * \boldsymbol{a}_X)'(\tfrac{1}{2}\boldsymbol{J} + \tfrac{1}{2}\boldsymbol{I}),$$

$$S_{34} = (\boldsymbol{m}_X - \mathrm{sgn}(\boldsymbol{w}) * \boldsymbol{a}_X)(\boldsymbol{n}_X + \mathrm{sgn}(\boldsymbol{w}) * \boldsymbol{b}_X)',$$

$$S_{44} = (\boldsymbol{n}_X + \mathrm{sgn}(\boldsymbol{w}) * \boldsymbol{b}_X)(\boldsymbol{n}_X + \mathrm{sgn}(\boldsymbol{w}) * \boldsymbol{b}_X)'(\tfrac{1}{2}\boldsymbol{J} + \tfrac{1}{2}\boldsymbol{I}),$$

where $\boldsymbol{1} = (1, \ldots, 1)'$, $\boldsymbol{0} = (0, \ldots, 0)'$ are vectors of length N, $\boldsymbol{I}$ is the identity matrix of size $N \times N$, $\boldsymbol{J}$ is the matrix of ones of size $N \times N$, i.e. $\boldsymbol{J} = [1]_{N \times N}$.

By (20) we get

$$\boldsymbol{w} = \sum_{i=1}^{N} \Big((\gamma_{1i} - \gamma_{1i}^*)\boldsymbol{m}_{X_i} + (\gamma_{2i} - \gamma_{2i}^*)\boldsymbol{n}_{X_i} + (\gamma_{3i} - \gamma_{3i}^*)(\boldsymbol{m}_{X_i} - \mathrm{sgn}(\boldsymbol{w}) * \boldsymbol{a}_{X_i})$$

$$+ (\gamma_{4i} - \gamma_{4i}^*)(\boldsymbol{n}_{X_i} + \mathrm{sgn}(\boldsymbol{w}) * \boldsymbol{b}_{X_i}) \Big),$$

and now the only unknown value to be found is $B = (m_B, n_B, a_B, b_B)$. By the Karush–Kuhn–Tucker theorem we obtain

$$\begin{cases} \gamma_{1i}(\varepsilon + \xi_{1i} - m_{Y_i} + \langle \boldsymbol{w}, \boldsymbol{m}_{X_i}\rangle + m_B) = 0, \\ \gamma_{1i}^*(\varepsilon + \xi_{1i}^* + m_{Y_i} - \langle \boldsymbol{w}, \boldsymbol{m}_{X_i}\rangle - m_B) = 0, \\ \gamma_{2i}(\varepsilon + \xi_{2i} - n_{Y_i} + \langle \boldsymbol{w}, \boldsymbol{n}_{X_i}\rangle + n_B) = 0, \\ \gamma_{2i}^*(\varepsilon + \xi_{2i}^* + n_{Y_i} - \langle \boldsymbol{w}, \boldsymbol{n}_{X_i}\rangle - n_B) = 0, \\ (C - \gamma_{1i})\xi_{1i} = 0, \\ (C - \gamma_{1i}^*)\xi_{1i}^* = 0, \\ (C - \gamma_{2i})\xi_{2i} = 0, \\ (C - \gamma_{2i}^*)\xi_{2i}^* = 0, \end{cases}$$

and

$$\begin{cases} m_B = m_{Y_i} - \langle \boldsymbol{w}, \boldsymbol{m}_{X_i}\rangle - \varepsilon \ \ \text{dla} \ \ \gamma_{1i} \in (0, C), \\ m_B = m_{Y_i} - \langle \boldsymbol{w}, \boldsymbol{m}_{X_i}\rangle + \varepsilon \ \ \text{dla} \ \ \gamma_{1i}^* \in (0, C), \\ n_B = n_{Y_i} - \langle \boldsymbol{w}, \boldsymbol{n}_{X_i}\rangle - \varepsilon \ \ \text{dla} \ \ \gamma_{2i} \in (0, C), \\ n_B = n_{Y_i} - \langle \boldsymbol{w}, \boldsymbol{n}_{X_i}\rangle + \varepsilon \ \ \text{dla} \ \ \gamma_{2i}^* \in (0, C). \end{cases}$$

Similarly as in case of triangular fuzzy numbers, to find a_B, b_B we should solve the following problems

$$\min_{a_B \geq 0} \sum_{i=1}^{N} \big(|m_{Y_i} - a_{Y_i} - (\langle \boldsymbol{w}, \boldsymbol{m}_{X_i}\rangle + m_B - \langle |\boldsymbol{w}|, \boldsymbol{a}_{X_i}\rangle - a_B)|_{\varepsilon} \big),$$

$$\min_{b_B \geq 0} \sum_{i=1}^{N} \big(|n_{Y_i} + b_{Y_i} - (\langle \boldsymbol{w}, \boldsymbol{n}_{X_i}\rangle + n_B + \langle |\boldsymbol{w}|, \boldsymbol{b}_{X_i}\rangle + b_B)|_{\varepsilon} \big).$$

6.3 Quadratic Loss Function

For the quadratic loss function all inequality constraint simplify to equality constraints. Hence we have to minimize

$$S(\boldsymbol{w}, \boldsymbol{\xi}_1, \boldsymbol{\xi}_2, \boldsymbol{\xi}_3, \boldsymbol{\xi}_4) = \frac{1}{2}||\boldsymbol{w}||^2 + \frac{C}{2}\sum_{k=1}^{4}\sum_{i=1}^{N}\xi_{ki}^2 \tag{22}$$

subject to

$$\begin{cases} m_{Y_i} - \langle \boldsymbol{w}, \boldsymbol{m}_X \rangle - m_B = \varepsilon + \xi_{1i}, \\ n_{Y_i} - \langle \boldsymbol{w}, \boldsymbol{n}_X \rangle - n_B = \varepsilon + \xi_{2i}, \\ (m_{Y_i} - a_{Y_i}) - (\langle \boldsymbol{w}, \boldsymbol{m}_{X_i} \rangle + m_B - \langle |\boldsymbol{w}|, \boldsymbol{a}_{X_i} \rangle - a_B) = \varepsilon + \xi_{3i}, \\ (n_{Y_i} + b_{Y_i}) - (\langle \boldsymbol{w}, \boldsymbol{n}_{X_i} \rangle + n_B + \langle |\boldsymbol{w}|, \boldsymbol{b}_{X_i} \rangle + b_B) = \varepsilon + \xi_{4i}, \end{cases} \tag{23}$$

where C is a soft margin parameter.

The Lagrangian function for the minimization of (22) subject to (23) looks as follows

$$L(\boldsymbol{w}, \boldsymbol{\xi}_1, \boldsymbol{\xi}_2, \boldsymbol{\xi}_3, \boldsymbol{\xi}_4) = \frac{1}{2}||\boldsymbol{w}||^2 + \frac{C}{2}\sum_{k=1}^{4}\sum_{i=1}^{N}\xi_{ki}^2$$

$$- \sum_{i=1}^{N}\gamma_{1i}\left(\varepsilon + \xi_{1i} - m_{Y_i} + \langle \boldsymbol{w}, \boldsymbol{m}_{X_i} \rangle + m_B\right)$$

$$- \sum_{i=1}^{N}\gamma_{2i}\left(\varepsilon + \xi_{2i} - n_{Y_i} + \langle \boldsymbol{w}, \boldsymbol{n}_{X_i} \rangle + n_B\right)$$

$$- \sum_{i=1}^{N}\gamma_{3i}\left(\varepsilon + \xi_{3i} - (m_{Y_i} - a_{Y_i}) + (\langle \boldsymbol{w}, \boldsymbol{m}_{X_i} \rangle + m_B - \langle |\boldsymbol{w}|, \boldsymbol{a}_{X_i} \rangle - a_B)\right)$$

$$- \sum_{i=1}^{N}\gamma_{4i}\left(\varepsilon + \xi_{4i} - (n_{Y_i} + b_{Y_i}) + (\langle \boldsymbol{w}, \boldsymbol{n}_{X_i} \rangle + n_B + \langle |\boldsymbol{w}|, \boldsymbol{b}_{X_i} \rangle + b_B)\right),$$

where $\gamma_{ki} \in \mathbb{R}$. After calculating the derivatives and comparing them with zero we get

$$\begin{pmatrix} 0 & 0 & 0 & 0 & 1' & 0' & 1' & 0' \\ 0 & 0 & 0 & 0 & 0' & 1' & 0' & 1' \\ 0 & 0 & 0 & 0 & 0' & 0' & 1' & 0' \\ 0 & 0 & 0 & 0 & 0' & 0' & 0' & 1' \\ 1 & 0 & 0 & 0 & S_{11} & S_{12} & S_{13} & S_{14} \\ 1 & -1 & 0 & 0 & S'_{12} & S_{22} & S_{23} & S_{24} \\ 1 & 0 & 1 & 0 & S'_{13} & S'_{23} & S_{33} & S_{34} \\ 1 & 0 & 1 & 0 & S'_{14} & S'_{24} & S'_{34} & S_{44} \end{pmatrix} \begin{pmatrix} m_B + \varepsilon \\ n_B + \varepsilon \\ a_B - \varepsilon \\ b_B + \varepsilon \\ \gamma_1 \\ \gamma_2 \\ \gamma_3 \\ \gamma_4 \end{pmatrix} = \begin{pmatrix} 0 \\ 0 \\ 0 \\ 0 \\ m_Y \\ n_Y \\ m_Y - a_Y \\ n_Y + b_Y \end{pmatrix},$$

with

$$S_{11} = m_X m'_X + I/C,$$
$$S_{12} = m_X n'_X,$$
$$S_{13} = m_X (m_X - \text{sgn}(w) * a_X)',$$
$$S_{14} = m_X (n_X + \text{sgn}(w) * b_X)',$$
$$S_{22} = n_X n'_X + I/C,$$
$$S_{23} = n_X (m_X - \text{sgn}(w) * a_X)',$$
$$S_{24} = n_X (n_X + \text{sgn}(w) * b_X)',$$
$$S_{33} = (m_X - \text{sgn}(w) * a_X)(m_X - \text{sgn}(w) * a_X)',$$
$$S_{34} = (m_X - \text{sgn}(w) * a_X)(n_X + \text{sgn}(w) * b_X)',$$
$$S_{44} = (n_X + \text{sgn}(w) * b_X)(n_X + \text{sgn}(w) * b_X)',$$

where $\mathbf{1} = (1, \ldots, 1)'$, $\mathbf{0} = (0, \ldots, 0)'$ are vectors of length N, I is the identity matrix of size $N \times N$. Finally, for a new observation X we get

$$Y(X) = \langle w, X \rangle + B = (\langle w, m_X \rangle + m_B, \langle w, n_X \rangle + n_B, \langle |w|, a_X \rangle + a_B, \langle |w|, b_X \rangle + b_B).$$

6.4 A Loss Function Based on the Trutschnig Distance

Similarly as for triangular fuzzy numbers our goal now is to solve the linear fuzzy regression problem utilizing the Trutschnig distance. Let $A = \widetilde{T}(m_A, n_A, a_A, b_A)$ and $B = \widetilde{T}(m_B, n_B, a_B, b_B)$ be two arbitrary trapezoidal fuzzy numbers. By (4) we get

$$t_\theta^2(A, B) = \int_0^1 \left([\text{mid}(A_\alpha) - \text{mid}(B_\alpha)]^2 + \theta[\text{spr}(A_\alpha) - \text{spr}(B_\alpha)]^2 \right) d\alpha.$$

Let us denote

$$I_1 := \int_0^1 [\text{mid}(A_\alpha) - \text{mid}(B_\alpha)]^2 d\alpha,$$

$$I_2 := \int_0^1 [\text{spr}(A_\alpha) - \text{spr}(B_\alpha)]^2 d\alpha.$$

For the trapezoidal fuzzy numbers we get

$$I_1 = \frac{1}{4} \Big((m_A + n_A - a_A + b_A - m_B - n_B + a_B - b_B)^2$$

$$+ (m_A + n_A - a_A + b_A - m_B - n_B + a_B - b_B)(a_A - b_A - a_B + b_B)$$

$$+\frac{1}{3}(a_A - b_A - a_B + b_B)^2)$$

$$= \frac{1}{4}\left(\left(m_A + n_A - \frac{1}{2}a_A + \frac{1}{2}b_A - m_B - n_B + \frac{1}{2}a_B - \frac{1}{2}b_B\right)^2 + \frac{1}{12}(a_A - b_A - a_B + b_B)^2\right),$$

$$I_2 = \frac{1}{4}((n_A - m_A + a_A + b_A - n_B + m_B - a_B - b_B)^2$$

$$+ (n_A - m_A + a_A + b_A - n_B + m_B - a_B - b_B)(a_B + b_B - a_A - b_A)$$

$$+\frac{1}{3}(a_B + b_B - a_A - b_A)^2)$$

$$= \frac{1}{4}\left(\left(n_A - m_A + \frac{1}{2}a_A + \frac{1}{2}b_A - n_B + m_B - \frac{1}{2}a_B - \frac{1}{2}b_B\right)^2 + \frac{1}{12}(a_B + b_B - a_A - b_A)^2\right).$$

Hence, the minimization of the distance $t_\theta^2(A, B)$ is equivalent to the minimization of

$$E_1 = \left(m_A + n_A - \frac{1}{2}a_A + \frac{1}{2}b_A - m_B - n_B + \frac{1}{2}a_B - \frac{1}{2}b_B\right)^2,$$

$$E_2 = (a_A - b_A - a_B + b_B)^2,$$

$$E_3 = \left(n_A - m_A + \frac{1}{2}a_A + \frac{1}{2}b_A - n_B + m_B - \frac{1}{2}a_B - \frac{1}{2}b_B\right)^2,$$

$$E_4 = (a_B + b_B - a_A - b_A)^2.$$

To find estimators of a vector w and fuzzy number $B = \tilde{T}(m_B, n_B, a_B, b_B)$, we have to minimize

$$S(w, \xi_1, \xi_2, \xi_3, \xi_4) = \frac{1}{2}||w||^2 + \frac{C}{2}\sum_{i=1}^{N}(\xi_{1i}^2 + \xi_{2i}^2 + \theta\xi_{3i}^2 + \theta\xi_{4i}^2) \quad (24)$$

subject to

$$\begin{cases} m_{Y_i} + n_{Y_i} - \frac{1}{2}a_{Y_i} + \frac{1}{2}b_{Y_i} - \langle w, m_{X_i}\rangle - m_B - \langle w, n_{X_i}\rangle - n_B \\ \qquad + \frac{1}{2}(\langle|w|, a_{X_i}\rangle + a_B) - \frac{1}{2}(\langle|w|, b_{X_i}\rangle + b_B) = \varepsilon + \xi_{1i}, \\ a_{Y_i} - b_{Y_i} - \langle|w|, a_{X_i}\rangle - a_B + \langle|w|, b_{X_i}\rangle + b_B = \varepsilon + \xi_{2i}, \\ n_{Y_i} - m_{Y_i} + \frac{1}{2}a_{Y_i} + \frac{1}{2}b_{Y_i} + \langle w, m_{X_i}\rangle + m_B - \langle w, n_{X_i}\rangle - n_B \\ \qquad - \frac{1}{2}(\langle|w|, a_{X_i}\rangle + a_B) - \frac{1}{2}(\langle|w|, b_{X_i}\rangle + b_B) = \varepsilon + \xi_{3i}, \\ \langle|w|, a_{X_i}\rangle + a_B + \langle|w|, b_{X_i}\rangle + b_B - a_{Y_i} - b_{Y_i} = \varepsilon + \xi_{4i}, \end{cases}$$

where C is a positive constant.

The Lagrangian function for the above problem is

$$L(w, \xi_1, \xi_2, \xi_3, \xi_4) = \frac{1}{2}\|w\|^2 + \frac{C}{2}\sum_{i=1}^{N}(\xi_{1i}^2 + \xi_{2i}^2 + \theta\xi_{3i}^2 + \theta\xi_{4i}^2)$$

$$-\sum_{i=1}^{N}\gamma_{1i}\left(\varepsilon + \xi_{1i} - m_{Y_i} - n_{Y_i} + \frac{1}{2}a_{Y_i} - \frac{1}{2}b_{Y_i} + \langle w, m_{X_i}\rangle + m_B\right.$$

$$+ \langle w, n_{X_i}\rangle + n_B - \frac{1}{2}(\langle |w|, a_{X_i}\rangle + a_B) + \frac{1}{2}\left(\langle |w|, b_{X_i}\rangle + \frac{1}{2}b_B\right)\right)$$

$$-\sum_{i=1}^{N}\gamma_{2i}\left(\varepsilon + \xi_{2i} - a_{Y_i} + b_{Y_i} + \langle |w|, a_{X_i}\rangle + a_B - \langle |w|, b_{X_i}\rangle - b_B\right)$$

$$-\sum_{i=1}^{N}\gamma_{3i}\left(\varepsilon + \xi_{3i} - n_{Y_i} + m_{Y_i} - \frac{1}{2}a_{Y_i} - \frac{1}{2}b_{Y_i} - \langle w, m_{X_i}\rangle - m_B\right.$$

$$+ \langle w, n_{X_i}\rangle + n_B + \frac{1}{2}(\langle |w|, a_{X_i}\rangle + a_B) + \frac{1}{2}\left(\langle |w|, b_{X_i}\rangle + \frac{1}{2}b_B\right)\right)$$

$$-\sum_{i=1}^{N}\gamma_{4i}\left(\varepsilon + \xi_{4i} - \langle |w|, a_{X_i}\rangle - a_B - \langle |w|, b_{X_i}\rangle - b_B + a_{Y_i} + b_{Y_i}\right).$$

After calculating all desired derivatives of the above function and comparing them with zero we get

$$\begin{pmatrix} 0 & 0 & 0 & 0 & 1' & 0' & 0' & 0' \\ 0 & 0 & 0 & 0 & 0' & 1' & 0' & 0' \\ 0 & 0 & 0 & 0 & 0' & 0' & 1' & 0' \\ 0 & 0 & 0 & 0 & 0' & 0' & 0' & 1' \\ 1 & 0 & 0 & 0 & S_{11} & S_{12} & S_{13} & S_{14} \\ 0 & 1 & 0 & 0 & S_{21} & S_{22} & S_{23} & S_{24} \\ 0 & 0 & 1 & 0 & S_{31} & S_{32} & S_{33} & S_{34} \\ 0 & 0 & 0 & 1 & S_{41} & S_{42} & S_{43} & S_{44} \end{pmatrix} \begin{pmatrix} m_B \\ n_B + \frac{3}{2}\varepsilon \\ a_B \\ b_B - \varepsilon \\ \gamma_1 \\ \gamma_2 \\ \gamma_3 \\ \gamma_4 \end{pmatrix} = \begin{pmatrix} 0 \\ 0 \\ 0 \\ 0 \\ m_Y \\ n_Y \\ a_Y \\ b_Y \end{pmatrix},$$

with

$$S_{11} = (m_X + n_X - \frac{1}{2}\,\mathrm{sgn}(w) * a_X + \frac{1}{2}\,\mathrm{sgn}(w) * b_X)m_X' + I/(2C),$$

$$S_{12} = (\mathrm{sgn}(w) * a_X - \mathrm{sgn}(w) * b_X)m_X' + I/(4C),$$

$$S_{13} = (n_X - m_X + \frac{1}{2}\,\mathrm{sgn}(w) * a_X + \frac{1}{2}\,\mathrm{sgn}(w) * b_X)m_X' - I/(2\theta C),$$

$$S_{14} = -(\mathrm{sgn}(w) * a_X + \mathrm{sgn}(w) * b_X)m_X' - I/(4\theta C),$$

$$S_{21} = (m_X + n_X - \frac{1}{2}\,\mathrm{sgn}(w) * a_X + \frac{1}{2}\,\mathrm{sgn}(w) * b_X)n_X' + I/(2C),$$

$$S_{22} = (\mathrm{sgn}(w) * a_X - \mathrm{sgn}(w) * b_X)n_X' + I/(4C),$$

$$S_{23} = (n_X - m_X + \frac{1}{2}\,\mathrm{sgn}(w) * a_X + \frac{1}{2}\,\mathrm{sgn}(w) * b_X)n_X' + I/(2\theta C),$$

$$S_{24} = -(\mathrm{sgn}(w) * a_X + \mathrm{sgn}(w) * b_X)n_X' + I/(4\theta C),$$

$$S_{31} = (m_X + n_X - \frac{1}{2}\,\mathrm{sgn}(w) * a_X + \frac{1}{2}\,\mathrm{sgn}(w) * b_X)(sgn(w) * a_X)',$$

$$S_{32} = (a_X - b_X)a_X' + I/(2C),$$

$$S_{33} = (n_X - m_X + \frac{1}{2}\,\mathrm{sgn}(w) * a_X + \frac{1}{2}\,\mathrm{sgn}(w) * b_X)(sgn(w) * a_X)',$$

$$S_{34} = -(a_X + b_X)a_X' - I/(4\theta C),$$

$$S_{41} = (m_X + n_X - \frac{1}{2}\,\mathrm{sgn}(w) * a_X + \frac{1}{2}\,\mathrm{sgn}(w) * b_X)(sgn(w) * b_X)',$$

$$S_{42} = (a_X - b_X)b_X' - I/(2C),$$

$$S_{43} = (n_X - m_X + \frac{1}{2}\,\mathrm{sgn}(w) * a_X + \frac{1}{2}\,\mathrm{sgn}(w) * b_X)(sgn(w) * b_X)',$$

$$S_{44} = -(a_X + b_X)b_X' - I/(2\theta C),$$

where $\mathbf{1} = (1, \ldots, 1)'$, $\mathbf{0} = (0, \ldots, 0)'$ are vectors of length N, I is the identity matrix of size $N \times N$.

Therefore,

$$w = \sum_{i=1}^{N} \gamma_{1i}\left(m_{X_i} + n_{X_i} - \frac{1}{2}\,\mathrm{sgn}(w) * a_{X_i} + \frac{1}{2}\,\mathrm{sgn}(w) * b_{X_i}\right)$$

$$+ \sum_{i=1}^{N} \gamma_{2i}\left(\mathrm{sgn}(w) * a_{X_i} - \mathrm{sgn}(w) * b_{X_i}\right)$$

$$+ \sum_{i=1}^{N} \gamma_{3i}\left(n_{X_i} - m_{X_i} + \frac{1}{2}\,\mathrm{sgn}(w) * a_{X_i} + \frac{1}{2}\,\mathrm{sgn}(w) * b_{X_i}\right)$$

$$- \sum_{i=1}^{N} \gamma_{4i}\left(\mathrm{sgn}(w) * a_{X_i} + \mathrm{sgn}(w) * b_{X_i}\right)$$

and for a new observation X we get

$$Y(X) = \langle w, X \rangle + B = (\langle w, m_X \rangle + m_B, \langle w, n_X \rangle + n_B, \langle |w|, a_X \rangle + a_B, \langle |w|, b_X \rangle + b_B).$$

7 Illustrative Examples and Comparisons

In this section we present two examples of SVM based fuzzy linear regression and compare the results with different approaches suggested in literature. Let $\{X_i, Y_i\}_{i=1}^N$ be a training dataset, where $X_i = (X_{i1}, \ldots, X_{ip})$ denotes a vector of fuzzy numbers and Y_i is a fuzzy number.

To evaluate the quality of each model we use the RMSE coefficient, i.e.

$$\text{RMSE} = \sqrt{\frac{1}{N} \sum_{i=1}^N d_2^2(\hat{Y}_i, Y_i)},$$

where $\hat{Y}_i$ is an estimator of Y_i and d_2 is the metric defined by (1).

We begin with imprecise data modeled by triangular fuzzy numbers.

Example 1 Let us consider a relationship between students' grades (Y_i) and their family incomes (X_i) using imprecise data given in Table 1 (see [7]).

For the above data we construct the following 5 models:

- L-SVM—SVM fuzzy linear model with the linear loss function,
- S-SVM—SVM fuzzy linear model with the quadratic loss function,
- T-SVM—SVM fuzzy linear model with the loss function based on the Trutschnig distance,
- Diamond—the Diamond's model described in [7],
- Kao-Chyyu—the Kao and Chyyu model described in [22].

The parameters for all considered SVM models were chosen so that RMSE would be minimal. Hence, for the SVM model with the linear loss function we obtained $C = 1000$; in the SVM model with the quadratic loss function we got $C = 1000$; in the SVM model with the loss function based on the Trutschnig distance we received $C = 2$ and $\theta = 0.11$. In all the models ε were equal to zero. Estimator of vector $\text{sgn}(\boldsymbol{w})$ was obtained by constructing a linear model with m_X considered as the independent variable and m_Y taken as the dependent variable.

Table 1 Fuzzy data on students' grades and their family incomes (Example 1)

$Y_i = (m_{Y_i}, a_{Y_i}, b_{Y_i})$	$X_i = (m_{X_i}, a_{X_i}, b_{X_i})$
(4.0, 0.60, 0.80)	(21.0, 4.20, 2.10)
(3.0, 0.30, 0.30)	(15.0, 2.25, 2.25)
(3.5, 0.35, 0.35)	(15.0, 1.50, 2.25)
(2.0, 0.40, 0.40)	(9.0, 1.35, 1.35)
(3.0, 0.30, 0.45)	(12.0, 1.20, 1.20)
(3.5, 0.53, 0.70)	(18.0, 3.6, 1.80)
(2.5, 0.25, 0.38)	(6.0, 0.60, 1.20)
(2.5, 0.50, 0.50)	(12.0, 1.80, 2.40)

Fig. 2 SVM based fuzzy regression model with the linear loss function

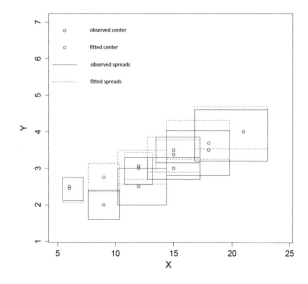

Fig. 3 SVM based fuzzy regression model with the quadratic loss function

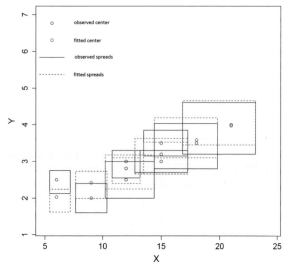

Figures 2, 3 and 4 show the observed data and their estimates. In particular, the black dots represent cores of the observed points, i.e. (m_{X_i}, m_{Y_i}), while the red dots represent their estimates, i.e. $(m_{X_i}, m_{\widehat{Y}_i})$. On the other hand, four corners of each solid box are given by $(m_{X_i} - a_{X_i}, m_{Y_i} - a_{Y_i})$, $(m_{X_i} + b_{X_i}, m_{Y_i} - a_{Y_i})$, $(m_{X_i} - a_{X_i}, m_{Y_i} + b_{Y_i})$ and $(m_{X_i} + b_{X_i}, m_{Y_i} + b_{Y_i})$, while four corners of each dotted box are defined by $(m_{X_i} - a_{X_i}, m_{\widehat{Y}_i} - a_{\widehat{Y}_i})$, $(m_{X_i} + b_{X_i}, m_{\widehat{Y}_i} - a_{\widehat{Y}_i})$, $(m_{X_i} - a_{X_i}, m_{\widehat{Y}_i} + b_{\widehat{Y}_i})$ and $(m_{X_i} + b_{X_i}, m_{\widehat{Y}_i} + b_{\widehat{Y}_i})$.

Looking at the results presented in Table 2 we see that the SVM model with the quadratic loss function and the model with the loss function based on the Trutschnig

Fig. 4 SVM based fuzzy regression model with the loss function utilizing the Trutschnig distance

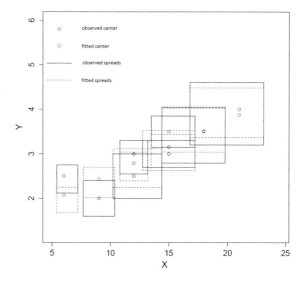

Table 2 Results of Example 1

Model	RMSE
L—SVM	0.66
S—SVM	0.54
T—SVM	0.55
Diamond	0.71
Kao-Chyyu	0.83

distance have nearly identical RMSE and smaller than the SVM model with the linear loss function. It is worth noting that all three SVM based models have smaller RMSE than both the Diamond and the Kao-Chyyu models.   □

Example 2 In this example we consider the trapezoidal fuzzy data shown in Table 3, discussed earlier in [1] for which we construct five regression models (four of them discussed also in Example 1):

• L-SVM—SVM fuzzy linear model with the linear loss function,
• S-SVM—SVM fuzzy linear model with the quadratic loss function,
• T-SVM—SVM fuzzy linear model with the loss function based on the Trutschnig distance,
• Diamond—the Diamond's model described in [7],
• Arabpour-Tata—the Arabpour and Tata model described in [1].

The model parameters were chosen as follows: $C = 500$ for the SVM model with the linear loss function, $C = 500$ for the SVM model with the quadratic loss function, $C = 0.3$ and $\theta = 0.08$ for the SVM model with the loss function based on the Trutschnig distance. In all the models ε was equal to zero. Estimator of

Table 3 Data used in Example 2

$Y_i = (m_{Y_i}, n_{Y_i}, a_{Y_i}, b_{Y_i})$	$X_i = (m_{X_i}, n_{X_i}, a_{X_i}, b_{X_i})$
(3.5, 3.75, 4.25, 4.5)	(1.5, 1.75, 2.25, 2.5)
(5.0, 5.25, 5.75, 6.0)	(3.0, 3.25, 3.75, 4.0)
(6.5, 7.00, 8.00, 8.5)	(4.5, 5.00, 6.00, 6.5)
(6.0, 6.25, 6.75, 7.0)	(6.5, 6.75, 7.25, 7.5)
(8.0, 8.25, 8.75, 9.0)	(8.0, 8.25, 8.75, 9.0)
(7.0, 7.50, 8.50, 9.0)	(9.5, 10.00, 11.00, 11.5)
(10.0, 10.25, 10.75, 11.0)	(10.5, 10.75, 11.25, 11.5)
(9.0, 9.25, 9.75, 10.0)	(12.0, 12.25, 12.75, 13)

Table 4 Results of Example 2

Model	RMSE
L—SVM	1.87
S—SVM	1.62
T—SVM	1.76
Diamond	3.05
Arabpour-Tata	4.79

vector sgn(w) was obtained by constructing a linear model with m_X considered as the independent variable and m_Y taken as the dependent variable.

Looking at the results given in Table 4 we may conclude that, as in Example 1, all three SVM based models have smaller RMSE than the Diamond and the Arabpour-Tata models. The smalles RMSE was received by the SVM model with the quadratic loss function. □

To sum up, the fuzzy regression models utilizing SVMs behaved nicely in our experiments, significantly better than the classical fuzzy regression tools. When comparing SVM based models with different loss functions it appeared that those with the quadratic loss and with the loss function based on the Trutschnig distance were better than the one with the linear loss. It seems that one of the advantages of the SVM fuzzy regression models is the presence of the parameters (especially parameter C) which enables to control the fit of the model by introducing something like a penalty for the estimation errors.

8 Conclusions

The growing need for the decision support in presence of imprecise data and perceptions entails in development of various fuzzy methods in data analysis, like fuzzy regression. This paper is dedicated to the new and promising approach to fuzzy regression based on the support vector machines. We have presented main ideas and some existing solutions for triangular fuzzy numbers. Moreover, we have also generalized the results known from the literature for the trapezoidal data and suggested the original new SVMs based fuzzy regression methods.

It is worth noting that in our examples the considered fuzzy regression methods utilizing SVMs turned out to behave better than the classical fuzzy regression approaches. However, more studies are still required and this encouraging conclusion should be confirmed in further experiments. It seems also that fuzzy regression method utilizing the loss function based on the Trutschnig distance should be further examined since it gives some freedom in choosing model parameters that might be possibly optimize.

One may ask whether trapezoidal fuzzy numbers deliver enough tools for modeling imprecise information in real-life situations. Of course, one can find examples where fuzzy numbers with nonlinear sides would fit better to data than trapezoidal fuzzy numbers. However, due to experience of many researchers and practitioners, trapezoidal fuzzy numbers in most cases appear as an optimal compromise between simplicity required for data processing and calculations and the accuracy in describing imprecise perceptions. This is also the reason that the trapezoidal approximation of fuzzy numbers is developing so rapidly. Therefore, going back to fuzzy regression, it seems that in the case of fuzzy data described by complicated membership functions the trapezoidal approximation of those fuzzy data is recommended before any method of fuzzy regression would be applied (for the theory and practical algorithms of the trapezoidal approximation we refer the reader, e.g., to [11, 12] and the literature cited there).

References

1. Arabpour, A.R., Tata, M.: Estimating the parameters of a fuzzy linear regression model. Iran. J. Fuzzy Syst. **5**, 1–19 (2008)
2. Bargiela, A., Pedrycz, W., Nakashima, T.: Multiple regression with fuzzy data. Fuzzy Sets Syst. **158**, 2169–2188 (2007)
3. Bertoluzza, C., Corral, N., Salas, A.: On a new class of distances between fuzzy numbers. Mathw. Soft Comput. **2**, 71–84 (1995)
4. Chang, Y.H.O., Ayyub, B.M.: Fuzzy regression methods-a comparative assessment. Fuzzy Sets Syst. **119**, 187–203 (2001)
5. Diamond, P.: Fuzzy least squares. Inf. Sci. **46**, 141–157 (1988)
6. Diamond, P.: Least squares and maximum likelihood regression for fuzzy linear models. In: Kacprzyk, J., Fedrizzi, M. (eds.) Fuzzy Regression Analysis, pp. 137–151. Wiley, New York (1992)
7. Diamond, P., Tanaka, H.: Fuzzy regression analysis. Fuzzy Sets Decis. Anal. Oper. Res. Stat. **1**, 349–387 (1988)
8. Dubois, D., Prade, H.: Operations on fuzzy numbers. Int. J. Syst. Sci. **9**, 613–626 (1978)
9. Fletcher T.: Support vector machines explained. www.cs.ucl.ac.uk/staff/T.Fletcher/
10. Grzegorzewski, P.: Metrics and orders in space of fuzzy numbers. Fuzzy Sets Syst. **97**, 83–94 (1998)
11. Grzegorzewski, P.: Trapezoidal approximations of fuzzy numbers preserving the expected interval-algorithms and properties. Fuzzy Sets Syst. **159**, 1354–1364 (2008)

12. Grzegorzewski, P.: Algorithms for trapezoidal approximations of fuzzy numbers preserving the expected interval. In: Bouchon-Meunier, B., Magdalena, L., Ojeda-Aciego, M., Verdegay, J.-L., Yager, R.R. (eds.) Foundations of Reasoning under Uncertainty, pp. 85–98. Springer, Berlin (2010)

13. Grzegorzewski, P., Mrwka E.: Linear regression analysis for fuzzy data, In: Proceedings of the 10th IFSA World Congress–IFSA 2003, Istanbul, Turkey, pp. 228–231. June 29-July 2, 2003

14. Grzegorzewski, P., Mrwka, E.: Regression analysis with fuzzy data. In: Grzegorzewski, P., Krawczak, M., Zadrony, S. (eds.) Soft Computing-Tools, Techniques and Applications, pp. 65–76. Warszawa, Exit (2004)

15. Grzegorzewski P., Pasternak-Winiarska, K.: Weighted trapezoidal approximations of fuzzy numbers. In: Carvalho, J.P., Dubois, D., Kaymak, U., Sousa, J.M.C (eds.), Proceedings of the Joint 2009 International Fuzzy Systems Association World Congress and 2009 European Society of Fuzzy Logic and Technology Conference, pp. 1531–1534, Lisbon

16. Grzegorzewski, P., Pasternak-Winiarska, K.: Bi-symmetrically weighted trapezoidal approximations of fuzzy numbers. In: Abraham, A., Benitez Sanchez, J.M., Herrera, F., Loia, V., Marcelloni, F., Senatore, S. (eds.), Proceedings of Ninth International Conference on Intelligent Systems Design and Applications, pp. 318–323. Pisa (2009)

17. Hastie, T., Tibshirani, R., Friedman, J.: The Elements of Statistical Learning. Springer, Berlin (2008)

18. Hong, D.H., Hwang, C.: Support vector fuzzy regression machines. Fuzzy Sets Syst. **138**, 271–281 (2003)

19. Hong, D.H., Hwang C.: Fuzzy Nonlinear Regression Model Based on LS-SVM in Feature Space, pp. 208–216. Springer, Berlin (2006)

20. Kacprzyk, J., Fedrizzi, M.: Fuzzy Regression Analysis. Omnitech Press, Heidelberg (1992)

21. Körner, R., Näther, W.: Linear regression with random fuzzy variables: extended classical estimates, best linear estimates, least squares estimates. Inf. Sci. **109**, 95–118 (1998)

22. Kao, C., Chyyu, C.L.: A fuzzy linear regression model with better explanatory power. Fuzzy Sets Syst. **126**, 401–409 (2002)

23. Näther, W.: Linear statistical inference for random fuzzy data. Statistics **29**, 221–240 (1997)

24. Redden, D.T., Woodall, W.H.: Properties of certain fuzzy linear regression methods. Fuzzy Sets Syst. **64**, 361–375 (1994)

25. Smola, A.J., Scholkopf, B.: A tutorial on support vector regression. Stat. Comput **14**, 199–222 (2004)

26. Tanaka, H.: Fuzzy data analysis by possibilistic linear models. Fuzzy Sets Syst. **24**(1987), 363–375 (1987)

27. Tanaka, H., Uegima, S., Asai, K.: Linear analysis with fuzzy model. IEEE Trans. Syst. Man Cybern. **12**, 903–907 (1982)

28. Tanaka, H., Watada, J.: Possibilistic linear systems and their application to the linear regression model. Fuzzy Sets Syst. **27**, 275–289 (1988)

29. Trutschnig, W., González-Rodríguez, G., Colubi, A., Gil, M.A.: A new family of metrics for compact, convex (fuzzy) sets based on a generalized concept of mid and spread. Inf. Sci. **179**, 3964–3972 (2009)

30. Vapnik, V.N.: The Nature of Statistical Learning Theory. Springer, Berlin (1995)

31. Vapnik, V.N.: Statistical Learning Theory. Wiley, New York (1998)

32. Zeng, W., Li, H.: Weighted triangular approximation of fuzzy numbers. Int. J. Approx. Reason. **46**, 137–150 (2007)

Quantified Quality Criteria of Contextual Bipolar Linguistic Summaries

Mateusz Dziedzic, Janusz Kacprzyk, Sławomir Zadrożny
and Guy De Tré

Abstract In our previous work we have proposed the concept of a contextual bipolar linguistic summary. It is an extension of the seminal concept of a linguistic summary proposed by Yager which is based on the application of Zadeh's calculus of linguistically quantified propositions to more intuitive and human consistent data mining. The original Yager's concept evolved over the years and our recent contribution to this theory is the inclusion of the concept of bipolarity of information and preferences. This enrichment of the notion of the linguistic summary calls for specialized measures of its quality, interestingness, etc. We further study this problem and in this paper we propose a new approach to assessing the quality of this type of summaries.

Keywords Linguistic summary · Context · Bipolarity · Measures of quality · Data mining

1 Introduction

More than 30 years have passed since the seminal work of Yager [33, 34] in which the concept of the linguistic summaries was introduced. Over all these years this concept

M. Dziedzic (✉)
Department of Automatic Control and Information Technology, Cracow University
of Technology, ul. Warszawska 24, 31-155 Kraków, Poland
e-mail: Mateusz.Dziedzic@ibspan.waw.pl

M. Dziedzic · G. De Tré
Department of Telecommunications and Information Processing, Ghent University,
Sint-Pietersnieuwstraat 41, 9000 Ghent, Belgium

G. De Tré
e-mail: Guy.DeTre@UGent.be

J. Kacprzyk · S. Zadrożny
Systems Research Institute, Polish Academy of Sciences, ul. Newelska 6,
01-447 Warszawa, Poland
e-mail: Janusz.Kacprzyk@ibspan.waw.pl

S. Zadrożny
e-mail: Slawomir.Zadrozny@ibspan.waw.pl

© Springer International Publishing Switzerland 2016
G. De Tré et al. (eds.), *Challenging Problems and Solutions
in Intelligent Systems*, Studies in Computational Intelligence 634,
DOI 10.1007/978-3-319-30165-5_7

has been further developed [12–14, 21, 31] and is enjoying still growing interest of the researchers; cf., e.g., [1, 4, 24, 29, 32]. A distinctive feature of Yager's approach is the use of natural language securing the intuitive character of the summaries, their easily interpretable and human-friendly format of presenting the discovered regularities, founded on Zadeh's *linguistically quantified propositions* [35]. According to Zadeh's paradigm, the trends and patterns found in data are presented in a formally strict and well defined, but at the same time easily comprehensible to an end-user form of (quasi-)natural language expressions. This concept of linguistic summaries has been further developed in our previous works, cf. [18, 20, 21], and here we continue this direction of research. Our approach to the linguistic summarization is closely related to our earlier work on *flexible fuzzy queries*. The essence of such queries is the use of natural language expressions to describe the data sought. Recently, a new line of research on flexible querying has emerged which is based on some psychological observations of the way humans are carrying out the evaluation of the broadly meant alternatives what is directly relevant to forming conditions on the data sought in the realm of a database querying. Namely, it is argued that humans separately consider the *positive* and the *negative aspects* of the considered alternatives. This triggered the research on queries with conditions of two types corresponding to the positive (desired) and negative (being avoided) features of data sought, respectively. Starting with the work of Dubois and Prade [5] such queries are referred to as *bipolar queries* and studying them gained significant recognition (cf., e.g., [2, 6–8, 17, 23, 30, 36, 39–41]). In the most popular approach those two assessments, represented by separate conditions, are not considered as equally important. To allow for an explicit aggregation of such bipolar conditions, among others, an explicit aggregation was investigated. As a result, two bipolar operators have been introduced, i.e. the "and possibly" and "or if impossible" [40, 42] operators, whose purpose is to model the aggregation of two conditions taking into account a possibility or impossibility (within the actual data) of satisfying one of them.

Following the peculiar association between flexible queries and linguistic summaries of data, the foregoing bipolarity concept has been introduced into linguistic summaries. After some necessary modifications, including introduction of a *context* of the bipolar query, we proposed concepts of a *contextual bipolar query* [42, 44] and a bipolar *contextual linguistic summaries* [9–11, 26]. The latter of the two is further developed in the present paper with a particular emphasis on summaries' quality criteria [9] and quantifiers used therein.

The following parts of the paper are organized as follows. In Sect. 2 we briefly introduce the flexible querying and linguistic summaries of data. The flexible queries are extended further in Sect. 3 with concepts of bipolarity and context, which leads to the introduction of contextual linguistic summaries in Sect. 4. Then, in Sect. 5, quality criteria of introduced contextual summaries are discussed and further developed. Finally, some conclusions and overall discussion are provided.

2 Flexible Fuzzy Queries and Linguistic Summaries

In our approach we identify a query with conditions that the data sought have to fulfill. Referring to the standard query language SQL we assume the notion of the query in this paper is basically equated with the `WHERE` clause of its `SELECT` instruction. This approach, despite simplified, preserves the most important aspects of a query which are essential for us.

In classical query languages preferences of users must be expressed precisely. However, due to the fact that their original form is a natural language expression, in many practical situations they are imprecise. For example, one (a non-DBMS expert) may be concerned primarily with the cost while looking for an apartment to rent and express his or her preference as:

$$\text{Find } cheap \text{ apartments for rent in Kraków.} \tag{1}$$

The above observation led to many studies on representing such imprecise statements in database queries [27, 37]. In an approach, referred here to as fuzzy linguistic queries, such imprecise terms as, e.g., *cheap* are represented by fuzzy sets defined in the domains of respective attributes.

Usually, a *dictionary* of linguistic terms is an important part of an implementation of systems supporting flexible fuzzy querying. Such a dictionary contains predefined linguistic terms and corresponding fuzzy sets, as well as terms defined by the users, used in queries; cf., e.g. [19]. Linguistic terms collected in such a dictionary form a perfect starting point to derive meaningful *linguistic summaries* of a database.

2.1 Linguistic Summaries of Data

In almost all approaches to the linguistic summarization of data (cf. [25]) as linguistic summaries we understand quasi natural language sentences describing some characteristic features present in data set under consideration. Those sentences are, as originally proposed by Yager [33, 34], represented by linguistically quantified propositions based on Zadeh's calculus of linguistically quantified propositions [35] as the underlying formalism. The statement representing a linguistic summary points out some properties shared by a number of data items and the proportion of these data items is expressed using a *linguistic quantifier*. That idea has been further developed by Kacprzyk and Yager [15], and Kacprzyk, Yager and Zadrożny [16, 18, 20].

Assuming $R = \{t_1, \ldots, t_n\}$ is a set of tuples (a relation) in a database, representing, e.g., a set of employees; $A = \{A_1, \ldots, A_m\}$ is a set of attributes defining schema of the relation R, e.g., salary, age, education_level, etc. in a database of employees ($A_j(t_i)$ denotes a value of attribute A_j for a tuple t_i), the linguistic summary of a set R is a linguistically quantified proposition which is an instantiation of the following generic form:

$$Q_{t \in R} \left(U(t), S(t) \right) \tag{2}$$

composed of the following elements:

- *Summarizer S* which is a fuzzy predicate representing, e.g., an expression "an employee is well-educated", formed using attributes of the set A;
- *Qualifier U* (optional) which is another fuzzy predicate representing, e.g., a set of "young employees", formally its absence is covered by (2) assuming $U = R$;
- *Linguistic quantifier Q*, e.g., "most" expressing the proportion of tuples satisfying the summarizer (optionally, among those satisfying a qualifier);
- *Truth (validity) T* of the summary, i.e. a number from $[0, 1]$ expressing the truth of a respective linguistically quantified proposition.

In what follows we often use an abbreviated form of this notation, i.e. $Q(U, S)$ or $Q(S)$ if $U = R$, dropping the t's where it does not lead to any misunderstanding and assuming the quantifier Q to refer to the whole relation R.

As already mentioned, in Yager's original approach the linguistic quantifiers are represented using Zadeh's calculus of linguistically quantified propositions. A *normal* and *monotonically non-decreasing* proportional linguistic quantifier Q is represented by a fuzzy set in $[0, 1]$ and $\mu_Q(x)$ states the degree to which the proportion of $100 \times x$ % of elements of the universe match the meaning of the quantifier Q. For example, $\mu_{most}(0.8) = 1.0$ means that the 80 % of tuples from the universe match the meaning of "most" to the degree of 1.0.

The truth degree of the linguistic summary (2) is computed as follows:

$$T(Q_{t \in R}(U(t), S(t))) = \mu_Q \left(\frac{\sum_{i=1}^{n} (\mu_U(t_i) \wedge \mu_S(t_i))}{\sum_{i=1}^{n} \mu_U(t_i)} \right). \tag{3}$$

where $\wedge$ denotes the minimum operator or, in general, a t-norm operator. In case the qualifier U is missing the formula (3) is simplified to:

$$T(Q_{t \in R} S(t)) = \mu_Q \left(\frac{1}{n} \sum_{i=1}^{n} \mu_S(t_i) \right). \tag{4}$$

As a remark, in what follows we will denote by T the truth of any linguistically quantified expressions, not only the summaries itself.

Linguistic quantifiers may be modeled in many various ways—cf., e.g., a recent survey by Delgado et al. [3]. These various approaches may be readily adopted in the framework of the linguistic summarization of data (cf., e.g. [10]). However, due to the computational simplicity and convenience we adopt in our investigations and in this paper the original Zadeh's calculus of linguistically quantified propositions mentioned above.

3 Bipolarity in Flexible Querying

As already mentioned, flexible fuzzy queries described above provide a very useful way of representation of users preferences, mainly allowing for a gradual, not only binary, character of them. Another view of modeling preferences, cf. [5], assumes that the decision maker often comes up with somehow independent evaluations of positive and negative features of alternatives in question.

This leads to a general concept of *bipolar query* [5] against a database, formed by two conditions representing the positive and negative assessments. In the approach most popular in recent studies those two assessments are not equally important: the negative condition, or rather its complement denoted C, is treated as a *constraint*, while the positive condition, P, is only *wished* to be satisfied.

Such a semantics of the bipolar queries, generically denoted as (C, P), can be traced back to the seminal work of Lacroix and Lavency, [28]. The evaluation of such a query results in two degrees corresponding to the satisfaction of the positive and negative condition(s), which are then aggregated by special operators.

In our previous works (e.g. [36, 38–40] and [42, 43], respectively) we proposed a pair of two asymmetric operators to combine both conditions, C and P:

1. Conjunction-like "and possibly" operator which aggregates their satisfaction degrees depending on the possibility of a simultaneous matching of both conditions, somehow favoring the (complement of the) negative condition C; and
2. Disjunction-like "or if impossible" operator which favors the positive condition P by taking into account the negative condition's complement C only if it is not possible to fulfill the positive one.

3.1 "And Possibly" Operator

The first type of proposed bipolar queries may be written in a general form of:

$$C \text{ and possibly } P \tag{5}$$

and would be denoted $C \wedge_p P$ in what follows. Such a query may be illustrated with an example:

$$\text{Find employees that are } \textit{young} \text{ "and possibly" earn a } \textit{high salary} \tag{6}$$

interpreted two fold:

1. If there is a tuple which satisfies (by satisfying a condition, here and in what follows, we understand satisfying it to a high degree) both conditions, then and only then it is actually *possible* to satisfy both of them and each tuple has to do so; or
2. If there is no such a tuple, then condition P can be ignored.

Thus, in the crisp framework, in the former case the query (5) reduces to the regular conjunction $C \wedge P$ while in the latter it simply reduces to condition C.

In the crisp case the matching degree of the $C \wedge_p P$ query against a tuple t may be formalized as [28]:

$$C(t) \wedge_p P(t) = C(t) \wedge (\exists_{s \in R} (C(s) \wedge P(s)) \Rightarrow P(t)) . \tag{7}$$

The execution of such a query in the *crisp* case boils down to a selection of tuples satisfying condition C followed by a selection, from among them, those satisfying condition P, if any. However, in a *fuzzy* case the situation is much more complicated and, at the same time, interesting. First of all, the formula (7) may be interpreted in various ways depending on the operators of the logical connectives. For the minimum, maximum and the Kleene–Dienes implication one obtains:

$$T\left(C(t) \wedge_p P(t)\right) = \min(\mu_C(t), \max(1 - \exists C P, \mu_P(t))), \tag{8}$$

where $\exists C P$ denotes $\max_{s \in R} \min(\mu_C(s), \mu_P(s))$.

3.2 "Or if Impossible" Operator

Recently we proposed a second type of bipolar operator [42] providing queries generalized as:

$$P \text{ or if impossible } C \tag{9}$$

in what follows denoted by $P \vee_i C$ and exemplified with a query:

$$\text{Find employees who earn a } \textit{high salary} \text{ "or if impossible" are } \textit{young} \tag{10}$$

interpreted as follows:

1. If there is no tuple (in the dataset) which satisfies positive condition P, then and only then it is actually *impossible* to satisfy it and the second condition C is taken into consideration; on the other hand
2. If there is such a tuple, then only condition P is considered.

Thus, in the crisp framework, in the former case the query (9) reduces to the condition C while in the latter case it reduces to P. This, for the crisp case, may be formally written as:

$$P(t) \vee_i C(t) = P(t) \vee (\neg \exists_{s \in R} P(s) \wedge C(t)) . \tag{11}$$

For fuzzy conditions P and C one may once again employ the minimum and maximum operators to interpret the conjunction and disjunction respectively, and obtain:

$$T\left(P(t) \vee_i C(t)\right) = \max(\mu_P(t), \min(1 - \exists P, \mu_C(t))), \qquad (12)$$

where $\exists P$ denotes $\max_{s \in R} \mu_P(s)$.

3.3 Context in Bipolar Queries

Both operators, the "and possibly" and "or if impossible", refer to the whole dataset/database when deciding on the possibility or impossibility of satisfying the conditions. Thus, the result of this calculation is the same for every tuple in the dataset. However, one can imagine a situation where the possibility/impossibility test should be rather performed only on a part of the dataset related to the tuple considered, e.g. in an employee database one may want to focus on employees of every department separately.

This tuple-related neighborhood is referred to here as a *context* of the tuple, or the query itself, and may be exemplified by:

> Find employees that are *young* "and possibly—with regards to the context of the colleagues *from the same department*—" earn a *high* (13) salary;

satisfied by an employee if:

1. He or she satisfies both conditions, i.e. is *young* and earns a *high salary*; or
2. He or she is *young* and there is no other employee *in (the context of) his or her department* who is both *young* and earns a *high salary*.

Such examples, constructed with any of the two operators led us to the concept of *contextual bipolar queries* (or, briefly, *contextual queries*, see e.g. [42, 44]) and further into the concept of *contextual bipolar linguistic summaries* (*contextual linguistic summaries*, see e.g. [10, 11, 26, 43]).

A generic form of the contextual bipolar query with the "and possibly" operator is formalized as:

$$C \text{ and possibly } P \text{ w.r.t. context } W \qquad (14)$$

in our previous works also noted as (C and possibly P with respect to W) and in what follows denoted as $C \wedge_p^W P$, where conditions C and P are interpreted as in (5). The predicate W denotes the context of a tuple t for which the query is evaluated, i.e., a part of the database that is related via W to the t and forms the context in which the possibility of fulfilling both C and P is calculated:

$$\text{Context}(t) = \{s \in R : W(t, s)\}. \qquad (15)$$

In general, the context defining predicate may be fuzzy, characterized by its membership function μ_W, and then the context is a fuzzy set of tuples characterized by the membership function:

$$\mu_{\text{Context}(t)}(s) = \mu_W(t, s) \tag{16}$$

The semantics of the query (14) may be formulated as:

$$C(t) \wedge_p^W P(t) = C(t) \wedge (\exists_{s \in R} (C(s) \wedge P(s) \wedge W(t, s)) \Rightarrow P(t)) \tag{17}$$

and for the minimum, maximum and Kleene-Dienes implication operators one obtains a fuzzified version:

$$T\left(C(t) \wedge_p^W P(t)\right) = \min(\mu_C(t), \max(1 - \exists CPW, \mu_P(t))), \tag{18}$$

where $\exists CPW$ denotes $\max_{s \in R} \min(\mu_C(s), \mu_P(s), \mu_W(t, s))$.

For the "or if impossible" operator the contextual bipolar query in the generic form is as follows:

$$P \text{ or if impossible } C \text{ w.r.t. context } W, \tag{19}$$

further denoted as $P \vee_i^W C$, and may be exemplified by a query:

Find employees that earn a *high salary* "or if impossible—with regards to (the context of) the colleagues *from the same depart-* *ment*—" are *young*; $\tag{20}$

satisfied by an employee if:

1. He or she satisfies the former condition, i.e. earns a *high salary*; or
2. He or she satisfies the latter condition, i.e. is *young*, and there is no other employee *in (the context of) the same department* who satisfies the former one, i.e. earns a *high salary*.

This semantics is formally expressed by a formula:

$$P(t) \vee_i^W C(t) = P(t) \vee (\neg \exists_{s \in R}(P(s) \wedge W(t, s)) \wedge C(t)) \tag{21}$$

and modeled for the fuzzy case (following (18)) as:

$$T\left(P(t) \vee_i^W C(t)\right) = \max(\mu_P(t), \min(1 - \exists PW, \mu_C(t))), \tag{22}$$

where $\exists PW$ denotes $\max_{s \in R} \min(\mu_P(s), \mu_W(t, s))$.

For the clarity, from now on we will use a notion of $\exists \Box$ (with a combination of C, P and W in the place of $\Box$) as a generic formula for the existence-quantified parts of formulas (7), (11), (17) and (21) (as, respectively, $\exists CP$, $\exists P$, $\exists CPW$ and $\exists PW$), not only for their fuzzified examples as above.

4 Contextual Linguistic Summaries

As mentioned in Sect. 3.3 the concept of a contextual linguistic summary, proposed by us in [11] and further developed and analyzed in [10, 26, 43], is strictly related to the concept of a contextual query in the same way as the "standard" linguistic summary is related to a flexible fuzzy query, as shown in [21].

In general, the contextual linguistic summaries follow the form of (2), the novelty lies in the use of the pair of logical operators "and possibly" and "or if impossible" (introduced in Sects. 3.1 and 3.2, respectively) in the summarizers.

4.1 Summaries with the $\wedge_p^W$ Operator

Contextual linguistic summaries, in the form consisting of the modified "and possibly" operator, in what follows shortly denoted as $Q\left(C \wedge_p^W P\right)$, describes regularities in the data exemplified by the pattern: Q of tuples possess *some property C* "and possibly" fulfill *some other property P*, whereas the mentioned *possibility* is considered within the context W of each of the individual tuples, as in the case of contextual queries.

Lets consider the following example over an employees dataset:

> *Most* employees are *young*, "and possibly—with regards to (the context of) the colleagues *from the same department*—" earn a *high* salary. (23)

The summarizer of the above summary is satisfied by an employee if he or she fulfills the former condition (C), i.e. is *young*, and one of the following occurs:

1. He or she fulfills also the latter condition, i.e. he or she is *young* and earn a *high salary*; or
2. He or she does not fulfill the latter one and there is no other employee within *(the context of) the same department* who fulfills both conditions, i.e. is *young* and earn a *high salary*. (24)

In the former case it is actually *possible* to satisfy both conditions within the specified context of the tuple considered, which makes both of them necessary to be fulfilled by every other tuple in this particular context to satisfy the summarizer. In the latter case, as there is no tuple in the specified context fulfilling both conditions, it is obviously *not possible* to fulfill them simultaneously, thus for every tuple in this specific context it is sufficient to satisfy only the former condition to satisfy the summarizer.

It is especially worth noting that both cases can occur together in one summary, but in separate contexts. Lets imagine a simple example of a company with two "same aged", but not equally paid departments:

- Dept. A, which employees are not only *young*, but also *highly paid*; and
- Dept. B, consisting solely of *young* employees receiving very *moderate salary* (i.e. *not high at all*).

For this company the summary (23) has truth degree $T = 1.0$ (with *most* modeled by an identity quantifier $\mu_{Q^{\mathbb{I}}}(x) = x$), as every employee of Department A satisfies both conditions, which results in fulfilling the first case of (24), and every employee from the Department B satisfies the former one and no one therein satisfy the latter, exemplifying the second case of (24). Note that, if both departments employ more or less the same number of employees (and are the only departments in the company, as assumed earlier) then, at the same time, a standard linguistic summary such as "*Most* employees are *young* and earn a *high salary*" will be true to the degree equal at most 0.5.

As showed above, the summaries $Q\left(C \wedge_p^W P\right)$ cover the tuples which satisfy the underlying contextual query. According to Sect. 4.1, the truth degree of the summarizer $C \wedge_p^W P$ for a tuple t is bounded from below and from above as follows:

$$T\left(C(t) \wedge P(t)\right) \leq T\left(C(t) \wedge_p^W P(t)\right) \leq T\left(C(t)\right). \tag{25}$$

The actual truth degree of such a summarizer depends on the truth degree of $\exists C P W$ for "*most* of t's", which is discussed in more detail in Sect. 5.

4.2 Summaries with the $\vee_i^W$ Operator

The modified version of "or if impossible" operator lets us formulate another form of contextual linguistic summaries, denoted as $Q\left(P \vee_i^W C\right)$, describing properties of the pattern: Q of tuples possess *some property P* "or if impossible" fulfill *some other property C*, whereas the mentioned *impossibility* is considered within the context W of each of the individual tuples, as in the case of contextual queries.

Following our previous example of an employees dataset and incorporating the $\vee_i^W$ operator one may get an exemplary summary:

> *Most* employees earn a *high salary*, "or if impossible—with regards to (the context of) the colleagues *from the same department*—" are *young*; (26)

where an employee satisfies the summarizer if:

1. He or she satisfies the former condition, i.e. earns a *high salary*; or
2. He or she meets the latter one, i.e. is *young*, and there is no employee (27) *within (the context of) the same department* who fulfills the former one.

As in the $\wedge_p^W$ operator example, both cases can occur together in one summary, but in separate contexts. As above, lets imagine a simple example of a company with two departments arranged as follows:

- Dept. A, which employees earn a *high salary* (their age is irrelevant); and
- Dept. B, consisting solely of *insufficiently paid*, yet *young* employees.

For this exemplary company the summary (26) has truth degree $T = 1.0$ (with *most* modeled by an identity quantifier $\mu_{Q^{\mathbb{I}}}(x) = x$), as every employee of Department A satisfies the former condition, which results in fulfilling the first case of (27), and every employee from the Department B satisfies the latter one and no one therein satisfy the former, which exemplifies the second case of (27).

Like in the $Q\left(C \wedge_p^W P\right)$ case, the summaries $Q\left(P \vee_i^W C\right)$ cover tuples which satisfy the underlying contextual query. As mentioned in Sect. 4.2, the truth degree of the summarizer $P \vee_i^W C$ for a tuple t is bounded from below and from above as follows:

$$T\left(P(t)\right) \leq T\left(P(t) \vee_i^W C(t)\right) \leq T\left(P(t) \vee C(t)\right). \tag{28}$$

The exact truth degree of such a summarizer depends on the truth degree of $\exists PCW$ of (21) for "*most* of t's", which is further discussed in Sect. 5.

5   Quality Criteria of Contextual Linguistic Summaries

In [10] we stated, for the "and possibly" operator case, that the quality of the context $W(t, s)$ of a linguistic summary $Q\left(C \wedge_p^W P\right)$ itself and of the premise of the implication occurring therein, i.e. $\exists CPW$, have to be taken into account when measuring the quality/interestingness of the contextual linguistic summary. As shown in [43], the same observation is valid for the second type of contextual linguistic summaries $Q\left(P \vee_i^W C\right)$, i.e. the one composed of the "or if impossible" operator, where the quality of the context $W(t, s)$ and the quality of $\exists PW$ has to be also taken into consideration.

Namely, for the first type of proposed summaries, i.e. $Q\left(C \wedge_p^W P\right)$, if the predicates C, P and W are such that the $\exists CPW$ in (18) is true to a very low or a very high degree for *most* of t's, then the summarizer $C \wedge_p^W P$ does not make much sense even if the truth value of a summary is high. As we showed in Sect. 3.3, this is due to the behavior of the contextual query $C \wedge_p^W P$ which turns locally, for a given tuple

t, into C or $C \wedge P$ query, respectively, when the truth degree of $\exists C P W$ for this particular t is close to 0 or close to 1, and thus if this happens for *most* of the tuples t then a standard linguistic summary $Q(C)$ or $Q(C \wedge P)$ is then more appropriate while the contextual linguistic summary is then redundant or even misleading.

The same observation applies to the second type of the proposed summaries, i.e. $Q\left(P \vee_i^W C\right)$, where if for *most* of t's the $\exists P W$ of (21) is true to a very high (close to 1) or very low (close to 0) degree the summarizer reduces to P or to $\neg P \wedge C$ respectively, following the interpretation for the contextual query $P \vee_i^W C$ shown in Sects. 3.3 and 4.2. In those situations it is, again, better to simplify the proposed summaries to $Q(P)$ and $Q(P \vee C)$, respectively. In fact, the latter summary, referring to the upper bound defined by (28), may take a more specific form, depending on some additional assumptions. For example, if W is reflexive, i.e., $\forall t \ \mu_W(t, t) = 1$ and, as assumed earlier, for *most* of t's the $\exists P W$ is true to a very low (close to 0) degree and the summary $Q\left(P \vee_i^W C\right)$ is true to a high degree then this summary should be replaced with $Q(\neg P \wedge C)$. This stems from the fact that $\exists P W$ low implies P low due to the assumed reflexivity of W.

From the above remarks an observation regarding the context W itself can be derived. Namely, if for "*most* of t's" there does not exist s such that $W(t, s)$, i.e., contexts of tuples defined by W are empty, then:

1. The premise of the implication in (17) is false and the $Q\left(C \wedge_p^W P\right)$ summary loses its bipolar character and its truth value is independent of the value of P;
2. The $\exists P W$ of the (21) is false and the $Q\left(P \vee_i^W C\right)$ summary analogously loses the bipolar character and the use of the $P \vee_i^W C$ summarizer is no longer meaningful.

In [10, 43] we proposed a solution to this problem in a form of additional quality measures (besides the summary truth value) expressed using the following linguistically quantified propositions:

Criterion 1:
$$Q_{t \in R} \exists_{s \in R \setminus \{t\}} W(t, s), \tag{29}$$

Criterion 2:
$$Q_{t \in R} \exists_{s \in R \setminus \{t\}} (C(s) \wedge P(s) \wedge W(t, s)), \tag{30}$$

Criterion 3:
$$Q_{t \in R} \exists_{s \in R \setminus \{t\}} (P(s) \wedge W(t, s)). \tag{31}$$

where quantifier Q, following our reasoning, is defined as for "*most* of t's", which in the simplest case may be represented by the identity quantifier $\mu_{Q^I}(x) = x$.

The truth values of the above linguistically quantified propositions represent the degree to which given criterion is satisfied. Thus, in what follows we will denote this degree as $T(Crit \ 1)$, $T(Crit \ 2)$ and $T(Crit \ 3)$, respectively.

As can be easily seen, Criterion 1 concerns only the quality of the context, thus it is meant to be applied to both types of contextual linguistic summaries—e.g., as a preliminary step in the summarization process, while the Criteria 2 and 3 are intended to evaluate the quality of whole contextual summaries $Q\left(C \wedge_p^W P\right)$ and $Q\left(P \vee_i^W C\right)$, respectively.

The basic idea is to let the end user to define thresholds of acceptance for all three Criteria. Other possible scenario is to accept contexts with value of Criterion 1 higher than a user-defined threshold and then take into account the values of Criterion 2 and 3 while selecting the best summaries to be presented to the user.

5.1 Quantifiers in Quality Criteria

After deeper insight into quality measures (29)–(31) we propose to extend them by using more sophisticated quantifiers (such as "*at least a few*", "*more than a half*" or even defined by fuzzy sets with trapezoidal membership functions) in place of the classical existential quantifier.

Criterion 1: In the Criterion 1, concerning only the context W itself, existential quantifier used so far allowed for contexts resulting in a high number of very small "groups". For example, for a toy dataset from Table 1a, where, if context (of the crisp character for clarity, but the following discussion could easily be extended to fuzzy contexts as well) is defined as:

$$W(t, s) = 1 \text{ iff } A(t) = A(s); \qquad 0 \text{ otherwise}, \tag{32}$$

then the Criterion 1 is fully satisfied owing to the use of the existential quantifier as every tuple has "at least one"—and to be precise, just one—other tuple within its own context. One can easily extend this situation to larger datasets, where this behavior may be undesirable.

Another drawback of the original form of Criterion 1 is that it allows for a contexts W that group all the tuples together (for an example simply replace all $A(t)$ values in Table 1a with 1.0), in which case the usage of contextual summaries is pointless.

To address this issue we propose to redefine Criterion 1 as follows:

Criterion 1:

(redefinition)

$$Q^{\text{I}}_{t \in R} Q^{\text{II}}_{s \in R \setminus \{t\}} W(t, s), \tag{33}$$

where both quantifiers, Q^{I} and Q^{II}, could be independently defined with fuzzy sets (although for Q^{I} we still assume the identity quantifier $\mu_{Q^{\text{I}}}(x) = x$). From now on by Criterion 1 we will mean the redefined formula.

Through replacing the existential quantifier with the Q^{II} defined by a fuzzy set with a trapezoidal membership function (cf. Fig. 1 for an exemplary quantifier $Q_{\text{trap}(x;0.0,0.2,0.6,0.8)}$) the above issue might be solved. There still exists a need for a proper definition of such a quantifier, but with some common sense in the background and an easily implemented participation of the end user this is not an issue.

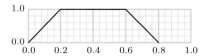

Fig. 1 An example of a quantifier $Q_{\mathrm{trap}(x;0.0,0.2,0.6,0.8)}$ defined with a trapezoidal membership function trap$(x; 0.0, 0.2, 0.6, 0.8)$

As a safe "rule of thumb", we propose to start with a previously mentioned quantifier $Q^{\mathrm{II}} = Q_{\mathrm{trap}(x;0.0,0.2,0.6,0.8)}$ to deal with both drawbacks: it will "accept" contexts W which, for "*most* tuples t", ensure the appropriate minimum and maximum sizes of context's groups. For our exemplary dataset (a) from Table 1a extended to 100 tuples, the value of the Criterion 1 would be 0.05, which clearly indicates, that this particular context W and all summaries based on it should be discarded. On the other hand, if we choose the context W so that every tuple (from the extended example, i.e. 100 tuples) has 9 other tuples in its "context group" (i.e. the value of $W(t, s) = 1.0$ within the groups) the value of Criterion 1 is 0.45, which should be acceptable.

Criteria 2 and 3: A similar reasoning can be carried out for the other two measures, i.e. Criterion 2 (30) and Criterion 3 (31). Through replacing the existential quantifier in their formulas one acquire more flexibility and "smoothness" in summary's evaluation—one can define quantifier Q^{II} to met his or her demands regarding the minimal proportion of tuples which have to satisfy simultaneously conditions C and P for summaries $Q\left(C \wedge_p^W P\right)$ and condition P for summaries $Q\left(P \vee_i^W C\right)$ to meet user's understanding of "possibility" or "impossibility" referred to in the contextual summaries.

Let us redefine the Criteria 2 and 3 as follows:

Criterion 2:

(redefinition)

$$Q^{\mathrm{I}}_{t \in R} Q^{\mathrm{II}}_{s \in R \setminus \{t\}} \left(C(s) \wedge P(s) \wedge W(t, s)\right), \tag{34}$$

Criterion 3:

(redefinition)

$$Q^{\mathrm{I}}_{t \in R} Q^{\mathrm{II}}_{s \in R \setminus \{t\}} \left(P(s) \wedge W(t, s)\right). \tag{35}$$

where the quantifier Q^{I}, as in our original propositions, is defined as for "*most* of t's"—what in the simplest case may be represented by the identity quantifier $\mu_{Q^{\mathrm{I}}}(x) = x$—and the quantifier Q^{II} will be discussed later on. In what follows we will denote by the Criterion 2 and Criterion 3 this new, redefined formulas.

For those two measures we propose to use as a Q^{II} a slightly less demanding quantifier, e.g. "*at least a few* t's", defined for example as $Q_{\mathrm{trap}(x;0.0,0.1,1.0,1.0)}$. To support this proposition let us analyze another toy example with a sample dataset (b) from Table 1b.

Table 1 Exemplary "datasets"

(a)

t	$A(t)$
t_1	1.00
t_2	1.00
t_3	0.98
t_4	0.98
t_5	0.96
t_6	0.96
t_7	0.94
t_8	0.94
t_9	0.92
t_{10}	0.92
...	...

(b)

t	$C(t)$	$P(t)$	$A(t)$
t_1	1.0	1.0	1.0
t_2	1.0	0.0	1.0
t_3	1.0	0.0	1.0
t_4	1.0	0.0	1.0
t_5	1.0	0.0	1.0
...	...	...	...
t_{50}	1.0	0.0	1.0
t_{51}	1.0	0.0	0.0
t_{52}	1.0	0.0	0.0
...	...	...	...
t_{96}	1.0	0.0	0.0
t_{97}	1.0	0.0	0.0
t_{98}	1.0	0.0	0.0
t_{99}	1.0	0.0	0.0
t_{100}	1.0	1.0	0.0

For the purpose of the analysis of the redefinitions of Criteria 2 and 3 let us assume that the data are divided by the "crisp" context (32) into two context groups (tuples t_1–t_{50} and tuples t_{51}–t_{100}, respectively). Please note that in both groups there is exactly one tuple with simultaneously satisfied conditions C and P (tuples t_1 and t_{100}, respectively).

The original formula of the Criterion 2, i.e. (30), will almost completely accept the context of a summary $Q\left(C \wedge_p^W P\right)$ as the value of this criterion is 0.98. However, it may be not in line with the semantics of the contextual "possibility" to satisfy $C \wedge P$ as only 2% of the tuples do satisfy conditions C, P and W simultaneously. This behavior is especially undesirable for the scenario of aggregation of the summaries' truth values and corresponding values of Criterion 2, where the latter have direct impact on the quality assessment of the whole summary.

When using the individual thresholds for all criteria, including Criterion 2 and 3, this impact is softened as even if the summary will be "let through" the Criterion 2 its low truth value T will directly indicate that it is not correct and will not be accepted in the final set of summaries. However, both approaches may be combined and the "virtual" high value of Criterion 2 can have direct influence on the quality measure of the summary, artificially overstating it, even when the thresholds will be used.

A similar motivation can be shown for the old formula of Criterion 3, i.e. (31), and the summaries $Q\left(P \vee_i^W C\right)$. Using the same example from Table 1b as above, the two tuples satisfying condition P will result in fulfilling the existential quantifier in $\exists P W$ for every other tuple and thus will give high values of Criterion 3 (also 0.98 to be precise) for the whole summary which might be seen as rather inappropriate as only 2% of tuples fulfill the conditions P and W simultaneously, and thus the possibility of fulfilling them both should be treated as low.

The redefined formulas for Criterion 2 and 3, i.e. (34) and (35) respectively, address the above issue. A proper definition of the quantifier Q^{II}, e.g. as a "*at least a few*s" with a trapezoidal membership function trap(x; 0.0, 0.1, 1.0, 1.0) may help to adjust the impact of the cardinality of a set of tuples fulfilling relevant conditions on the values of additional criteria, e.g., by specifying that "at least around 10% of tuples" have to met them to state the possibility or impossibility of meeting the bipolar conditions.

5.2 Note on Quantifiers in "and Possibly" and "or, if Impossible" Operators Itself

Following our approach for summaries' quality criteria one may incorporate more sophisticated quantifiers than the existential one into the bipolar operators (14) and (19) itself. As a motivational example let us recall the dataset (b) from Table 1b, for which the truth values of the summaries $Q\left(C \wedge_p^W P\right)$ and $Q\left(P \vee_i^W C\right)$ is equal 0.02 due to the fact that there are only two tuples (out of 100) actually fulfilling both conditions C and P simultaneously (for the former summary) or the condition P (for the latter). Even for such a small dataset it feels like the influence of the single tuples is excessive, and the existential quantifier in (17) and (21) projects this influence to datasets of larger sizes.

Through replacement of the existential quantifier with a softer one, e.g. "*at least a few t*'s" as in Criteria 1–3, one achieves that the larger proportions of tuples have to meet the appropriate conditions to confirm the possibility or impossibility of fulfilling them in the "overall view" of the contexts as satisfied. Formally, the contextual summaries may be redefined as follows:

$$Q_t \left(C(t) \wedge_p^W P(t) \right)$$
$$= Q_{t \in R}^{\mathrm{I}} \left(C(t) \wedge \left(Q_{s \in R}^{\mathrm{II}} \left(C(s) \wedge P(s) \wedge W(t, s) \right) \Rightarrow P(t) \right) \right) \tag{36}$$

and

$$Q_t \left(P(t) \vee_i^W C(t) \right) = Q_{t \in R}^{\mathrm{I}} \left(P(t) \vee \left(\neg Q_{s \in R}^{\mathrm{II}} (P(s) \wedge W(t, s)) \wedge C(t) \right) \right), \tag{37}$$

where quantifier Q^{I} is a part of the summary (i.e. is chosen from a set of available quantifiers) and quantifier Q^{II} serves as the above mentioned "*at least a few t's*" quantifier, exemplified, e.g., as $Q_{\mathrm{trap}(x; 0.0, 0.1, 1.0, 1.0)}$, which seems to be a good point of departure for more sophisticated tuning, including an end-user's contribution.

6 Conclusions

In this paper we have further developed the concept of the contextual linguistic summaries and their derivation. In particular, we have proposed some extensions to the earlier proposed quality measures regarding the contexts itself as well their combinations with the predicates of the summary. The main point of the proposed extensions is the idea to replace the existential quantifier what provides for the more flexibility in defining the possibility and impossibility of the satisfaction of the predicates involved. We illustrate the proposed concepts on some simple examples to better, in a more intuitive way, present their essence.

What concerns further research, it may include the more in-depth study of the relation between the discussed quality measures and the properties of the context, in the spirit of our earlier papers [40, 44] as well as a more thorough experimental verification of the usefulness of the proposed solutions. Moreover, an important direction seems to be the analysis of the new contextual linguistic summaries from the point of view of their comprehensiveness as proposed by Kacprzyk and Zadrożny [22].

Acknowledgments Mateusz Dziedzic contribution is supported by the Foundation for Polish Science under International PhD Projects in Intelligent Computing. Project financed from The European Union within the Innovative Economy Operational Programme (2007–2013) and European Regional Development Fund. This work was also partially supported by the National Science Centre (NCN) under Grant No. UMO-2012/05/B/ST6/03068.

References

1. Castillo-Ortega, R., Marín, N., Sánchez, D., Tettamanzi, A.G.: Quality Assessment in Linguistic Summaries of Data. In: Greco, S., Bouchon-Meunier, B., Coletti, G., Fedrizzi, M., Matarazzo, B., Yager, R.R. (eds.) Advances in Computational Intelligence, communications in computer and information, vol. 298, science edn, pp. 285–294. Springer, Berlin (2012)
2. De Tré, G., Zadrożny, S., Bronselaer, A.J.: Handling bipolarity in elementary queries to possibilistic databases. IEEE Trans. Fuzzy Syst. **18**(3), 599–612 (2010)
3. Delgado, M., Ruiz, M.D., Sánchez, D., Vila, M.A.: Fuzzy quantification: a state of the art. Fuzzy Sets Syst. **242**, 1–30 (2014)
4. Donis-Diaz, C.A., Muro, A., Bello-Pérez, R., Morales, E.: A hybrid model of genetic algorithm with local search to discover linguistic data summaries from creep data. Expert Syst. Appl. **41**(4), 2035–2042 (2014)
5. Dubois, D., Prade, H.: Bipolarity in flexible querying. In: Andreasen, T., Motro, A., Christiansen, H., Larsen, H.L. (eds.) FQAS 2002, LNAI, vol. 2522, pp. 174–182. Springer, Berlin (2002)
6. Dubois, D., Prade, H.: Gradualness, uncertainty and bipolarity: Making sense of fuzzy sets. Fuzzy Sets Syst. 192, 3–24 (2012). http://linkinghub.elsevier.com/retrieve/pii/S0165011410004598
7. Dubois, D., Prade, H.: On various forms of bipolarity in flexible querying. In: Proceedings of the 8th conference of the European Society for Fuzzy Logic and Technology (Eusflat) (2013). http://www.atlantis-press.com/php/paper-details.php?id=8414
8. Dubois, D., Prade, H.: Modeling and if possible and or at least: different forms of bipolarity in flexible querying. Flexible Approaches in Data, Information and Knowledge Management, pp. 3–19. Springer International Publishing, Berlin (2014)
9. Dziedzic, M., Kacprzyk, J., Zadrożny, S.: On some quality criteria of bipolar linguistic summaries. In: Ganzha, M., Maciaszek, L.A., Paprzycki, M. (eds.) Proceedings of the 2013 Federated Conference on Computer Science and Information Systems (FedCSIS), pp. 643–646. Kraków, Poland (2013)
10. Dziedzic, M., Kacprzyk, J., Zadrożny, S.: Bipolar linguistic summaries: A novel fuzzy querying driven approach. In: Joint IFSA World Congress and NAFIPS Annual Meeting, IFSA/NAFIPS, Edmonton, Alberta, Canada. pp. 1279–1284, IEEE, June 24–28 2013
11. Dziedzic, M., Zadrożny, S., Kacprzyk, J.: Towards bipolar linguistic summaries: a novel fuzzy bipolar querying based approach. In: Proceedings of 2012 IEEE International Conference on Fuzzy Systems (FUZZ-IEEE), pp. 1–8. IEEE (2012)
12. George, R., Srikanth, R.: Data summarization using genetic algorithms and fuzzy logic. In: Herrera, F., Verdegay, J. (eds.) Genetic Algorithms and Soft Computing, pp. 599–611. Springer, Heidelberg (1996)
13. Kacprzyk, J., Yager, R.R., Zadrożny, S.: A fuzzy logic based approach to linguistic summaries of databases. Int. J. Appl. Math. Comput. Sci. **10**, 813–834 (2000)
14. Kacprzyk, J., Ziółkowski, A.: Database queries with fuzzy linguistic quantifiers. IEEE Trans. System, Man Cybern. 16(3), 474–479 (1986)
15. Kacprzyk, J., Yager, R.R.: Linguistic summaries of data using fuzzy logic. Int. J. Gen. Syst. **30**, 33–154 (2001)
16. Kacprzyk, J., Yager, R.R., Zadrożny, S.: A fuzzy logic based approach to linguistic summaries of databases. Int. J. Appl. Math. Comput. Sci. **10**, 813–834 (2000)
17. Kacprzyk, J., Zadrożny, S.: Compound bipolar queries: combining bipolar queries and queries with fuzzy linguistic quantifiers. In: Pasi, G., Montero, J., Ciucci, D. (eds.) Proceedings of the 8th conference of the European Society for Fuzzy Logic and Technology (EUSFLAT-13). pp. 848–855. Atlantis Press (2013)
18. Kacprzyk, J., Zadrożny, S.: On a fuzzy querying and data mining interface. Kybernetika **36**, 657–670 (2000)
19. Kacprzyk, J., Zadrożny, S.: Computing with words in intelligent database querying: standalone and internet-based applications. Inf. Sci. **134**(1–4), 71–109 (2001)

20. Kacprzyk, J., Zadrożny, S.: Data mining via linguistic summaries of databases: an interactive approach. In: Ding, L. (ed.) A New Paradigm of Knowledge Engineering by Soft Computing, pp. 325–345. World Scientific, Singapore (2001)
21. Kacprzyk, J., Zadrożny, S.: Linguistic database summaries and their protoforms: towards natural language based knowledge discovery tools. Inf. Sci. **173**(4), 281–304 (2005)
22. Kacprzyk, J., Zadrożny, S.: Comprehensiveness and interpretability of linguistic data summaries: A natural language focused perspective. In: Proceedings of 2013 IEEE Symposium on Computational Intelligence for Human-like Intelligence, CIHLI 2013, Singapore, pp. 33–40. IEEE, April 16–19 2013. http://dx.doi.org/10.1109/CIHLI.2013.6613262
23. Kacprzyk, J., Zadrożny, S.: Hierarchical bipolar fuzzy queries: towards more human consistent flexible queries. In: Proceedings of 2013 IEEE International Conference onFuzzy Systems (FUZZ), pp. 1–8 (2013)
24. Kacprzyk, J., Zadrożny, S.: Computing with words, protoforms and linguistic data summaries: towards a novel natural language based data mining and knowledge discovery tools. J. Autom. Mobile Robot. Intell. Syst. **8**(3), 52–58 (2014)
25. Kacprzyk, J., Zadrożny, S.: Linguistic summaries of time series: A powerful and prospective tool for discovering knowledge on time varying processes and systems. In: Seising, R., Trillas, E., Kacprzyk, J. (eds.) Towards the Future of Fuzzy Logic, Studies in Fuzziness and Soft Computing, vol. 325, pp. 65–77. Springer International Publishing (2015)
26. Kacprzyk, J., Zadrożny, S., Dziedzic, M.: A novel view of bipolarity in linguistic data summaries. In: Kczy, L.T., Pozna, C.R., Kacprzyk, J. (eds.) Issues and Challenges of Intelligent Systems and Computational Intelligence, Studies in Computational Intelligence, vol. 530, pp. 215–229. Springer International Publishing (2014)
27. Kacprzyk, J., Zadrożny, S., Tré, G.D.: Fuzziness in database management systems: half a century of developments and future prospects. Fuzzy Sets Syst. 281, 300–307 (2015). http://dx.doi.org/10.1016/j.fss.2015.06.011
28. Lacroix, M., Lavency, P.: Preferences: Putting more knowledge into queries. In: Stocker, P.M., Kent, W., Hammersley, P. (eds.) Proceedings of the 13th International Conference on Very Large Data. pp. 217–225. Morgan Kaufmann Publishers Inc. (1987)
29. Marín, N., Sánchez, D.: On generating linguistic descriptions of time series. Fuzzy Sets Syst. **1**, 1–25 (2015)
30. Matthé, T., De Tré, G., Zadrożny, S., Kacprzyk, J., Bronselaer, A.: Bipolar database querying using bipolar satisfaction degrees. Int. J. Intell. Syst. **26**(10), 890–910 (2011)
31. Rasmussen, D., Yager, R.: SummarySQL–a fuzzy tool for data mining. Intell. Data Anal. **1**(1–4), 49–58 (1997)
32. Wilbik, A., Keller, J.M.: A fuzzy measure similarity between sets of linguistic summaries. IEEE Trans. Fuzzy Syst. **21**(1), 1282–1288 (2013)
33. Yager, R.: A new approach to the summarization of data. Inf. Sci. **28**, 69–86 (1982)
34. Yager, R.: On linguistic summaries of data. In: Frawley, W., Piatetsky-Shapiro, G. (eds.) Knowledge Discovery in Databases, pp. 347–363. AAAI/MIT Press, Cambridge (1991)
35. Zadeh, L.A.: A computational approach to fuzzy quantifiers in natural languages. Comput. Math. Appl. **9**, 149–184 (1983)
36. Zadrożny, S.: Bipolar queries revisited. In: Torra, V., Narukawa, Y., Miyamoto, S. (eds.) Modelling Decisions for Artificial Intelligence (MDAI 2005). LNAI, vol. 3558, pp. 387–398. Springer, Berlin (2005)
37. Zadrożny, S., De Tré, G., De Caluwe, R., Kacprzyk, J.: An overview of fuzzy approaches to flexible database querying. In: Galindo, J. (ed.) Handbook of Research on Fuzzy Information Processing in Databases, pp. 34–53. Information Science Reference, New York (2008)
38. Zadrożny, S., Kacprzyk, J.: Bipolar queries and queries with preferences. In: Proceedings of 17th International Workshop on Database and Expert Systems Applications (DEXA '06). pp. 415–419 (2006)
39. Zadrożny, S., Kacprzyk, J.: Bipolar queries: a way to deal with mandatory and optional conditions in database querying. Uncertainty Approaches for Spatial Data Modeling and Processing, pp. 117–132. Springer, Berlin (2010)

40. Zadrożny, S., Kacprzyk, J.: Bipolar queries: an aggregation operator focused perspective. Fuzzy Sets Syst. **196**, 69–81 (2012)
41. Zadrożny, S., Kacprzyk, J., De Tré, G.: Bipolar queries in textual information retrieval: a new perspective. Inf. Proc. Manag. **48**(3), 390–398 (2012)
42. Zadrożny, S., Kacprzyk, J., Dziedzic, M.: Contextual bipolar queries: "or if impossible" operator case. In: Proceedings of the 2015 Conference of the International Fuzzy Systems Association and the European Society for Fuzzy Logic and Technology (IFSA-EUSFLAT 2015). Atlantis Press (2015)
43. Zadrożny, S., Kacprzyk, J., Dziedzic, M.: On a new type of contextual queries and linguistic summaries of a bipolar type. In: Proceedings of the 2015 IEEE International Conference on Fuzzy Systems (FUZZ-IEEE 2015). IEEE (2015)
44. Zadrożny, S., Kacprzyk, J., Dziedzic, M., De Tré, G.: Contextual bipolar queries. In: Jamshidi, M., Kreinovich, V., Kacprzyk, J. (eds.) Advance Trends in Soft Computing, Studies in Fuzziness and Soft Computing, vol. 312, pp. 421–428. Springer International Publishing, Berlin (2014)

Part III
Multi-agent Based Technologies

Microgrids and Management of Power

Weronika Radziszewska and Zbigniew Nahorski

Abstract The advancements in technology, changes in power usage patterns and the pressure on the renewable technologies are forcing changes in the electric power grids and the electric infrastructure. The new challenges appear and also new ways of dealing with problems. The concepts of prosumer and microgrid emerged. To make these feasible and safe, the management of power usage and production is required to maintain the balance of power. Demand side management and production side management consider techniques to deal with cost-effective power balancing problems. In this article, the concept of complex energy management system is presented; a specific case study for a research and education center is considered. This limits the scope of the management, as the demand side management should not limit the users in performing their professional duties, this restriction is less present in the case of households. The outline of the system is presented with the short description of its elements.

Keywords Energy management system · Demand side management · Production side management · Task scheduling · Short-time balancing · Multi-agent system

1 Introduction

Renewable power sources are perceived as a solution that can help fight climate change. The renewable power sources are sources that produce energy from natural processes, such as irradiance, water, wind, waves, etc. These power sources have the advantage of having non exhaustible fuel. The disadvantage lies in the unpredictability of production and sometimes short lifespan of the devices. The existence of such

W. Radziszewska (✉) · Z. Nahorski
Systems Research Institute, Polish Academy of Sciences,
Newelska 6, 01-447 Warsaw, Poland
e-mail: Weronika.Radziszewska@ibspan.waw.pl

Z. Nahorski
e-mail: Zbigniew.Nahorski@ibspan.waw.pl

© Springer International Publishing Switzerland 2016
G. De Tré et al. (eds.), *Challenging Problems and Solutions
in Intelligent Systems*, Studies in Computational Intelligence 634,
DOI 10.1007/978-3-319-30165-5_8

sources boosted the changes in the power grid, introducing the bi-directional power flow, which lead to introduction of prosumers.

Prosumer is a concept that was originally defined in economy as a junction between words professional and consumer. It was adapted by the energy sector and changed its origin to words 'producer' and 'consumer'. A prosumer is a unit (connected to the grid) that internally produces and consumes energy. Usually the term prosumer is used regarding small structures (households, city districts, villages) with microsources or renewable sources connected. Such configuration has a rational economical explanation: the cost of construction of such facility is smaller than expected revenue. As the production and consumption of the power within the prosumer grid do not always balance, a prosumer can be seen by the external grid as a source that delivers energy to the grid or as a load that consumes it, depending on a current power flow. Due to the small production and consumption abilities such prosumer would be mostly exchanging very limited amounts of power with external power grid.

Usually prosumers are small energy units and individually they do not impact a lot on an overall balance of the grid, but in numbers they can improve the state of the power grid by avoiding using power during peak hours. To do this, the prosumer has to be aware of the situation and, if necessary, actively limit its usage or shift it to different time. At the moment, there is a lot of research done to develop system that would help manage the power production (if possible) and optimize the usage of power, see e.g. [3, 4, 12, 20, 31].

In this article the problem of management of the prosumer microgrid is considered (the concept of microgrids is described in the following section). The problem of power balancing is presented in Sect. 3. The approaches for power management, both from consumption and production point of view are presented in Sects. 4 and 5. A complex energy management system elaborated for specific microgrid is presented in Sect. 6. It takes into consideration the specific requirements and limitations of the microgrid. Described in Sect. 6.1, it is a research and education center equipped with renewable and non-renewable power sources. The last section concludes the article.

2 Microgrid

A microgrid is a group of consumers, producers, prosumers, or energy storage devices located on small area, that can operate autonomously. The microgrid usually constitutes a low (400/230 V) or medium (1–60 kV) voltage network. Very often, microgrids are equipped with power sources (e.g. gas microturbines, micro wind turbines, photovoltaic panels) or power storage (e.g. batteries, flywheels). Such infrastructure poses a big challenge to the management of energy, as the small changes in power production or consumption have big influence on the state of microgrid. In spite of that, microgrids have a number of advantages, especially when equipped with small energy sources (renewables or not) and when there is a possibility to store power, even in limited amount. A characteristic feature of a microgrid is that it can be treated as one entity from the point of view of the larger network. This work considers the power issues inside a microgrid, for discussion of additional advantages of using microgrids see [16]. Some realized projects connected to microgrid research, are described in [10].

A microgrid can work in 'synchronous mode', meaning that it is connected to a larger grid and exchanges power with it. However, microgrid can work disconnected from the main grid; this is a so called 'island mode' operation. Microgrid can be in such state when the external grid is unavailable or if it is possible to perfectly balance production and consumption of electricity internally.

A microgrid is not just a smaller version of a macrogrid. The issue has been discussed in detail in [13]. The physical effects in low-voltage grids are different than in high voltage grids, which are less vulnerable to small fluctuations of power. Moreover, a possible autonomous (island) operation of a microgrid requires solving of additional problems. For example, subsistence of the frequency, which is normally controlled by the external grid, has to be solved. In the island operation mode a microgrid often does not have enough power to support a usual load all the time; there should be a mechanism of switching off the loads with lower priorities.

3 Power Balancing

Power produced by some renewable sources (especially micro sources, which might lack the ability to manage their current production level) fluctuates dynamically due to sudden changes in e.g. wind and solar irradiance. Predictors, to some extent, can forecast the production and help minimizing the imbalances, but the predictions are not perfect. Consumption of energy is also very changeable and often unpredictable, especially in small microgrids, where a single device can make a noticeable difference in overall power usage. This means that the actions of a single person can make a noticeable disruption from a typical daily power usage profile.

The power in the microgrid has to have good properties, which means that the parameters of the power as e.g. phase angle and voltage have to be within certain limits, to ensure that the microgrid is in balance. The microgrid is in balance when the amount of power produced or delivered to the microgrid equals the usage of this power plus the possible loses of power.

Balancing should make the amount produced:

$$s(\Delta t_k) = \int_{t \in \Delta t_k} s(t)dt$$

for a given time period (Δt_k), equal to the amount consumed:

$$d(\Delta t_k) = \int_{t \in \Delta t_k} d(t)dt$$

at that time. The real energy balancing is a continuous process, but from the operational point of view it can be quantified to a number of short time periods Δt.

$$\sum_{i=0}^{n} s_i(\Delta t_k) = \sum_{j=0}^{m} d_j(\Delta t_k) + L(\Delta t_k), \ \Delta t_k \in T$$

where $n \in N$ is the number of active producers and $m \in M$ is the number of active consumers. The losses of power during transmission ($L(\Delta t_k)$) are relatively small for microgrids, their amount depends on network structure. While their absence does not influence the theoretical solution, it does allow for a simplification of the model.

A microgrid in general can consist of producers, consumers and prosumers. Each of these can be considered a controllable or uncontrollable unit. Uncontrollable devices are those which are not manageable by the grid or by a management system. To this category belong most of the power consuming devices and small renewable power sources (in which power production depends on weather conditions). It is important to note that controllable/uncontrollable in this context are considered in relation to a management system: a lamp is controllable by a person, but as we do not want the system to decide on switching it on or off, for the system it is uncontrollable. The balancing problem reverts to a decision problem of setting the operating point of controllable devices in the microgrid, so that supply and demand are equal. To simplify a model, all uncontrollable devices can be aggregated to a single value: this value is either 0 (perfect balance within the group of uncontrollable devices), positive (behaves as producer) or negative (behaves as consumer), but this aggregated value is not constant over time.

The power sources have physical limitations: a minimal and maximal operating point, a time necessary for changing the operation point, etc. Managing a controllable power source means deciding if the device will be active in the next time period ($s_i(\Delta t_k)$), and if so, determine the amount of power it will provide.

The time for achieving balance is limited, and this time does not increase for a bigger microgrid. This is a problem that cannot only be solved by adding more computing power, but requires a more intelligent approach. Distributed optimizations, such as provided by agent systems, are often proposed, as their behavior tends to provide a good enough solution within the time constraints set.

There have been many papers dealing with problem of power balancing, see for example [14]. However, as pointed out in [38], due to a dynamic generation and demand of the electric power, and need to obtain the power balance, the grids with renewable energy sources require application of even more complex control systems. They are usually called the energy management systems (EMS). General architectures of energy management systems might be found in [27, 28, 38]. These systems often include such modules as a control module oriented to optimization of the grid operating costs, a module cooperating with the distribution grid operator, and a module ensuring reliable supply of energy. In most real life installations, a microgrid is connected to an external power grid, which can provide or absorb a large amount of power. In large power grids, a constant reserve of production power is kept in order to cover occurring imbalances.

The problem of power balancing is slightly different on each level of the power grid. Balancing power in the high voltage network can benefit from big aggregation of consumption. There the daily and weekly cycles dominate [15] and the inertia of the grid is much larger. In microgrids, the changes in consumptions still have visible cycles, but the random behavior plays a bigger role and the inertia of devices is smaller. This requires fast decision making regarding change of the operation point of sources and consumers in the grid. That poses a computation challenge, especially when the number of nodes is large and an energy management system has to balance the energy in all nodes, considering also all the physical limitation of the devices within a defined time period.

Effective balancing requires some kind of schema of cooperation between the producers of energy. The most straightforward schema is the centralized management: it is then possible to have one predictor of demand (e.g. that which gives the smallest errors), based on which the plan for production is made and the system distributes the power production. Centralized systems offer the possibility of, earlier mentioned, optimal production distribution [36], possibly considering multi-criteria decision making. Centralized systems unfortunately have a number of different disadvantages: sensitivity to central controller failure, poor scalability, and requirement of full control over the sources. Full control may not be a problem in microgrids with a single owner, but may be unacceptable in a more general situation. A centralized system might also not be able to consider specific preferences of the source owners or might give unacceptable results when a source owner happens to actively make decisions on its own (although that should not happen in a well designed system).

Non-centralized solutions have been also developed and showed promising results. Among them agent-based power balancing systems are quite a popular approach. Due to the intrinsic characteristics of the agents, these system are distributed. A classification of different energy management schemes for agent-based systems can be found in [29]. Agents can represent single devices, nodes in the power grid, subsets of nodes or even single microgrids. The presented categories of management schemes are: central-hierarchical control structure, distributed-hierarchical control structure, and decentralized control structure (peer-to-peer relation). The hierarchical organization of agents introduces an order and defines agent's functions in optimization and decision making. This can speed up the processing of the data, by dividing and distributing the tasks for calculation. The hierarchy can handle power distribution in a similar way as centralized systems. Completely decentralized control structures are extremely robust to failures and can quickly adapt to changing conditions. However, because there is a larger exchange of data and more intensive communication, such systems tend to operate slower, which might be the cause of imbalances not being resolved in time.

The last group of control systems are the ones based on market structures. The market is the central element of the balancing process, but the participants decide what kind of offer is placed on the market. In such approaches, money and cost functions play the role of ordering the power from most desired sources (i.e. cheapest and most efficient) down to the sources that are used only in emergency (i.e. more expensive power producers). And vice versa, if loads to be powered are considered, the more important (with higher price) are chosen first. Presentation of market based energy control systems can be found in [21, 37, 38].

The information shared between producers is a property of a used communication scheme, which can depend on the level of cooperation, the size of the microgrid or other factors, such as cost of power production, ownership, regulations, etc. In microgrids with one owner, there can be full cooperation with full flow of information; allowing for central balancing to be used. When competition of producers is present, the flow of information may be constrained to the minimal level that is necessary for the balancing process.

4 Demand Side Management and Demand Response

With the development of smart grids, the ideas for optimizing energy consumption went even further: to ensure the stable parameters for the current and to ensure rational prices for the power, the consumers could actively take part in managing the energy usage. A new interdisciplinary research area called Demand Side Management (DSM) emerged. DSM has several main goals: to convince people to take part in energy optimization (also energy saving), to find the best way to communicate them the current status of the network and to develop appliances that would optimize power usage without the human intervention.

The first problem lies in explaining the problems and making users realize that they can make a difference by actively managing the energy usage. However, such actions require that people adjust their lifestyle to the current situation. A peak in demand can be caused by many people doing the same thing at the same time (e.g. switching on home appliances, cooking lunch), which usually requires additional power sources. Getting people to shift their energy consumption lowers the peaks, and lower peaks are easier and cheaper to maintain (lower requirements to the additional power sources to cover the short time peaks). To convince people to make such effort, they should be clearly informed about the status of the power grid. The most popular way of informing people is by introducing different prices. When there is a peak of consumption, the price of energy is high, and it is lower when there is an excess of energy. That idea was behind introducing peak and off-peak tariffs [32], which currently is done by providing the consumers with fixed intervals where peaks are known to occur. Financial incentives are the most common one and easily understandable, but as Cialdini shows in [5] an even better incentive is the feeling of being in competition (e.g. between neighbors).

The action of changing the prices of energy more dynamically to influence the demand has its own name: demand response. This is a very promising technology that can be introduced in near future. It spans over the innovations in automatic responsiveness of the devices, managing of reserves, market strategies and introducing different incentives. A review of currently ongoing research in demand response and partly in DSM is presented in [2].

To simplify the consumption management, there is the idea to create intelligent appliances that would actively delay or modify their operation cycles to reduce the power peak. Such devices exist (e.g. some washing machines), but they are still very unpopular due to lack of trust of people—they do not like the feeling that something is happening outside of their knowledge—which partly is caused by the lack of understanding of how such systems work and what they do. What is more, such devices have limited applicability, as most of the usage is human-driven. In [4] the problem of interactive and background power usage is discussed and a method to control background loads (e.g. refrigerator, dehumidifier) to reduce the power peaks is presented.

The biggest obstacle for introducing DSM technologies, is the lack of preparation of the legislation that would allow introducing a market mechanism for this communication. The legislation also lacks clear rules about exchange of information from the smart grid, and it should introduce simpler rules for the installation of microsources (both renewable and not).

DSM is somewhat linked to optimizing energy efficiency. While purely optimizing energy efficiency is not an active DSM, it may be the result of e.g. price incentives. Lowering energy usage can not only be achieved by using more energy efficient devices, but also by optimizing the activities of the devices present in the infrastructure. One example is the network infrastructure; the authors in [18, 19], describe algorithms to route network traffic differently, in order to minimize the amount of network hardware, and thus power, used.

5 Production Side Management

Production side management deals with deciding which of the available power sources and power storage units are used. This can be done for many purposes, e.g. balancing, maintaining good quality of power, or minimizing cost. While balancing adds a time aspect, thus adding speed of adjustment as a criterion, even the other purposes of production side management already pose problems. A common factor in these problems is the presence of uncertainty as neither the production of other power units nor the exact consumption is known.

The biggest cost for owners of uncontrollable microsources (like wind turbines, water turbines, photovoltaic panels) is the installation of devices and maintenance. The exploitation cost, except for repairs, are negligible, so the best strategy for the owner is to produce as much as possible. The power that is overproduced must be used or send to the power grid (assuming the microgrid is not in the island mode).

The owners of controllable power microsources (like micro gas turbines, reciprocating engines on biogas or cogenerations units) are in different situation. In their case producing power or even operating in the idle state means using the fuel, which has to be produced or purchased. Considering that switching on or off of the power source takes time (depending on the characteristic of the device), the first decision of the micro power source owner is when to switch the source on, and then at what operating point. Similar decisions problem arouse when the source can be switched off. If the microgrid operates in the island mode, the source owners decisions are crucial for proper operation of the microgrid. There are a number of methods and strategies to solve this issue. Some of them are centralized, treating the power production as multi-criteria optimization problem (considering cost, fairness and special requirements of the owners of the microsources), other solutions consider more distributed approach where owners have to cooperate or compete to reach the balance of power. Because the solution presented in this work is distributed, a deeper analysis of such an approach will be presented.

In a distributed approach, the amount of public information and what information are being exchanged is an important issue. For various reasons, as e.g. safety, competition, willingness to make profit, the producers tend to keep certain information private. The lack of information exchange can make it impossible to perform balancing. Such situation was considered in [1], where a method to deal with such ill-defined problem is suggested. It was called the El Farol Bar Problem. The extension of this problem, called the Potluck Problem, considers the supply and demand balancing with almost lack of information exchange [7]. Discussion about this topic has been presented in [24].

6   Complex Management of Power in the Microgrid

6.1   Case Study

The concept of the EMS presented in this work considers both demand side management and production side management, but within the limits given by the considered model of the microgrid. The microgrid used in the research is based on the original plans for the Research Center of the Polish Academy of Sciences 'The Conversion of Energy and Renewables' in Jabłonna. The simplified schema of the microgrid is presented in Fig. 1. The diagram shows all production and storage nodes in the different buildings. The consumption nodes are not presented in the figure. The Research Center is located in a group of five buildings equipped with power sources: renewable (photovoltaic panels, micro wind turbines, micro water power plant) and non-renewable (reciprocating engine, gas micro combined heat and power plant).

The research center was conceived to study applications of renewable energy and new technologies. As such, in the description, the center aims to research energy generation, energy storage but also energy in the form of heat and transfer of heat. The EMS focuses on the electrical energy side of the research center. The main source of heat are condensing gas boilers (CGB), installed in four of the buildings. Electricity and heating are interdependent: air-conditioning and ventilation use power to operate and impact temperature, while on the other hand a reciprocating engine and a gas

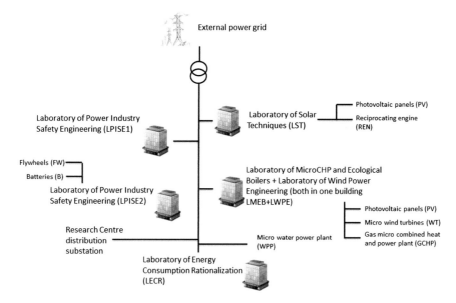

Fig. 1 The microgrid is composed of five buildings and a water power plant, connected together and sharing one connection to the external grid (top of the figure)

microturbine are combined heat and power (CHP) units, the heat from exhaust gases is recuperated and used for heating. Temperature change and water heating is much slower process than the flow of electricity. Heating management is simpler and well recognized; consequently this research focuses solely on electrical energy.

This work was done in the context of a common project, in which, among others, the group from Warsaw University of Technology and the present authors participated. The models of the grid and its devices were designed by a group from Warsaw University of Technology, described in [22, 39]. All the technical and economical aspects of the grids according to the Polish Energy Law [34] have been taken into account to make the model as realistic as possible. Also current developments such as integration of distributed energy resources [13], AC-coupled hybrid systems [6] and island mode operation [30] has been considered.

All of the loads are divided into two groups: the ones that have to be powered in any conditions—unconditionally supplied (unconditionally reserved) and the ones that can be switched off under power deficit—conditionally supplied (conditionally reserved). All equipment that have priority in receiving power, such as e.g. controllers of sources and power network, important computer equipment and emergency lights, are connected to the unconditionally supplied group. Those devices that can be switched off in the event of power deficit are connected to the conditionally supplied group. This differs from the classification to the controllable and uncontrollable devices. The criterion here is necessity of getting power at any cost or not. In [39], authors suggest even that the unconditional supplying section should be outside of the control of energy management system. The EMS should not disturb the operation of unconditionally supplied equipment so it does not manage them. The monitoring of such nodes is still available and they also should be registered in the EMS as all remaining nodes.

In total, there are 128 nodes (production and consumption nodes together), of which 54 are unconditionally reserved.

6.2 Model of the Energy Management System

Due to the complexity of the management problem described here, a two-phase optimization of the power in the grid is applied. The general structure of the system is presented in Fig. 2. The system comprise of a set of specialized systems. The model of the microgrid was described in [21, 27], the simulators of production in [26], the simulators of consumption in [25]. The energy management system is composed by three main systems: Planner, Short-Time Energy Balancing and Energy trading system. In the next sections the parts of EMS are described with more details.

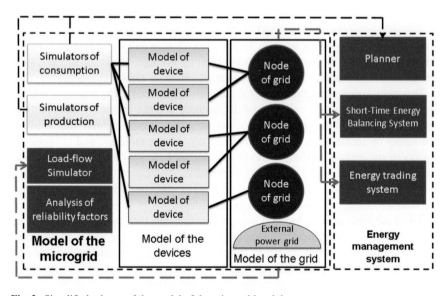

Fig. 2 Simplified schema of the model of the microgrid and the energy management system

6.3 Planner

The Planner realizes the first stage: optimization of power usage and production using the predictions and data available (realizes demand side management). It uses the planned tasks of activities and forecasts of uncontrollable production and consumption of energy in the grid. The Planner suggests the best time for the realization of submitted events, taking into account limitations defined by the user. The Planner also schedules the optimal operating point of the power generators in the microgrid and helps defining the conditions of long term deals with the external power provider. The Planner uses a simulated annealing, which is a heuristic algorithm, to find the best scheduling of events. The aim is minimization of the cost of operating the research center. The model and algorithms of the Planner were published in [8, 9].

Because the research and conference center considered in the project has specific aim and purpose, the aspects of demand side management had to be adjusted to the characteristics of the microgrid. The priority of the research center is to perform research, educate and organize scientific events. While demand side aspects can be taken into consideration, the research and operation of the research center has to take priority over power efficiency. Scheduling activities and knowledge on recurring tasks can contribute to adjustment of demand without disturbing daily operations. The Planner collects data about the events via the room and equipment reservation system. A user reserving the conference room has to give constraints about the time of the event (lecture, conference, etc.), the system shows the best time for the event, when the energy can be provided in a cheapest way. If the time does not fit the user, she/he has to insert different constraints and request the Planner to recalculate the schedule.

The Planner can determine the usage of power for a given schedule and for each moment calculate the optimal (for balancing power) operating point of the controlled power sources. Knowing these operating points, the Planner can calculate the cost of the schedule, which is then minimized using a metaheuristic method.

The Planner can never be totally accurate, especially due to the long term schedule calculation. The temporary deviations from the schedule are taken care of by the Short-Time Energy Balancing system. The Short-Time Energy Balancing system is responsible for automatically adjusting the operating point of the controllable devices, in order to equalize supply and demand of power in the microgrid. The power should be balanced at any point in time, requiring that the system should work almost real-time. Due to the architecture of the system, it is not possible to formally prove that the system is a real-time system, consequently the operation of it will be referred to as a on-line balancing.

6.4   Short-Time Energy Balancing System

The Short-Time Energy Balancing system is the second component in the two stage optimization performed by the EMS. In this work, the Short-Time Energy Balancing system is a multi-agent system in which each node is represented by an autonomic agent. A node groups a set of devices, it can contain either energy consumers, uncontrollable energy sources (e.g. photovoltaic panels), controllable sources (e.g. micro gas turbine) or energy storage (e.g. batteries). This approach to short-time energy balancing is a distributed approach, where every node communicates with others and makes a decision. The multi-agent paradigm was considered to be very useful in this situation: agents can be given the necessary intelligence to make a decision, and communication between agents is one of the main aspects in a multi-agent system.

The balancing mechanism is initiated by a node connected with an energy consumer or an uncontrollable source, whose operating point changes: a light gets switched on or off, or the output of a photovoltaic panel increases or decreases. When the operating point of a consumer increases, the effect on the system is the same as if the operating point of an uncontrollable source decreases: there will be a deficit of power (a negative imbalance), and additional power needs to be supplied. The reverse happens when the operating point of a consumer decreases or that of an uncontrollable source increases. At this point, the node that causes the imbalances signals to all other devices that an imbalance occurs and requests offers from devices to solve this imbalance. The only devices that are possibly capable of dealing with the imbalance are the controllable sources, they answer the request for offers: for a negative imbalance each controllable source provides the amount of power it can supply and its cost, for a positive imbalance this will be the amount of power they can decrease and the profit. From this list of offers, the node that caused the imbalance chooses the preferred option (e.g. with the lowest cost or highest profit) potentially selecting multiple devices to cover the imbalance.

6.5   Energy Trading System

The microgrid defined in the project cannot be fully self-sustainable, as the planned production abilities are much smaller than the possible peak consumption. The estimated amount of power that can be produced in perfect weather conditions and with all power sources activated, is around 250 kW. The possible consumption can reach twice that value (in the project the sum of maximum consumption for each node was estimated to around 950 kW, which is nearly impossible to happen). The real difference between production and consumption of power is expected to vary between a possibly small overproduction (in which case power has to be sold to the power grid) and a possibly large deficit of power that has to be compensated by the power from the external network.

This implies that the microgrid will have to trade with the external power network. Under the present law, the only way the microgrid can trade power is by making deals with the power provider, without the possibility of negotiating short term changes. Current deals of small magnitudes are priced according to fixed tariffs and the presented microgrid would fall in this category, and as such, active participation in the market would not be possible.

Due to the properties of electric current, its trading on a level of high and middle voltages is realized using different types of deals and levels of markets. Electric power is a special type of commodity: it cannot be easily and losslessly stored, supply and demand have to balance and the availability of power is changing in time. To manage the balancing problem, the power is traded on different levels: there is a long-term market, where bilateral deals are made; there is a real-time market, where power is traded on a stock market; and there is a balancing market where the occurring imbalances are settled. The detailed description of power markets can be found in [17, 23, 35]. Energy markets were opened not long ago [17] and are still under subsequent development in many countries. Challenges with new technologies give incentives to redefine the power electricity market, e.g. in [11] authors suggest treating electric energy markets as multicommodity markets.

Even though real markets are not yet prepared for trading with the microgrids, in this project some level of price negotiation was assumed. First, the long-term deals for power are considered available. This possibility is supported by the Planner, in its schedule the amount of power that should be purchased or sold to the external power grid is defined. The market for such small amounts of power does not exist so the way of determining the fluctuation of prices had to be designed. It was assumed that the microgrid offers too little power to influence the structure and prices on the market. In this case, the prices on the market were modeled using neural networks and the price of the power from the external power provider more or less fluctuated according to this pattern. The system was described in details in [33].

The Short-Time Energy Balancing system is trading small amounts of energy with external power grid. At the moment, short-time trading is realized using fixed prices of power per kW, but the extension of using varied prices is easily added. In that case, the system could have a different goal: instead of minimizing the power exchange between external power grid and microgrid, it would optimize the price of power. Then, a situation can be possible when the microgrid lowers the operating point of its microsources to buy more cheaper energy from the external power grid.

6.6   Example

The example of the Planner operation is schematically outlined in Fig. 3. The exemplary tasks in the figure are the aggregation of a number of nodes needed for the tasks and these tasks represent various events in the microgrid. For example, the Tasks D and E are the general (averaged) consumption of the groups of nodes (such as lights, socket usage, ventilation). Tasks A, B and C represent short lasting events,

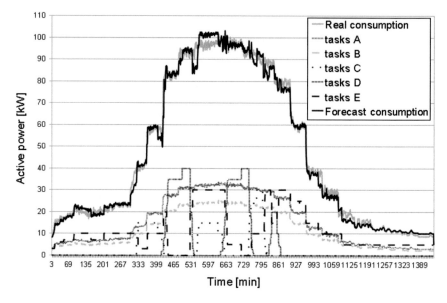

Fig. 3 Example of the planner schedule and forecast of consumption

such as a meeting, a lesson and operation cycles of some laboratory equipment. The forecasted consumption in the figure is the sum of the aggregated consumption of a number of tasks. This consumption profile is generated by the Planner. However, in reality, the consumption profile deviates from this forecasted consumption. Solving this deviation is the task of the Short-Time Energy Balancing system, which deals with the on-line changes.

In Fig. 4 the outcome of the Short-Time Energy Balancing system is presented. The simulation was done without the consideration of the Planner preset operating points of the controllable devices. The system managed to balance power and the figure presents the changes in operating point of the devices. As can be seen, the system first uses the power from the cheapest source and when it cannot produce more power, more expensive sources are used. In the last minutes of the experiment, the consumption of power is below the minimal operating point of the controllable power sources, so the microgrid needs to send the excess power to the external power grid. This means that the microgrid for this short period sells power to the network. The amount sold is small, which limits the possibility to negotiate prices on the power market.

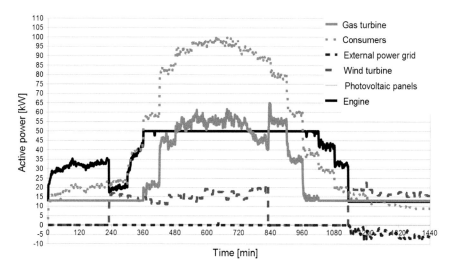

Fig. 4 Example of the management of controllable power sources by the Short-Time Energy Balancing system

7 Conclusion

The presented here case study is an example of energy management in a microgrid. The solution described outlines the issues that might be solved by the new technologies. The character of the research and education center limits the use of demand side management, while allows for more advanced production side management.

The developed EMS consists of a number of modules that work together in order to achieve energy balancing. Models of the devices and of the microgrid are the base to perform power optimization. The system for energy trading is a basic attempt to model a still non-existent market for power trading in microgrids. The Planner realizes the scheduling of the defined tasks to realize partial demand side management and also to collect data about the power usage. It works based on the data given by the users and requires their cooperation. The Planner is equipped with a number of predictors to estimate the background consumption level and forecast production from renewable power sources.

The Short-Time Balancing system is a multi-agent, distributed system that manages the controllable sources (including the connection of external power grid) to balance the power almost real-time. In this system every node (grouping one or more devices) in the microgrid is represented by an agent. In literature, the scenarios without controllable sources dominate (like in [37]), in which case the agents of the consumption devices either request a load-level, or are imposed a load level. Contrary to those approaches, the presented approach includes controllable energy sources which also take part in the negotiation process: the consumption devices change their load level, send a signal that they cause an imbalance, after which it will be decided in negotiations which production device(s) will change its output to compensate for it.

The work presented provides a basis for further research. One of the problems that should be considered is the more advanced management of power storage units. In the current scenario, the power storage units are mainly reacting to imbalances when they occur, they could however take part in a more long-term planning. The possibility of scheduling charging and discharging batteries by the Planner might help to decrease imbalances and provide a cheaper solution. Another interesting aspect is the introduction of electric cars which, when connected to the grid, may also impact policies for energy balancing.

Acknowledgments The research of W. Radziszewska was supported by the Foundation for Polish Science under International PhD Projects in Intelligent Computing. Project financed from The European Union within the Innovative Economy Operational Programme 2007–2013 and European Regional Development Fund. The results have been partially obtained in the Polish Ministry of Science and Higher Education grant number N N519 580238.

References

1. Arthur, W.B.: Complexity in economic theory: inductive reasoning and bounded rationality. Am. Econ. Rev. **84**(2), 406–411 (1994)
2. Balijepalli, V.S.K.M., Pradhan, V., Khaparde, S.A., Shereef, R.M.: Review of demand response under smart grid paradigm. In: Innovative Smart Grid Technologies - India (ISGT India), 2011 IEEE PES, pp. 236–243 (2011)
3. Barker, S.K., Mishra, A.K., Irwin, D.E., Cecchet, E., Shenoy, P., Albrecht, J.: Smart*: an open data set and tools for enabling research in sustainable homes. In: Proceedings of the 2012 Workshop on Data Mining Applications in Sustainability (SustKDD 2012) (2012)
4. Barker, S.K., Mishra, A.K., Irwin, D.E., Shenoy, P.J., Albrecht, J.R.: Smartcap: Flattening peak electricity demand in smart homes. In Giordano, S., Langheinrich, M., Schmidt, A. (eds.) PerCom, pp. 67–75. IEEE (2012)
5. Cialdini, R., Schultz, W.: Understanding and motivating energy conservation via social norms. Technical report, William and Flora Hewlett Foundation (2004)
6. Cramer, G., Reekers, J., Rothert, M., Wollny, M.: The future of village electrification - More than two years of experiences with AC-coupled hybrid systems. In: Proceedings of 2nd European PV-Hybrid and Mini-Grid Conference, Kassel, Germany, pp. 1–6 (2003)
7. Enumula, P.K., Rao, S.: The potluck problem. Econ. Lett. **107**(1), 10–12 (2008)
8. Gorczyca, M., Krysiak, T., Lichtenstein, M.: Przegląd i analiza możliwości zastosowania metod szeregowania zadań związanych z poborem energii w ośrodku badawczym. Technical report, Systems Research Institute PAS (2011)
9. Gorczyca, M., Krysiak, T., Lichtenstein, M.: Power economic dispatch problem with modern cost functions - the complexity and approximation algorithms (2013). Manuscript
10. Hatziargyriou, N.D., Asano, H., Iravani, R., Marnay, Ch.: Microgrids: An overview of ongoing research, development, and demonstration projects. IEEE Power Energy Mag. **5**(4), 78–94 (2007)
11. Kaleta, M., Traczyk, T. (eds.) Modeling Multi-commodity Trade: Information Exchange Methods. Advances in Intelligent and Soft Computing, vol. 121. Springer (2012)
12. Kwak, J., Varantham, P., Mahesvaran, R., Tambe, M., Jazizadeh, F., Kavulya, G., Klein, L., Becerik-Gerber, B., Hayes, T., Wood, W.: Saves: A sustainable multiagent application to conserve building energy considering occupantants. In: Conitzer, V., Winikoff, M., Padgham, L.,

van der Hoek, W. (eds.) Proceedings of the 11th International Conference on Autonomous Agents and Multiagent Systems - Innovative Applications Track (AAMAS 2012) (2012)
13. Lasseter, R., Akhil, A., Marnay, Ch., Stephens, J., Dagle, J., Guttromson, R., Meliopoulous, A.S., Yinger, R., Eto, J.: White paper on integration of distributed energy resources: The certs microgrid concept. Technical report, CERTS (2002)
14. Linnenberg, T., Wior, I., Schreiber, S., Fay, A.: A market-based multi-agent-system for decentralized power and grid control. In: Mammeri, Z. (ed.) Proceedings of 2011 IEEE 16th Conference on Emerging Technologies & Factory Automation ETFA 2011, pp. 1–8. Paul Sabatier University, Toulouse (2011)
15. Lovins, A., Odum, M., Rowe, J.W.: Reinventing Fire: Bold Business Solutions for the New Energy Era. Chelsea Green Publishing Company, White River Junction (2011)
16. Lovins, A.B., Datta, E.K., Feiler, T., Rabago, K.R., Swisher, J.N., Lehmann, A., Wicke, K.: Small is profitable: the hidden economic benefits of making electrical resources the right size. Rocky Mountain Institute (2002)
17. Murray, B.: Power Markets and Economics: Energy Costs, Trading. Wiley, Emissions (2009)
18. Niewiadomska-Szynkiewicz, E., Sikora, A., Arabas, P., Kamola, M., Mincer, M., Kołodziej, J.: Dynamic power management in energy-aware computer networks and data intensive computing systems. Future Gener. Comput. Syst. 37, 284–296 (2014)
19. Niewiadomska-Szynkiewicz, E., Sikora, A., Arabas, P., Kołodziej, J.: Control system for reducing energy consumption in backbone computer network. Concurr. Comput.: Prac. Exp. 25(12), 1738–1754 (2013)
20. Nistor, S., Wu, J., Sooriyabandara, M., Ekanayake, J.: Cost optimization of smart appliances. In Innovative Smart Grid Technologies (ISGT Europe). In: 2011 International Conference and Exhibition on 2nd IEEE PES, pp. 1–5 (2011)
21. Pałka, P., Radziszewska, W., Nahorski, Z.Z.: Balancing electric power in a microgrid via programmable agents auctions. Control Cybern. 4(41), 777–797 (2012)
22. Parol, M., Wasilewski, J., Wójtowicz, T., Nahorski, Z.: Low voltage microgrid in a research and educational center. In: CD Proceedings of the Conference "Elektroenergetika ELEN 2012", p. 15 (2012)
23. Piekut, S., Skoczek, S., Dąbrowski, Ł.: Raport o rynku energii elektrycznej w Polsce. RWE Stoen (2012)
24. Radziszewska, W., Kowalczyk, R., Nahorski, Z.: El farol bar problem, potluck problem and electric energy balancing - on the importance of communication. In: Paprzycki, M., Ganzha, M., Maciaszek, L. (eds.) Proceedings of the 2014 Federated Conference on Computer Science and Information Systems. Annals of Computer Science and Information Systems, vol. 2, pages 1515–1523. IEEE (2014)
25. Radziszewska, W., Nahorski, Z.: Simulation of energy consumption in a microgrid for demand side management by scheduling. In: Paprzycki, M., Ganzha, M., Maciaszek, L. (eds.) Proceedings of the 2013 Federated Conference on Computer Science and Information Systems, pp. 679–682. IEEE (2013)
26. Radziszewska, W., Nahorski, Z.: Modeling of power consumption in a small microgrids. In: 28th International Conference on Informatics for Environmental Protection: ICT for Energy Efficiency, EnviroInfo 2014, Oldenburg, Germany, pp. 381–388 (2014)
27. Radziszewska, W., Nahorski, Z., Parol, M., Pałka, P.: Intelligent computations in an agent-based prosumer-type electric microgrid control system. In: Kóczy, L.T., Pozna, C.R., Kacprzyk, J. (eds.) Issues and Challenges of Intelligent Systems and Computational Intelligence. Studies in Computational Intelligence, vol. 530, pp. 293–312. Springer (2014)
28. Ramchurn, S.D., Vytelingum, P., Rogers, A., Jennings, N.R.: Putting the 'smarts' into the smart grid: a grand challenge for artifitial intelligence. Commun. ACM 55(4), 86–97 (2012)
29. Rohbogner, G., Fey, S., Hahnel, U.J.J., Benoit, P., Wille-Haussmann, B.: What the term agent stands for in the smart grid definition of agents and multi-agent systems from an engineer's perspective. In: 2012 Federated Conference on Computer Science and Information Systems (FedCSIS), pp. 1301–1305 (2012)

30. Rua, D., Pereira, L.F.M., Gil, N., Lopes, J.A.P.: Impact of multi-microgrid communication systems in islanded operation. In: 2011 2nd IEEE PES International Conference and Exhibition on Innovative Smart Grid Technologies (ISGT Europe), pp. 1–6 (2011)
31. Schaerf, A., Shoham, Y., Tennenholtz, M.: Adaptive load balancing: A study in multi-agent learning. J. Artif. Intell. Res. **2**, 475–500 (1995)
32. Schleich, J., Klobasa, M.: How much shift in demand? Findings from a field experiment in Germany. In: ECEEE summer study proceedings, pp. 1919–1926. Fraunhofer-Gesellschaft (2013)
33. Stańczak, J.: Podsystem handlu energią elektryczną z operatorem sieci zewnętrznej. Metoda generowania cen przez operatora. Technical report, Systems Research Institute PAS (2013)
34. The energy law act dated 10 Apr 1997 (Polish)
35. Toczyłowski, E., Kaleta, M., Kacprzak, P., Pałka, P.: Modelowanie rynków energii elektrycznej wybrane zagadnienia. Technical report, Systems Research Institute PAS (2007)
36. Tsikalakis, A.G., Hatziargyriou, N.D.: Centralized control for optimizing microgrids operation. IEEE Trans. Energy Convers. **23**(1), 241–248 (2008)
37. Vasirani, M., Ossowski, S.: A collaborative model for participatory load management in the smart grid. In: Proceedings of the 1st International Conference on Agreement Technologies, pp. 57–70. CEUR (2012)
38. Vogt, H., Weiss, H., Spiess, P., Karduck, A.P.: Market-based prosumer participation in the smart grid. In: 4th IEEE International Conference on Digital Ecosystems and Technologies (DEST), pp. 592–597. IEEE (2010)
39. Wasilewski, J., Parol, M., Wojtowicz, T., Nahorski, Z.: A microgrid structure supplying a research and education centre - Polish case. In: 2012 3rd IEEE PES International Conference and Exhibition on Innovative Smart Grid Technologies (ISGT Europe), pp. 1–8 (2012)

Transaction Protocol and Mechanisms for Adaptive Management of Long-Running Tasks

Marcin Stępniak

Abstract Execution of real-world services can lead to many unexpected events that need to be handled. So that failures of tasks composed of such services frequently occur. Mechanisms for automated task accomplishment and failure handling in open and heterogeneous systems are proposed. They are based on general protocols derived from the well known OASIS Web Services Transaction standard WS-TX for business transactions. The protocols and mechanisms are implemented in the prototype Autero multi-robot system in which the robots cooperate so as to accomplish complex tasks.

Keywords Business transaction · Business service · Automated task realization

1 Introduction

The term *service* can be defined as *a valuable action, deed, or effort performed to satisfy a need or to fulfill a demand* [23]. In this paper, the business services will refer to services that are specific activities carried out by service providers for clients. A client who uses a business service has adequate resources required for its realization (the most common is money). The client also expects the service to be performed in a given time period.

Tasks commissioned by a client can also be performed by a composition of business services as a business process. The business process may also include the services realized only in computer systems (they need to have a common representation).

It is necessary to distinguish a business service from a traditional Web service executed solely within a computer system. A weather service which returns specific data (e.g. temperature, humidity, wind speed) for provided inputs (e.g. the location and date) may be an example of a Web service. Room painting, designing a website, and translation of a text are examples of business services. Each service can be classified as a certain type (category of action). As a result, one type can be assigned to multiple

M. Stępniak (✉)
Systems Research Institute, Polish Academy of Sciences, Warsaw, Poland
e-mail: martinus.st@gmail.com

© Springer International Publishing Switzerland 2016   179
G. De Tré et al. (eds.), *Challenging Problems and Solutions*
in Intelligent Systems, Studies in Computational Intelligence 634,
DOI 10.1007/978-3-319-30165-5_9

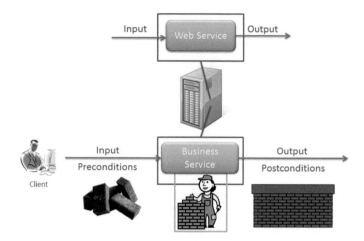

Fig. 1 The difference between a Web service and a business service

services with different parameters. The Fig. 1 visualizes the concept of a business service. Bricklayer requires a supply of materials for the construction (preconditions). From these materials he builds a wall ordered by a client (postconditions).

A client, who wants a task to be accomplished, is usually interested only in the results. The details of the task realization, and the process management can be an unnecessary burden. For that reason, task automating mechanisms are proposed.

Realization of a task is called a transaction. In Information Technology the term *transaction* means "the interaction and managed outcome of a well-defined set of tasks" [11]. The history of a transaction management can be viewed from a temporal perspective from the Stone Age to modern times [22]. The Stone Age is the time period without transactions. In classic history people realized the need for mechanisms to ensure the proper operation of applications that support multiple users working concurrently. Thus, the first transaction mechanisms saw the light. In the Middle Ages business applications grew more complex and hence the requirements for heterogeneous distributed systems and databases increased. Consequently, distributed transactions were developed. The Renaissance brought a combination of process control systems with transaction support, which led to the development of many transaction models suited to such systems. In the Renaissance, main emphasis was on reliability in processes execution. In modern times, we see the emergence of new application domains, in which the Internet plays a major role. To allow the proper operation of business processes in such an environment, transaction management must be adjusted accordingly. The processes are still a key component in this era, but greater emphasis is placed on a cooperation of independent parties, usually represented by loosely coupled web services. The solutions presented in this paper are part of a current trend, but also take into account future needs. The growing importance of the Internet and automation of processes and services may result in an increase in the number of clients interested in commissioning tasks. This creates a need to develop universal mechanisms and protocols to manage such processes, which will allow independent parties to cooperate in order to realize complex tasks.

Conventional transaction processing in tightly-coupled systems ensures ACID properties (Atomicity, Consistency, Isolation, Durability). It requires a close relationship between a transaction coordinator and participants. Ensuring all of these properties for long-running business processes may be impossible or at least impractical.

Transactions for business processes will be referred to as business transactions. Business transaction is defined as a set of actions carried out by services leading to the task accomplishment. Special transaction mechanism for handling failures is designed that has the following properties.

1. Failed services may be replaced with other services during task realization.
2. General plan may be changed.
3. The transaction ends after successful completion of the task, or inability to complete the task, or cancellation of the task.

2 Service Definition and Realization Phases

Since services cooperate in open, dynamic and heterogeneous environments, there is a need for a common representation of the domain. Therefore, the representation of the environment is realized in a form of an ontology to enable collaboration and semantic compatibility of communication in such environment. The ontology consists of concepts and relations between them, i.e. objects, object attributes, and the relations between objects. Each object is of a certain, pre-defined type. The object type is defined by its attributes, and by the internal (hierarchical) structure, i.e. object may consist of sub-objects and relations between these sub-objects. An elementary type has no internal structure, so it is defined only by attributes. A complex type has a hierarchical structure of subtypes. The ontology is defined as a hierarchical collection of types of objects (see [2]). Primitive attributes and relations are the key elements for constructing the types. The object itself, as an instance of its type, is defined by assigning specific values to its attributes and by specifying relations.

The ontology is common for all services and components of the system, and it is also used to specify tasks and define types of services called service interfaces consisting of the following elements.

- Type of service, i.e. type of action that the service performs.
- Specification of the inputs and outputs of the service.
- The conditions required for input of the service (preconditions), and the effects of its execution (postconditions) specifying the output. These conditions are expressed as relations between objects in the environment (ontology).
- Service attributes.

Service attributes contain information about the static features of a service and are used during planning and arrangement, for example, operation range for a transport service, and an average realization time. Ideas of using ontologies for service interfaces definitions can be easily found in the literature [13]. Service and task definitions can be expressed in various languages, e.g. OWL-S [16], Web Service Modeling Ontology [21], and Entish [1].

The way a service is executed often depends on client requirements, as well as the current state of the environment. Therefore, it is important to establish these conditions before the actual execution of the service. This operation is called an **arrangement**. In an arrangement phase, a request with client requirements is sent to the service. The request contains the information about available resources and the expected time of completion, as well as the current state of the environment and the specific requirements of other services in the business process. Service provider determines whether it can perform the task in accordance with the received parameters. If so, it sends a description of the conditions which must be met in order to execute it. If service provider does not agree to the client requirements, it sends a refusal. This process is repeated for all the services of the given type. Then client selects the best offer. Requirements sent to the service provider in the arrangement phase will be referred as an **intention**, and the response as a **commitment**.

Penalties for service providers and clients for violating the agreed contract should also be set during the arrangement phase. This could be a penalty for the client resignation, for withdrawing from the contract by the service provider or violation of other terms of the contract (failing to meet a contractual obligation in a timely fashion, poor quality, etc.).

During the execution phase, the service is executed in accordance with the agreement made in the arrangement phase. After execution phase, the service provider notifies the client about results of the service execution.

3   The System Architecture

The system architecture is an extension of the SOA paradigm (see [5, 12]). The additional elements are: Planer, Arrangement Module and Ontology. This architecture is depicted in Fig. 2.

Task Manager (TM for short)—represents a client, and provides a GUI for the client to define tasks and monitor their realization. It uses Planning Module for business process composition and Arrangement Module in the arrangement phase.

Service Manager (SM)—is a service provider interface for providing its services for an external client, in this case, TM is the client. SM controls the execution of the subtask delegated by TM.

Arrangement Module (AM)—is responsible for carrying out the arrangement phase. May implement various service selection mechanisms (e.g. a genetic algorithm [8]). AM may be implemented to be able to automatically select the best offer.

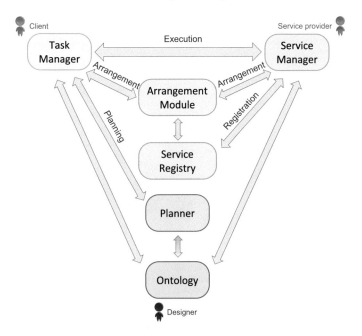

Fig. 2 The system architecture

Service Registry (SR)—stores information about services currently available in the system. Each service, so as to be available, must be registered by its manager in Service Registry. SR may filter the registered services taking into account their attributes.

Planner—is responsible for generating a set of abstract plans based on a client's task. These plans are transformed by Task Manager into a business processes.

Ontology—model containing definitions of types of services and related types of objects. It is a specific semantics of the domain. In fact, this element will be a module or application that enables a creation of a model (representation) of services environment, manages the ontology, and provides access to it by the other system components.

4 The Transaction Protocol

Communication between Task Manager and Service Manager is done according to the transaction protocol which defines the states of services and messages used to change them (see Fig. 3). It allows Task Manager to initialize particular phases of service invocation, monitor their progress, and perform additional actions, e.g. compensation. Service uses messages of the protocol to notify Task Manager

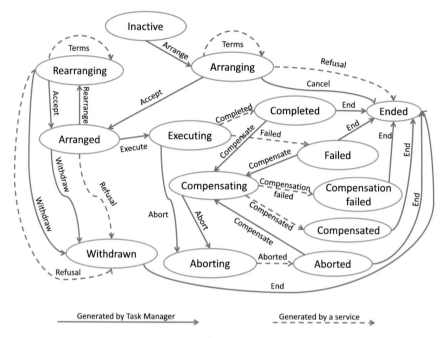

Fig. 3 Transaction protocol state transition diagram

about successful and unsuccessful events in the process of its realization. Each state is described in more detail below.

- **Inactive**—state that occurs only on the side of the Task Manager. It means that the node representing the service is part of the process, but the arrangement phase has not started yet.
- **Arranging**—on the Task Manager side no service is yet assigned to the node in the process. The service is in an arrangement phase.
- **Arranged**—state after accepting a commitment. From this point a particular service is associated with a corresponding node of the business process.
- **Rearranging**—rearranging conditions of the already arranged service. In this state, the service provider generates a new commitment to amended intentions of the client.
- **Executing**—the service performs its job.
- **Completed**—the service has successfully done its job.
- **Failed**—state after an unsuccessful execution of the service.
- **Compensating**—in this state, the service performs actions to compensate for previously performed actions.
- **Compensated**—state after successful completion of a compensation.
- **CompensationFailed**—if a compensation process fails, the service proceeds to this state.
- **Aborting**—in this state the service terminates the job.

- **Aborted**—state after termination of the execution.
- **Withdrawn**—state indicating that the service was withdrawn after completion of the arrangement phase.
- **Ended**—this state indicates that the client (Task Manager) will not request any additional actions. The service in this state can remove the information about the context of the transaction.

All necessary data required for a task execution is a part of the transaction protocol message (see Table 1). This method ensures the greater consistency of the system state. During the task execution, messages are sent according to the specific sequences. They can create different combinations, but a set of possible messages in a given state of the service is strictly defined in the transaction protocol.

The actual situation may include information about the state of the environment related to the service. It can be also a fragmentary data generated by the service. This information can be helpful for Task Manager to restore a proper realization of the task.

To ensure a consistent system state, the protocol messages must be submitted in accordance with the specified sequence. In addition, the parties must save information about whether the message has been acknowledged by the other party, as well as the message sent time. This information is used to decide if and when sender should retry sending the message. The Fig. 4 is a sequence diagram illustrating the process.

The diagram on Fig. 5 shows the complete sequence of the messaging protocol for the single-step business process. There are three services registered in Service Registry. All of them agree to perform services according to the specified parameters and send their offers (commitments). Task Manager chooses one offer and rejects the rest with the Cancel message. Then the selected service is executed and after the successful completion Task Manager is notified by sending a Completed message. End message ends the transaction. After receiving the message of this type the service does not have to store the transaction context.

Table 1 Data sent with messages

Message	Data
Arrange	Intention
Rearrange	New intention
Terms	Commitment
Execute	Input data
Completed	Output data
Failed	Failure description and actual situation
CompensationFailed	Failure description and actual situation
Compensated	Actual situation
Aborted	Actual situation

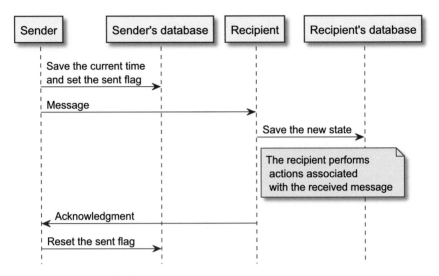

Fig. 4 Transaction protocol message sending sequence diagram

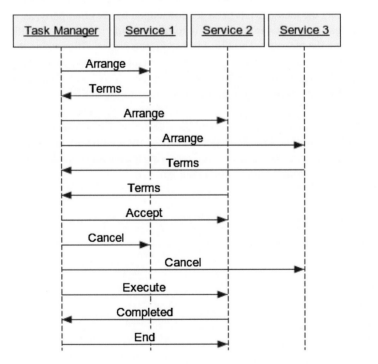

Fig. 5 Sequence diagram of an example transaction

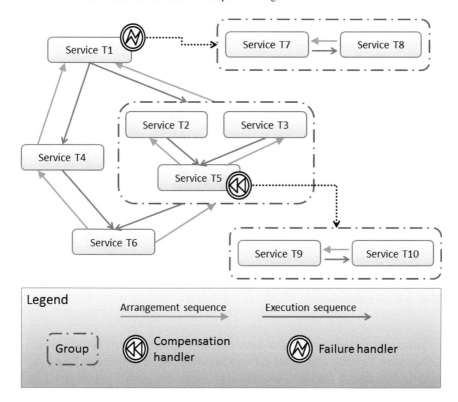

Fig. 6 An example of a plan

5 Task Realization

A client must define the task to be accomplished. The way of defining tasks and the format of their description depends on the specific implementation of Task Manager. While specifying the tasks, objects and relations defined in an ontology are used.

For a given task, Planner returns abstract plans that when arranged and executed, may realize the specified task. An abstract plan is represented as a directed graph where nodes are service types and edges correspond to a causal relationship between the output of one service and the input of the second service. The relationships determine the order of arrangement and then the execution of a concrete plan that has also a form of a directed graph (called business process) however, its nodes are concrete arranged services. Sometimes it is not possible to arrange the whole business process before the start of execution phase. In this case, the business process includes the unassigned nodes for which the arrangement is carried out later on.

In a concrete plan its node may represent a composed service (as a subprocess) consisting of already arranged services. Plan may also include handlers responsible for a compensation and failure handling. Figure 6 shows an example of a plan in a graphical form.

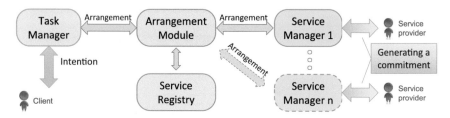

Fig. 7 An arrangement phase of a single service

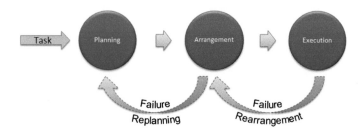

Fig. 8 Task realization phases

During the arrangement phase Task Manager passes a prepared intention to Arrangement Module. AM sends it to all registered services (Service Managers). Service Managers, for each of the obtained intentions, prepare commitment and send it back to AM. Once this phase of collecting commitments is complete, AM selects the best commitment or forward the set of commitments to TM, where a specific commitment can be selected automatically or by the user. This procedure is shown in the diagram in Fig. 7.

In the execution phase Task Manager invokes a service sending a complete input data to appropriate Service Manager. SM sends a response with the output data, being a service product or a confirmation of successful completion (e.g. changing situation in the environment).

Task Manager can stop the service execution before its completion. This may be caused by the task cancellation by the client, a failure during execution of parallel services in the plan (that cannot be replaced), or by changes in the environment making the current plan infeasible.

The general flow of task realization is shown in Fig. 8. The arrows pointing right show the normal flow, while the arrows pointing left illustrate the procedure in case of inability to perform a particular phase.

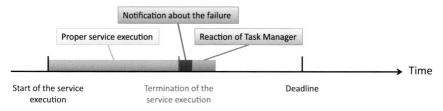

Fig. 9 Time diagram of a service failure with notification

Fig. 10 Time diagram of service a failure without notification

6 Failure Handling

Services provided in the real world are exposed to the occurrences of unexpected situations that prevent their proper execution. Therefore, the fault tolerance is an important aspect for systems using such services (see [19]). Task realization should be possible to complete after a failure of some services participating in it. Task Manager is responsible for handling such situations.

If the service cannot be executed properly, SM informs the Task Manager about the failure. TM can take appropriate actions at that moment. The Fig. 9 shows the sequence of events on a time-line.

Task Manager may not obtain notification about the service failure. In this case, this situation is detected after expiration of the time allocated for the service execution (see. Fig. 10).

Task Manager is equipped with a failure handling mechanism based on the simple algorithm shown in Fig. 11. Set S contains services included in the sub-process, which will be replaced. Set N contains services that are connected by execution order to the services from the set S. If a failure of a service performing a subtask occurs (fail message), the rearrangement procedure is invoked for searching for other services of the same type that can perform the subtask. If such services are available, the best offer will be chosen, and the subtask realization will be continued. Otherwise, the failed service is added to the set S. Then, preconditions and postconditions (PPC) are determined for services from the set S. For these conditions, a new abstract plan is generated by the Planner, and arranged into a sub-process attached to the main process (replacing the failed services from set S). If Planner does not return any plans, then the set N is created that contains all the services that are direct successors (in the execution order) of the failed services in set S. Then, set N is added to set S. The union of S and N becomes the new set S. The preconditions and postconditions

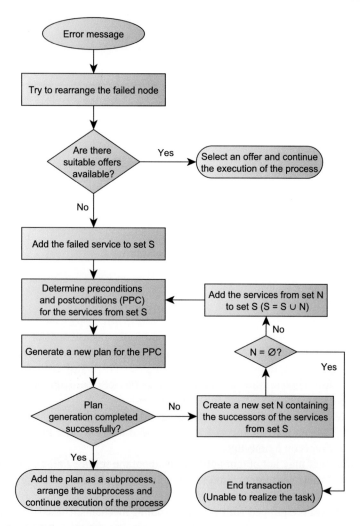

Fig. 11 Service failure handling algorithm

are determined for the new set S. Then the procedure is repeated. It ends either with the successful generation of a new plan or when the set N is empty. Empty set N means inability to carry out the task.

7 Prototype Implementation of the System

The designed transaction protocol and mechanisms have been implemented in the Autero system. It is a software platform for delegating tasks (by human users) to be accomplished in a multi-robot system. The system architecture is shown in Fig. 12

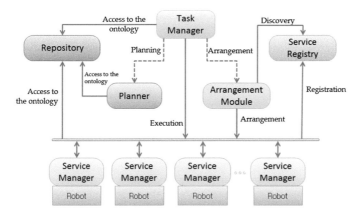

Fig. 12 The Autero architecture

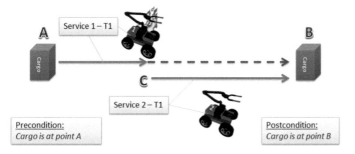

Fig. 13 Service replacement—Service 2 completes the task after the failure of the service 1

(for details, see [4]). Services are provided by heterogeneous robots. Repository stores ontology and provides access to it by the other system components. It also has a graphical user interface (GUI) for developing the ontology, and its management.

The system has been tested in a universal simulated environment implemented in Unity 3D. Simulation allows analyzing implemented algorithms for different scenarios. The scenarios with unexpected situations during a task realization are discussed below.

To illustrate the mechanisms of handling service failures in the simulated environment, a simple task of cargo transportation from point A to point B is used as an example. The original plan consists of one service that can accomplish the task.

Figure 13 shows the first step of the failure handling algorithm, i.e. replacing the failed service by a service of the same type. In the arrangement phase service 1 (of the robot 1) was chosen. The robot transported object only to point C, then it reported the failure and sent information to Task Manager that the transported object is at point C. Task Manager selected a service of the same type (robot 2) that transported the object from point C to destination point B accomplishing the task (Fig. 14).

Fig. 14 An example
scenario carried out in the
simulation—Robot 2
completes the task after the
failure of the robot 1
(damaged wheel)

Other unexpected situations may also occur and cause failures, for example, a robot may face an unknown obstacle impossible to bypass for the robots of that type. To recover from this failure, a service of other type is needed. In this situation, a new plan should be generated. The Fig. 15 shows the described situation and a possible solution. To complete the task, two services (of type T2 and T3) were selected, which will replace service 1 in order to transport the object from point C to point B. In the arrangement phase services 2 and 3 were selected. Service of type T2 can overcome the obstacles impossible for service of type T1, but also has a limited range. Therefore, it transports cargo only to the point D. The final step is accomplished by using service of type T3, which transports cargo by air, that is, a quad-copter robot can be used. All services are performed within a transaction that contains a dynamic set of participants. The transaction does not require all participants to successfully complete their tasks. The failure of a single service does not require the termination of the transaction. Termination is only necessary when it is not possible to continue the task.

Compensation is performed after an abortion of a subtask execution or the occurrence of a failure that interrupts the execution phase. It restores the state of the environment before start of the execution. Since restoring that situation is sometimes impossible, the compensation may change the situation resulting from the failure to

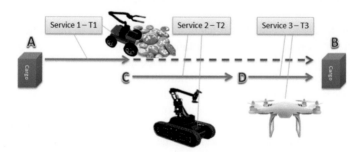

Fig. 15 Replacement of a failed service with a new plan

Fig. 16 Compensation procedure

a situation from which the task realization can be continued. Note that even for such simple tasks (that seem to be trivial) a universal failure recovery mechanism and corresponding compensations are not easy to design and implement.

The Fig. 16 shows a compensation procedure for the above task. Robot (transporting a cargo from A to B) received a message from TM ordering SM to compensate the performed actions. The compensation should be implemented as returning of the cargo back to the starting point A. It should be noted that in this case the internal service compensation procedure is used. This does not mean, however, that the same robot will be used to transport the cargo back. SM is responsible for controlling robots, and it may choose another robot to perform the compensation.

Task realization plan may contain predefined procedures for a failure handling and compensations. The Fig. 17 shows a simulated scenario of such a procedure. After a failure (spilled granules—C) of a granule transport service performed by a robot (A), TM invokes a proper failure handling procedure. In the above example the procedure contains a cleaning robot service (B) that collects the spilled granules.

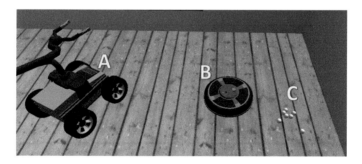

Fig. 17 Predefined failure handling procedure

8 Related Work

There are many protocols and mechanisms to handle transactions for loosely coupled distributed services. This section will discuss the most important of them and their limitations.

The ATOGS [15] (Adaptability in Transaction Oriented Grid Service) model enables transaction oriented services to execute in an adaptive way so as to handle failure problems (such as hardware faults, communication faults, software faults, and service expiry faults) using fault-tolerance techniques (i.e., check-pointing and replication). This model, however, is only suitable for atomic transactions.

Two-phase commit protocol is used for short-term transactions. It is not suitable for use with the long-term transaction because it blocks the resources until the completion of the execution of all operations in the transaction. It can be used only for arrangement phase to make it feasible to arrange the whole business process at once. However, the arrangement method described in this paper is usually sufficient, and the introduction of 2PC would benefit only in a few cases. If necessary, it is possible to expand the proposed protocol.

The TIP [14] protocol, like the 2PC protocol ensures atomicity condition for the transaction which entails limitations discussed above. Although the TIP protocol is designed to support business services, they are not represented as an electronic services. TIP transactions support only selected phases of a business service realization.

The main purpose of Tentative Hold Protocol [20] is non-blocking reservation of services (mainly their resources) prior to execution phase. This functionality can be easily implemented by Service Managers. After sending commitment SM may pre-book resources, but do not block them entirely. After receiving the commitment acceptance from a client, the resources would be blocked for the particular client, intended to be used in the execution phase. The use of THP gives no warranty of the service realization. The service provider can at any time withdraw the reservation.

WS-Coordination [7] together with the WS-Business Activity [9] protocol can be used to manage long-running tasks. The designed transaction protocol derives from these specifications. However, due to their limitations and different assumptions it has been significantly changed and the present state does not comply with those specifications. According to WS-Coordination specification, services must register to participate in the task. For this reason, they must have information about the task in which they participate.

Specifications included in the WS-CAF [6] are very similar to those in the WS-TX, as both are based on OTS [17] extensions. For this reason, they have the limitations discussed above. WS-CAF specifications are also less popular than competitive solutions.

BTP [18] protocol can be compared to the arrangement phase. The essence of this solution is getting commitment to perform a particular task. Supported transactions can be long-running, but this only applies to the duration of reservation or duration of a contract. It is not possible to control an execution phase, which is a significant limitation, because the execution phase usually lasts longer and generates the actual costs. The BTP protocol enables the operation of compensation, but this only applies to the reservation.

Agent Based Transaction (ABT) [10] allows replacement of individual participants in the transaction for their functional equivalents, which can increase chances of success and avoid costs of failure. ABT also allows services to create a hierarchy of transactions. ABT has been developed for traditional web services, and uses a simple 2PC protocol which is not suitable for business services performed in the real world.

9 Conclusions

A business transaction protocol and recovery mechanisms were proposed for tasks accomplishment carried out in the real world. They use an adequate representation of services and reflect real phases of a business service realization. The proposed protocol has a richer set of states (compared to WS-TX), so that it also supports the arrangement phase and allows for better management of a service execution. The recovery mechanisms of transactions adapt the task realization process to an actual state of the environment. This is especially important in a dynamic and open environments where services are provided by independent service providers, and it is not possible to guarantee their proper execution.

Early versions of the protocol and mechanisms have been implemented in the SOA-enT [3] system developed under the IT-SOA project. The current version has been significantly changed and improved, but still can be used in that system, yielding better task execution management.

The prototype system Autero verified that the proposed mechanisms of transactions are useful in the systems of heterogeneous robots in a simulated environment, and may improve their reliability. Tests performed in a real environment are always limited by the devices (robots) and their limited range and capabilities. From the point of view of the proposed information technology (the protocols) the fact that the environment is simulated is irrelevant.

Acknowledgments Autero has been developed within the project "RobREx—Autonomy for rescue and exploration robots", grant NRDC no PBS1/A3/8/2012. Marcin Stępniak was partially supported by the Foundation for Polish Science under International PhD Projects in Intelligent Computing.

References

1. Ambroszkiewicz, S.: Entish-a simple language for web service description and composition. Internet Technologies, Applications and Societal Impact, pp. 289–306. Springer, New York (2002)
2. Ambroszkiewicz, S., Bartyna, W., Faderewski, M., Terlikowski, G.: Multirobot system architecture: environment representation and protocols. Bull. Pol. Acad. Sci.: Tech. Sci. **58**(1), 3–13 (2010)
3. Ambroszkiewicz, S., Ambroszkiewicz, A., Bartyna, W., Baranski, M., Faderewski, M., Kulma, P., Mikulowski, D., Pilski, M., Ryzko, A., Stępniak, M., Terlikowski, G., Vojteshenko, I.: A platform for development of electronic markets of sophisticated business services. Advanced SOA Tools and Applications, pp. 73–124. Springer, New York (2012)
4. Ambroszkiewicz, S., Bartyna, W., Skarzynski, K., Szymczakowski, M., Stępniak, M.: Architecture of an autonomous robot at the it level. J. Autom. Mob. Robot. Intell. Syst. **9**(1), 34–40 (2015)
5. Bloor, R., Baroudi, C., Kaufman, M., et al.: Service Oriented Architecture for Dummies. Wiley, Hoboken (2007)
6. Bunting, D., Chapman, M., Hurley, O., Little, M., Mischkinsky, J., Newcomer, E., Webber, J., Swenson, K.: Web services composite application framework (WS-CAF) ver1. 0. Arjuna Technologies Limited, Fujitsu Software, IONA Technologies PLC, Oracle Corp and Sun Microsystems, Technical report (2003)
7. Cabrera, F., Copeland, G., Freund, T., Klein, J., Langworthy, D., Orchard, D., Shewchuk, J., Storey, T.: Web services coordination (WS-coordination). Joint specification by BEA, IBM, and Microsoft (2002)
8. Ding, Z., Liu, J., Sun, Y., Jiang, C., Zhou, M.: A transaction and QoS-aware service selection approach based on genetic algorithm. IEEE Trans. Syst. Man Cybern. **45**(7), 1035–1046 (2015)
9. Freund, T., Little, M.: Web services business activity (WS-businessactivity) version 1.1 (2007)
10. Jin, T., Goschnick, S.: Utilizing web services in an agent based transaction model. Extending Web Services Technologies, pp. 273–291. Springer, New York (2004)
11. Kaye, D.: Loosely Coupled: The Missing Pieces of Web Services. RDS Strategies LLC, Marin County (2003)
12. Krafzig, D., Banke, K., Slama, D.: Enterprise SOA: Service-Oriented Architecture Best Practices. Prentice Hall Professional, Indianapolis (2005)
13. Kuropka, D., Tröger, P., Staab, S., Weske, M.: Semantic Service Provisioning. Springer Science & Business Media, Berlin (2008)
14. Lyon, J., Evans, K., Klein, J.: Transaction internet protocol version 3.0. Technical report (1998)
15. Mahato, D., Umrao, L., Singh, R.: Adaptability in transaction oriented grid service. In: International Conference on Parallel, Distributed and Grid Computing (PDGC), pp. 239–244 (2014)
16. Martin, D., Paolucci, M., McIlraith, S., Burstein, M., McDermott, D., McGuinness, D., Parsia, B., Payne, T., Sabou, M., Solanki, M., Srinivasan, N., Sycara, K.: Bringing semantics to web services: the owl-s approach. In: Cardoso, J., Sheth, A. (eds.) Semantic Web Services and Web Process Composition. Lecture Notes in Computer Science, vol. 3387, pp. 26–42. Springer, Berlin (2005)
17. OMG: CORBA services: common object services specification, chapter. Object Transaction Service (1997)
18. Potts, M., Cox, B., Pope, B.: Business transaction protocol primer. OASIS Committee Supporting Document (2002)
19. Randell, B.: Fault tolerance in decentralized systems. IEICE Trans. Commun. **83**(5), 903–907 (2000)
20. Roberts, J., Srinivasan, K.: Tentative hold protocol-part 1. White Paper, sl: W3C (2001)
21. Roman, D., Keller, U., Lausen, H., de Bruijn, J., Lara, R., Stollberg, M., Polleres, A., Feier, C., Bussler, C., Fensel, D., et al.: Web service modeling ontology. Appl. Ontol. **1**(1), 77–106 (2005)

22. Wang, T., Vonk, J., Kratz, B., Grefen, P.: A survey on the history of transaction management: from flat to grid transactions. Distrib. Parallel Databases **23**(3), 235–270 (2008)
23. WebFinance: what is a service? definition and meaning (2014). http://www.businessdictionary. com/definition/service.html (Online; access: 12–10-2014)

A Hybrid Approach to Parallelization of Monte Carlo Tree Search in General Game Playing

Maciej Świechowski and Jacek Mańdziuk

Abstract In this paper, we investigate the concept of a parallelization of Monte Carlo Tree Search applied to games. Specifically, we consider General Game Playing framework, which has originated at Stanford University in 2005 and has become one of the most important realizations of the multi-game playing idea. We introduce a novel parallelization method, called Limited Hybrid Root-Tree Parallelization, based on a combination of two existing ones (Root and Tree Parallelization) additionally equipped with a mechanism of limiting actions available during the search process. The proposed approach is evaluated and compared to the non-limited hybrid version counterpart and to the Tree Parallelization method. The advantages over Root Parallelization are derived on a theoretical basis. In the experiments, the proposed method is more effective than Tree Parallelization and also than non-limited hybrid version in certain games.

Keywords Monte Carlo Tree Search · Upper Confidence Bounds Applied for Trees · General Game Playing · Parallelization · Parallel Computing

M. Świechowski (✉)
Ph.D. Studies at Systems Research Institute,
Polish Academy of Sciences, Warsaw, Poland
e-mail: m.swiechowski@mini.pw.edu.pl

J. Mańdziuk
Faculty of Mathematics and Information Science,
Warsaw University of Technology, Warsaw, Poland
e-mail: j.mandziuk@mini.pw.edu.pl

J. Mańdziuk
School of Computer Engineering, Nanyang Technological University,
Singapore, Singapore

© Springer International Publishing Switzerland 2016
G. De Tré et al. (eds.), *Challenging Problems and Solutions in Intelligent Systems*, Studics in Computational Intelligence 634,
DOI 10.1007/978-3-319-30165-5_10

1 Introduction

Monte Carlo Tree Search (MCTS) [5] is renowned for being the state-of-the-art algorithm of searching a game tree in a variety of domains. It is particularly useful in complex games with high branching factors such as Go [12], Hex [2], Havannah [27] or Arimaa [26] where no good evaluation function exists. Since the introduction of MCTS to a domain of General Game Playing (GGP) [13] in 2007, it has also become a backbone of almost all the strongest players [25]. GGP deals with creating autonomous agents capable of playing many games with a high level of competence. The term was proposed by Stanford Logic Group in 2005, together with the introduction of the General Game Playing Competition as the official world championships. Our player, called MINI-Player [22, 24], has been our annual entry to the competition since 2012.

Parallelization is understood as making the program run simultaneously on many computers and/or processing cores. In this paper, we are concerned with the parallelization of the MCTS algorithm in the GGP framework, specifically. The MCTS is relatively easy to parallelize compared to any classic Depth-First Search algorithm [19] (e.g., alpha-beta, min–max, MTD-f) because it contains fewer synchronization points. The synchronization can be less frequent because there is a simulation phase in MCTS, which is usually very time-consuming and isolated, i.e., the game tree does not need to be accessed in this phase. However, running multiple simulations in parallel steps away from the original idea of MCTS where for each iteration of the algorithm there is one simulation. Therefore, parallelization is not only an implementation issue but also a design choice.

The main motivation behind parallelization is to increase implementation efficiency of the Monte Carlo Tree Search algorithm. Naturally, the more simulations are performed the more accurate statistics of actions are collected. In GGP, each game is written in a universal logic language which is very slow to interpret compared to any game-dedicated representation. Utilizing many CPU cores is especially important in the GGP Competition scenario, where every participant runs the program on their own computational facilities. Parallelization in GGP area brings an additional difficulty stemming from the fact that the approach needs to be universally good, i.e. suitable for a variety of games.

This paper is organized as follows. In the next section, we briefly introduce the GGP competition setting and describe our MINI-Player and the MCTS algorithm in more details. In Sect. 3, related parallelization methods are presented which are the entry points to our method. The following section contains description of the proposed novel method applied in MINI-Player. The last two sections are devoted to results and conclusions, respectively.

2 Background

2.1 General Game Playing

As already mentioned in the introduction, General Game Playing refers to the area of creating autonomous multi-game playing agents. Each game in the GGP framework is defined in the Game Description Language (GDL) [15], which allows for describing any finite, synchronous, deterministic and perfect-information game. Apart from these constraints, games have no limitation, e.g., they can feature any number of players (including only one), can be of various genres (not only board games) can be cooperative, competitive or partially both.

GDL is a first-order logic language based on Datalog [1] and Prolog [4]. It is relatively easy to convert a GDL description to a Prolog program. A complete game state of any game is defined by a set of facts (predicates) which hold true. When designing methods of parallelization, one must keep in mind that the GDL description must be sent to all remote units in the system. Moreover, many methods require game states to be sent frequently. Although certain compression of the state is possible (e.g., conversion of strings to numbers), the representation is still less compact, in a general case, than a game-specific one [23]. Therefore, not only simulations are slower but also the communication overhead is higher in GGP.

In GGP, the communication takes place through messages sent via HTTP protocol. Players do not communicate with each other directly, but with a special component called the **Game Manager** (GM). The GM can be started locally or provided by the organizers of the GGP Competition. The defined messages are **START**, **PLAY**, **STOP**, **ABORT**, **PING** and **INFO**.

The **START** message is sent by the GM to each player when a game starts. It may look as follows: (*START match1 white* ((*role white*) (*role black*)....) *40 10*). The message contains the keyword "START", an identifier of the played match (to differentiate between matches), a role for the player (which must exactly match one of the roles defined in a game), the complete GDL description of a game and a pair of clocks (in seconds). The *START-clock* defines the initial preparation measured from sending the rules until the game starts. Players are expected to respond with "(ready)" before the *START-clock* expires. The *PLAY-clock* defines time available for players to make a move. After the successive *PLAY-clock*, PLAY messages are sent and the game state is updated by the GM. In the GGP competition, the *START-clock* is usually set to somewhere between 20 and 120 s whereas the *PLAY-clock* to 10–30 s. These relatively low settings significantly limit the possible approaches to creating and parallelizing a GGP agent.

The START message, which is sent only once, is followed by a number of **PLAY** messages at even intervals equal to the *PLAY-clock*. Before each *PLAY-clock* expires, each player has to send their chosen move. When it expires, the GM sends the PLAY message that contains moves chosen by all players (called a joint move) to all the players so they can update their game state accordingly. The GM is also responsible for checking whether the submitted moves are legal in the game. If not, it chooses random ones instead and/or disqualifies the illegally playing players. A play message may look as follows: (*PLAY match1* (*move a 1*) *noop*).

Technically, the **STOP** message is the last PLAY message with the additional meaning that the game has reached the conclusion. Players should update their state for the last time and check the outcome of the game. The **ABORT** message means that game has been terminated prematurely—either by manual intervention or a failure at the GM side.

The **PING** and **INFO** messages are used to check whether players are online. They should respond with "(available)" or "(busy)".

2.2 MINI-Player

MINI-Player [22, 24] has been designed and implemented as part of the Ph.D. thesis [20] and with the aim of taking part in the official GGP championships. All of parallelization methods investigated in this paper were implemented in MINI-Player. The key features of the program are:

1. **Monte Carlo Tree Search**. MCTS, which is the main routine of the player, is described in Sect. 2.3.
2. **Simulation strategies**. MINI-Player uses seven policies to bias the search in the simulation phase of the MCTS algorithm. The policies are Random, Approximate Goal Evaluation (greedily trying to fulfill a goal condition), History Heuristic (exploiting past good actions), Mobility (maximizing the relative number of available actions), Exploration (prioritizing novel game states during search), Statistical Symbols Counting [16] (dynamically constructed material-inspired evaluation function) and Score (detection of explicitly defined scoring condition in the GDL rules).
3. **Adaptive mechanism of selection of strategies**. The strategies are evaluated dynamically based on their empirical performance. A strategy is assigned to guide a simulation based on its evaluation and a confidence of the evaluation. The better or the less simulated the strategy the higher the probability of choosing it.
4. **Fast GDL interpreter**. We developed a dedicated GDL interpreter [21] which is faster than known Prolog distributions.
5. **Three-time (2012–2014) participation in GGP competition**. In 2012 and 2014, MINI-Player achieved the 7–8th place.

2.3 Monte Carlo Tree Search

The MCTS algorithm is an iterative simulation-based approach to searching the game tree. Each iteration consists of four phases depicted in Fig. 1.

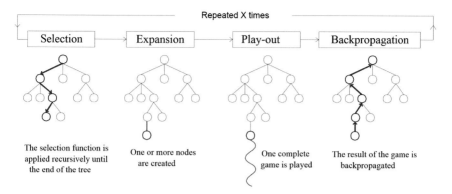

Fig. 1 Four phases of the MCTS algorithm. The figure is reproduced from [8]

Selection: start the search from the root node. Traverse the tree down until reaching a leaf node. In each node, choose the child node with the highest average score. More sophisticated approaches replace the average score with a specialized formula for determining the node to choose.

Expansion: if a state in the leaf node is not terminal choose a continuation which falls out of the tree and allocate a new child node. Typically, in the basic version of the method, just one new node is added per each expansion step.

Simulation: starting from a state associated to the newly expanded node, perform a full game simulation (i.e. to the terminal state).

Backpropagation: fetch the result of the simulated game. Update statistics (scores, visits) of all nodes on the path of simulation starting from the newly expanded node up to the root.

When the time budget is up, an action leading to the highest average score is chosen.

In the selection phase, MINI-Player uses the Upper Confidence Bounds Applied for Trees (UCT) [14] method. The purpose of the algorithm is to maintain a balance between exploration and exploitation in the selection step. Instead of sampling each action uniformly or choosing always the best action so far, the selection of the best action is made as follows:

$$a^* = arg \max_{a \in A(s)} \left\{ Q(s, a) + C \sqrt{\frac{ln\,[N(s)]}{N(s, a)}} \right\} \qquad (1)$$

where a—is an action; s—is the current state; $A(s)$—is a set of actions available in state s; $Q(s, a)$—is an assessment of performing action a in state s; $N(s)$—is a number of previous visits to state s; $N(s, a)$—is a number of times an action a has been sampled in state s; C—is a coefficient defining a degree to which the second component (exploration) is considered.

The first remarkably successful application of MCTS in games referred to Go. A majority of the strongest programs, e.g., MoGo [11] or CrazyStone [9] use variants of MCTS. In contrast to all the variations of min–max alpha-beta search, the MCTS is an aheuristic method. The aheuristic property means that there is no game-specific knowledge (heuristics) required so the method can be applied to a wide selection of problems.

3 Related Methods of Parallelization in GGP

There have been three major methods proposed for the task of parallelizing an MCTS-based player, i.e. Leaf Parallelization (LP) [6], Root Parallelization (RP) [6, 17] and Tree Parallelization (TP) [6, 7, 18]. These methods can also be called At-the-leaves Parallelization, Single-Run Parallelization and Multiple-Runs Parallelization, respectively. Our method is based on two of them, so we decided to dedicate a section for a short review of all the methods.

3.1 Leaf Parallelization

In Leaf Parallelization (LP), there is a single game tree with exclusive access from the master process (a distinguished thread on one of the machines). The MCTS/UCT algorithm is performed by the master process. When it reaches a leaf node, the state associated to that node is sent to all remote processes. Then, a simulation starting from that state is executed in parallel in each of these processes. The master process waits for the simulations to finish and collects the results. In Leaf Parallelization, only simulation and back-propagation phases differ in comparison to the original (sequential) MCTS formulation. The simulation phase consists of multiple independent simulation starting from the same state, whereas the back-propagation phases updates statistics aggregated from these multiple runs instead of one.

A simple idea and easy implementation are the biggest advantages of the Leaf Parallelization method. However, although many simulations performed from the same node increase the confidence of its statistics, it is not the most effective approach. The original MCTS/UCT algorithm chooses the node to simulate after each simulation, because every time the statistics are updated, a new node might be more suitable for the next simulation. In the LP approach, even the meaningless state has to be simulation at least K times, where K is the chosen number of remote processes. Moreover, the tree does not grow quicker than in the single-threaded MCTS. Figure 2 shows how Leaf Parallelization works.

Fig. 2 An illustration of the Leaf Parallelization method with 3 parallel processes. The parallel simulations start from the marked node and the MCTS algorithm waits until all of them are completed

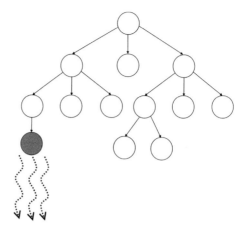

3.2   Root Parallelization

The main idea of Root Parallelization (RP) is to expand the game trees on each remote process individually and synchronize them occasionally. In the most typical approach, the master process passes raw messages (*START, PLAY, STOP, ABORT*) received from the GM to all connected machines. Each machine acts as a regular GGP player with the following two exceptions: (1)—they respond to the master node instead of to the GM; (2)—instead of the chosen action, they send MCTS statistics (total score, visits) stored in the top-most level of the game. The top-most level of the game tree contains the nodes related to all possible actions in the current state. The master process performs synchronization by aggregating the gathered statistics to its own game tree to compute the global average scores for actions. The most frequently chosen method of aggregation is a sum but certain other are possible, e.g., a sum of the best K or majority voting.

The biggest advantage of the method is a minimal communication overhead. The statistics gathered from the trees expanded by independent UCT algorithms are more confident but they may also be too overlapping (similar). In overall, the method scales relatively well and was successfully used by a GGP player called Ary [17].

Three variants of aggregating the results were tested: *Best* (select the best evaluated move from a distributed node), *Sum* (sum total scores and total visits and compute a global average scores), *Sum10* (*Sum* performed only for the top ten best evaluated moves) and *Raw* (send only average scores of moves from nodes without weighting by the number of total visits). The best results were reported for *Sum* and *Sum10* with no significant difference between them. Figure 3 shows how Root Parallelization works.

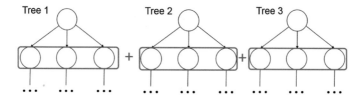

Fig. 3 An illustration of the Root Parallelization method with 3 parallel processes. Each process maintains an independent tree constructed in a sequential fashion. Statistics of the root's children are aggregated during a synchronization point

3.3   Tree Parallelization

Tree Parallelism (TP) resembles the idea of the sequential MCTS/UCT as closely as possible. There is a single game tree expanded by the master process. Whenever the MCTS/UCT algorithm schedules a new simulation, the master process checks for the first available process which is either a local thread (the preferred case) or a remote process. Once the master process sends the simulation request it proceeds to the next iteration of the MCTS/UCT. The simulation count (visits) of the previously selected node is incremented immediately but the scores are updated when the simulation is actually completed (therefore, until it happens, it is a virtual loss for all players). If no parallel unit is available, the system will wait or place the simulation request into a waiting queue, depending on the implementation. The advantage of Tree Parallelization over both Root and Leaf Parallelization is that more unique states are visited. The MCTS algorithm converges faster in the Leaf Parallelization approach. However, the communication overhead is significantly higher than in Root Parallelization. Figure 4 shows how Tree Parallelization works.

Fig. 4 An illustration of the Tree Parallelization method with 3 parallel processes. When a simulation is finished, statistics are updated in the tree and a new simulation is scheduled based on the currently observed statistics

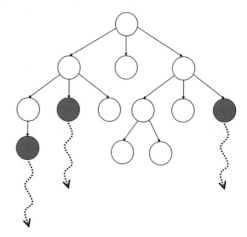

4 Our Hybrid Approach: Limited Root-Tree Parallelization

The idea of our method is to combine the best of both words. Within one machine we use Tree Parallelization which is then mixed by Root Parallelization between the machines. TP can be relatively efficiently performed on a single computer, because of the benefits gained from a shared memory. TP on remote machines, however, not only requires network communication but also serialization and deserialization of states which simulations start from. With the shared memory available in one program's instance, all the necessary synchronization can be put into a critical section or even a lock-free fashion can be pursued. The lock-free version [10] assumes that some synchronization errors (often called faulty updates) are possible. The idea is to detect such cases and ignore result from the simulation if a fault was encountered. In our implementation, the safer approach with the lock is used.

Although the shared game tree has to be accessed twice in one iteration of the MCTS algorithm (first during Selection/Expansion and next during Back-Propagation), both phases can reside within the same lock when a smart reordering of the MCTS phases is applied. Figure 5 shows how Hybrid Parallelization works.

A unique feature of the MINI-Player's parallelization approach is that each remote node operates on a subset of the tree to increase the variety of search. Each remote node (a computer) is assigned a unique identifier $ID \in [1, N]$. The maximum number (N) is globally available to all nodes. On the first possible level in the tree, where MINI-Player has more than one available action we narrow the search of each remote machine to the portion of the total moves defined by:

$$\left[\frac{ID - 1}{N}, \frac{ID + 1}{N} \right] \tag{2}$$

These intervals overlap to keep some redundancy in case a machine disconnects. The last interval is wrapped back to the beginning. For instance, if $N = 5$, the intervals are: $[0, \frac{2}{5}][\frac{1}{5}, \frac{3}{5}][\frac{2}{5}, \frac{4}{5}][\frac{3}{5}, 1]$ and "$[\frac{4}{5}, \frac{6}{5}]$" $= [\frac{4}{5}, 1] \cup [0, \frac{1}{5}]$.

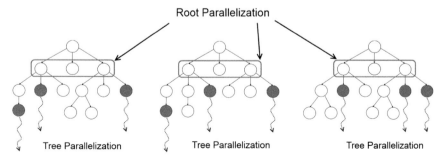

Fig. 5 An illustration of the Hybrid Parallelization method with 3 parallel nodes and 3 parallel processes in each node

These intervals are used on remote nodes to constrain the available actions to the UCT algorithm in the selection phase. If M is the total number of moves, then only actions with indices from the range as computed by Formula 3 are available to choose.

$$\left[\left\lfloor M * \frac{ID - 1}{N}\right\rfloor, \left\lfloor M * \frac{ID + 1}{N}\right\rfloor\right] \tag{3}$$

The redundancy caused by overlapping intervals can be further strengthened in two ways:

1. By increasing the overlapping margin i.e. $\left[\frac{ID-k}{N}, \frac{ID+k}{N}\right]$, where $k > 1$.
2. By having multiple machines assigned the same ID. For instance, with 8 machines, N can be set to 4 and each value of ID can be assigned to a pair of machines. In such a case, the same subset of actions will be checked twice due to redundancy. Since the MCTS algorithm is non-deterministic, repeated simulations may increase confidence of the statistics.

Actions of MINI-Player are limited only **in the first node** on each path coming from the **root** in which our player has more than 1 action available. The root is included, so if MINI-Player has more than 1 action in the root then it is the only node with limited actions. There are no limitations of actions chosen for the simulated opponents. There are two sample trees shown in Fig. 6 where the mechanism is illustrated. The set of allowed moves is limited only for MINI-Player because in this approach the assessment of actions in distributed programs is fair by means of being evaluated against all possible opponents' actions. If the search was limited to certain subsets of moves for all players, then some of them could have too promising assessment if they happened to be simulated together with only weak actions of the opponents. To solve this issue, each combination of players' actions would have to be simulated by some remote node what vastly increases complexity of the problem.

In the proposed parallelization approach, three types of nodes are used. The setup is presented in Fig. 7. Each physical computer is running one node at a time which can be the Master, a Hub or a Worker Node.

Fig. 6 An illustration of how MINI-Player's actions are limited in the UCT algorithm. The nodes are labeled by the number of legal moves for MINI-Player. *Circles* marked in *green* are the first nodes on each path from the root with more than 1 action available, so the limiting takes place there. Branches ending with *three dots* are not relevant to the algorithm's presentation

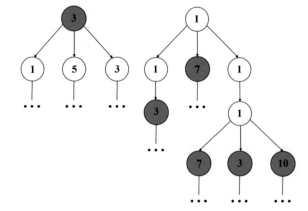

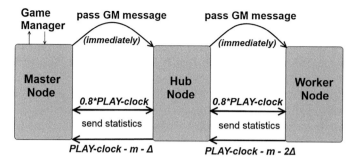

Δ - time margin for message gathering
m - time margin for the response to Game Manager

Fig. 7 A data flow in the proposed Hybrid Parallelization algorithm

1. **Master Node**—the main program, which communicates with the *Game Manager*. It passes *START, PLAY, STOP* and *ABORT* messages without any processing further to other types of nodes. The MCTS/UCT algorithm is used here to construct the game tree. Within a scope of the machine, the algorithm uses Tree Parallelization. Before a *PLAY-clock* clapses, the statistics are gathered and aggregated as in Root Parallelization method.

2. **Hub Node**—an intermediate layer passing messages back and forth between the Master Node and Worker Nodes. Hub Nodes are used only in situations, where there are so many computers available that maintaining connections between the Master Node and each of them would be infeasible. Each Hub Node gathers statistics from a certain group of Worker Nodes and sends them, already aggregated, to the Master Node. A high number of connections to the main machine active at same time (just before a move is made) renders it vulnerable to time-outs, which in turn can lead to game losses. This is important in the GGP Competition scenario. However, with the intermediate layer the responsibility of communicating with Worker Nodes is shared among Hub Nodes and the main machine is not overloaded.

3. **Worker Node**—a procedure of work is similar to the Master Node's one. The difference is only that Workers do not pass messages any further and respond with the statistics of level one nodes (just below the root) instead of the action to play.

4.1 Aggregation of Statistics

Worker Nodes send statistics computed for all currently available actions of all players by local MCTS/UCT algorithms. Because the current state in the game is stored in a tree root, the nodes which contain these statistics are the root's children. The total scores computed for all players and the number of visits are sent twice per each *PLAY-clock* interval. The first synchronization is when 80 % of the *PLAY-clock* elapses whereas the second one occurs right before it elapses. The Master Node sums the obtained data for every node which is present in its own game tree. This operation affects the average score which is now updated. Since the nodes are close to the root, the only situation in which they are not present is when a game has an enormously huge branching factor or the *PLAY-clock* is extremely small. Please recall that the UCT algorithm will prioritize actions which were never visited before, so K children of the root are sure to be visited in the first K simulations.

After the first aggregation, the Master Node propagates the aggregated statistics back to Hub Nodes which are further passed to Worker Nodes. Statistics on each machine become synchronized in the first levels of the game trees. From now on, until the *PLAY-clock* expires, the remote nodes are allowed to sample all actions, i.e., the search narrowing algorithm is switched off. In the remaining portion of the *PLAY-clock*, all game trees have the chance to expand in direction of the highest evaluated actions. If the narrowing mechanism was not switched off at any point, the machines which did not have access to the eventually chosen move, would have to build the game tree starting from the root only after every PLAY message. If only there are no more actions in a current state than available machines, Formula 3 guarantees that each action (the played action, in particular) is available to exactly two machines, therefore, all but two would have a degenerated game tree at each step.

4.2 Disadvantages of Using Intermediate Layers

A huge advantage of the Hybrid Parallelization method is that it can utilize a lot of threads and computers relatively easily. As depicted in Fig. 5, it has a hierarchical structure, where the Root component is higher in the hierarchy. In comparison to Tree Parallelization method, where a communication cost quickly becomes an issue, it can harvest significantly more power due to less frequent need of sending data. In TP, each request for a simulation generates a message, which can count to thousands per step between a single pair of machines. In HP only two messages are exchanged per step between a pair of machines. The utilization of computers can be even increased by adding more intermediate, i.e., Hub Node, layers. There is no difference whether a Hub Node passes messages to a Worker Node or another Hub Node. A downside of adding intermediate layers is that each such a layer has to include a small time margin—Δ—for communication. To avoid time-outs, the Master Nodes responds with the chosen action with a time margin—m. An ith layer Hub Node sends the gathered statistics to the Master Node with a margin equal to $m + i * \Delta$. A Worker Node sends the final statistics with a margin of $m + (K + 1) * \Delta$, where K denotes the number of intermediate layers.

4.3 Advantages over Root Parallelization

In comparison with Root Parallelization, the hybrid method scales better because the aggregation mechanism treats one computer, instead of one thread as in the original RP, as one parallel unit. For instance, if quad-core computers are considered, HP can employ approximately four times more machines with the same communication overhead. Moreover, game trees constructed by the Hybrid Parallel algorithm are of higher quality than in the Root Parallel one.

A predominance of the hybrid method over RP can be derived theoretically. For a given pair (game, parallelization method) there always exists a saturation point beyond which adding more threads does not increase the performance. Eventually, the synchronization and/or communication cost will be even detrimental to the overall player's strength. Let us denote this boundary number of parallel instances for (game1, RP) by B. For a given game, the Root component of Hybrid Parallelization will stop scaling around approximately the same number of units. However, the communication costs imposed by this component are completely unaffected by the synchronization costs from the Tree components running on each remote unit. In the case of a classic RP, the remote units are simply threads. In the case of HP, the remote units are k-core machines parallelized via TP. This concludes, that Hybrid Parallelization scales asymptotically better up to k times.

Root Parallelization has a chance to be more effective only with a small numbers of computers, when it is not yet saturated. In general, it is more valuable to compare HP and TP methods than HP and RP methods. The former are simply more different whereas the latter use the same parallelization idea when analyzed top-down (Figs. 3 and 5).

5 Results

We performed two experiments in the following hardware setup: 16 identical machines equipped with *Intel(R) Core i7-2600* processors (4 physical, 8 logical cores), 8GB of RAM and *Windows 7* 64-bit operating system. In both experiments, two players parallelized on 32 threads (8 machines × 4 cores), play a series of 100 matches against each other. The number of matches is limited to 100 due to significant amount of time required to complete the experiment.

In GDL, the obtained scores for players are defined by integer numbers from the [0, 100] interval. These scores, called goals, are valid only in terminal states. In order to make comparison of results from various games easier, we define the concepts of a win, draw and loss. One player wins the match if it achieves a higher score than the opponent it is tested against. In such a case, the latter loses the game. A draw occurs if both tested agents achieve equal scores. In both experiments, the total score for each player is calculated as shown in Eq. 4. The number of their wins is summed with a half of the number of draws. This method of calculating scores has been widely used in GGP, e.g., in [3], where multiple games are involved in an experiment. After 50 repetitions, the roles in the game are swapped to avoid any bias related to a starting position.

$$Score = (|WINS| + |DRAWS| * 0.5) \tag{4}$$

The first experiment was aimed at testing Limited Root-Tree Parallelization (**Limited-RT**) against the regular Root-Tree Parallelization (**RT**). In such a case, we can measure the impact of the proposed actions limiting mechanism and verify its usefulness.

Nine significantly different games, including chess-like games, connection games, market-inspired games, of various complexity (in terms of the numbers of possible unique states and actions) were chosen. For these selected games, Table 1 presents two game properties which are very important from the parallelization point of view: the average time required to perform one Monte Carlo simulation and the average number of possible moves (branching factor).

Table 2 presents the experimental results. Using 75 % confidence intervals, the total result (measured in wins-draws-losses) is 4-1-4 whereas within 95 % confidence intervals, the outcome is 3-3-3. The average score is slightly higher for the Limited Root-Tree variant, but it is not significant by margins with at least 75 % confidence. We can conclude that both methods are similarly strong but suitable for different games. Limiting actions on a machine seems to work well for Farming Quandries, because in this game usually the last actions computed by the GDL interpreter are the best, so the last two machines can take full advantage of that. Good performance of this method in Farmers and Pentago is probably caused by high average branching factors in that games, around 1000 and 100 respectively. Both games feature some universally good moves such as the buy/sell actions near the start/end, respectively, in Farmers or placing stones in the middle of four sub-boards in Pentago. Hex, having a relatively high branching factor too, features no universally good moves in the

Table 1 The average time required to complete one game simulation in our framework (on a single thread) and the approximate average branching factor of each game

Game	1-sim time (ms)	Branching factor
Breakthrough	4.6	20*
Checkers	20.7	2.8
Connect4	0.9	4
Farmers	1.7	1000
Farming quandries	1.5	8*
Hex 9×9	0.8	79*
Othello	298.5	10
Pentago	13.9	97.3
Pilgrimage	20.3	9.2*

The branching factors marked with a * were computed empirically based on massive simulations in large UCT trees

Table 2 Evaluation of Root-Tree Parallelization (**Limited-RT**) against the plain Root-Tree Parallelization (**RT**)

Game	Clock (s)		Limited-RT versus RT	75 % conf. interval	95 % conf. interval
Breakthrough	45	8	42.00	[4.16] ✓	[9.67]
Checkers	60	10	**51.00**	[3.99]	[9.30]
Connect4	40	5	37.50	[3.64] ✓	[8.47] ✓
Farmers	45	5	**57.00**	[3.86] ✓	[8.98]
Farming quandries	60	8	**77.00**	[3.49] ✓	[8.13] ✓
Hex 9×9	60	10	40.00	[4.12] ✓	[9.60] ✓
Othello	90	15	39.00	[4.06] ✓	[9.46] ✓
Pentago	45	8	**63.00**	[3.93] ✓	[9.15] ✓
Pilgrimage	90	10	**61.00**	[3.79] ✓	[8.83] ✓
Average			**51.94**		

The results above 50, which are shown in bold, are in favor of the first player appearing in the table, which is Limited-RT in this case. Values in column *Clock* represent the time alloted for *START-clock* (the left value) and *PLAY-clock* (the right value). There are both 75 and 95 % confidence intervals put in square brackets. Confident wins by either player are marked with ticks next to confidence intervals

above-mentioned sense. Connect Four has the second lowest branching factor, and there are no move patterns like in Farming Quandries, what is the reason for the plain Hybrid Method being stronger there. The same (low branching factor and no move ordering patterns), is true for Othello. In a game-dependent scenario, both parallelization methods should be initially verified (before choosing the right one), because neither is universally stronger.

In the second experiment (see Table 3), the proposed parallelization method (**Limited-RT**) was evaluated against a player using Tree Parallelization (**Tree**). The total result (wins-draws-losses) is 6-1-2 in favour of the former method when

Table 3 Evaluation of Root-Tree Parallelization (**Limited-RT**) against Tree Parallelization (**Tree**)

Game	Clock (s)		Limited-RT versus Tree	75 % conf. interval	95 % conf. interval
Breakthrough	45	8	**60.00**	[4.12] ✓	[9.60] ✓
Checkers	60	10	**62.50**	[3.88] ✓	[9.02] ✓
Connect4	40	5	**57.00**	[3.47] ✓	[8.08]
Farmers	45	5	**83.50**	[2.67] ✓	[6.21] ✓
Farming quandries	60	8	**86.00**	[2.80] ✓	[6.51] ✓
Hex 9×9	60	10	**64.00**	[4.04] ✓	[9.41] ✓
Othello	90	15	24.00	[3.60] ✓	[8.37] ✓
Pentago	45	8	**52.00**	[4.21]	[9.79]
Pilgrimage	90	10	38.00	[3.91] ✓	[9.10] ✓
Average			**58.60**		

See the description of Table 2 for the interpretation of results the 75 % confidence intervals are applied and 5-2-2 with the 95 % confidence. Moreover, the average score of **Limited-RT**—58.60—is significantly higher than that of **Tree**. The **Tree** approach appears to be generally better in slowly simulated games (with the exception of checkers). Please notice, that **RT** as well as **Limited-RT**, contain **Tree** as a subsystem limited to 4 CPU cores (in our case). The biggest gain of parallelization measured after adding an additional thread is noticed with the lowest number of threads. The more cores (threads), the worse scalability of **Tree**, so the hybrid methods are much more promising in maintaining scalability.

6 Conclusions

In this paper, a new method of parallelization of the MCTS algorithm was presented. The method was applied to a GGP program called MINI-Player. Its underpinning ideas are a combination of Root and Tree Parallelization and narrowing the search to only certain subsets of actions on remote units. We proposed a mechanism of data aggregation coming from the remote machines. The combination of the two mentioned parallelization methods appears to be more suitable for GGP scenario than Tree Parallelization alone (shown empirically) and Root Parallelization alone (discussed theoretically). The mechanism of narrowing actions is beneficial in certain class of games, especially with high branching factor and repetitive move patterns (e.g., when the last moves with respect to specific ordering are stronger in average).

A dynamic detection of the most suitable parallelization method based on game rules analysis in the *START-clock* time is one of the future goals. We also plan to perform further tests with a higher number of games using the proposed methods. In general, parallelization of the MCTS algorithm is a robust and promising way to increase its performance. The methods discussed in this paper in the scope of GGP should be easily adapted to any game or even beyond the game domain where the MCTS algorithm is used.

Acknowledgments M. Świechowski was supported by the Foundation for Polish Science under International Projects in Intelligent Computing (MPD) and The European Union within the Innovative Economy Operational Programme and European Regional Development Fund.

This research was financed by the National Science Centre in Poland, based on the decision DEC-2012/07/B/ST6/01527.

References

1. Abiteboul, S., Hull, R., Vianu, V.: Foundations of Databases. Addison-Wesley, Boston (1995)
2. Arneson, B., Hayward, R.B., Henderson, P.: Monte Carlo tree search in hex. IEEE Trans. Comput. Intell. AI Games **2**(4), 251–258 (2010)
3. Björnsson, Y., Finnsson, H.: CadiaPlayer: a simulation-based general game player. IEEE Trans. Comput. Intell. AI Games **1**(1), 4–15 (2009)

4. Bratko, I.: Prolog Programming for Artificial Intelligence. International Computer Science Series. Addison Wesley, Boston (2001)
5. Browne, C.B., Powley, E., Whitehouse, D., Lucas, S.M., Cowling, P.I., Rohlfshagen, P., Tavener, S., Perez, D., Samothrakis, S., Colton, S.: A survey of Monte Carlo tree search methods. IEEE Trans. Comput. Intell. AI Games **4**(1), 1–43 (2012)
6. Cazenave, T., Jouandeau, N.: On the parallelization of UCT. Proc. CGW07, 93–101 (2007)
7. Cazenave, T., Jouandeau, N.: A parallel Monte-Carlo tree search algorithm. In: Computers and Games, pp. 72–80. Springer, Berlin (2008)
8. Chaslot, G., Winands, M.H., Szita, I., van den Herik, H.J.: Cross-entropy for Monte-Carlo tree search. ICGA J. **31**(3), 145–156 (2008)
9. Coulom, R.: Efficient selectivity and backup operators in Monte-Carlo tree search. In: Computers and Games, pp. 72–83. Springer, Berlin (2007)
10. Enzenberger, M., Müller, M.: A lock-free multithreaded Monte-Carlo tree search algorithm. In: Advances in Computer Games, pp. 14–20. Springer, Berlin (2010)
11. Gelly, S., Silver, D.: Achieving Master Level Play in 9 x 9 Computer Go. In: AAAI, vol. 8, pp. 1537–1540 (2008)
12. Gelly, S., Kocsis, L., Schoenauer, M., Sebag, M., Silver, D., Szepesvári, C., Teytaud, O.: The grand challenge of computer Go: Monte Carlo tree search and extensions. Commun. ACM **55**(3), 106–113 (2012)
13. Genesereth, M.R., Love, N., Pell, B.: General game playing: overview of the AAAI competition. AI Mag. **26**(2), 62–72 (2005)
14. Kocsis, L., Szepesvári, C.: Bandit based Monte-Carlo planning. In: Proceedings of the 17th European Conference on Machine Learning. ECML'06, pp. 282–293. Springer, Berlin (2006)
15. Love, N., Hinrichs, T., Haley, D., Schkufza, E., Genesereth, M.: General Game Playing: Game Description Language specification. http://games.stanford.edu/readings/gdl_spec.pdf (2008)
16. Mańdziuk, J., Świechowski, M.: Generic heuristic approach to general game playing. In: Bieliková, M., Friedrich, G., Gottlob, G., Katzenbeisser, S., Turán, G. (eds.) SOFSEM. Lecture Notes in Computer Science, vol. 7147, pp. 649–660. Springer, Berlin (2012)
17. Méhat, J., Cazenave, T.: A parallel general game player. Künstliche Intell. **25**(1), 43–47 (2011)
18. Méhat, J., Cazenave, T.: Tree parallelization of ary on a cluster. In: Proceedings of the IJCAI-11 Workshop on General Game Playing (GIGA'11), pp. 39–43 (2011)
19. Plaat, A., Schaeffer, J., Pijls, W., de Bruin, A.: Best-first fixed-depth minimax algorithms. Artif. Intell. **87**(1–2), 255–293 (1996)
20. Świechowski, M.: Adaptive simulation-based meta-heuristic methods in synchronous multi-player games. Ph.D. thesis, Systems Research Institute, Polish Academy of Sciences, Warsaw, Poland (2015) in review
21. Świechowski, M., Mańdziuk, J.: Fast interpreter for logical reasoning in general game playing. J. Log. Comput. (2014). doi:10.1093/logcom/exu058
22. Świechowski, M., Mańdziuk, J.: Self-adaptation of playing strategies in general game playing. IEEE Trans. Comput. Intell. AI Games **6**(4), 367–381 (2014)
23. Świechowski, M., Mańdziuk, J.: Specialized vs. multi-game approaches to AI in games. In: Angelov, P., Atanassov, K., Doukovska, L., Hadjiski, M., Jotsov, V., Kacprzyk, J., Kasabov, N., Sotirov, S., Szmidt, E., Zadrożny, S. (eds.) Intelligent Systems'2014. Advances in Intelligent Systems and Computing, vol. 322, pp. 243–254. Springer International Publishing (2015)
24. Świechowski, M., Mańdziuk, J., Ong, Y.S.: Specialization of a UCT-based general game playing program to single-player games. IEEE Trans. Comput. Intell. AI Games (2015) (accepted for publication)
25. Świechowski, M., Park, H., Mańdziuk, J., Kim, K.: Recent advances in general game playing. Sci. World J. 2015. http://dx.doi.org/10.1155/2015/986262 (2015)
26. Syed, O., Syed, A.: Arimaa - a new game designed to be difficult for computers. ICGA **26**, 138–139 (2003)
27. Teytaud, F., Teytaud, O.: Creating an upper-confidence-tree program for Havannah. In: Advances in Computer Games, pp. 65–74. Springer, Berlin (2010)

Part IV
Intelligent Computing in
Decision Support Systems

A Consensus Reaching Support System for Multi-criteria Decision Making Problems

Dominika Gołuńska and Janusz Kacprzyk

Abstract We present an extension of a consensus reaching support system presented in our previous works to additionally accommodate a multi-criteria evaluation of options and importance weights of all criteria given by each agent (individual). The multi-criteria setting implies a need for some modification of concepts, tools and techniques proposed in our previous works with a single criterion. To improve the efficiency of the process we use some additional suggestions/hints provided for the moderator in the form of linguistic summaries, modified to the multi-criteria setting. We present an application for a real world problem which involves reaching of a sufficient agreement in the small group of human agents. The results obtained are intuitively appealing, and promising in terms of time and costs of opinion changes to reach a sufficient consensus.

Keywords Multi-criteria decision making · Consensus reaching process · Group decision support system · Linguistic data summaries · Degree of consensus · Linguistic preferences · Moderator · Measures of agreement

1 Introduction

Consensus can be meant in various ways—cf. [1, 12–14, 22, 29, 32–36], and the same concerns group decision making—cf., e.g., [2, 4, 14, 15, 22]. Basically, consensus—as meant in many constructive approaches, also here—aims at reaching an agreement of the participants of a decision making process by accommodating views and opinions of all agents (decision makers) involved to attain a decision, usually to choose an option or a course of action, that will yield what will be beneficial to the entire group. Therefore, consensus is based on a cooperation of the agents involved and involves

D. Gołuńska (✉) · J. Kacprzyk
Warszawa, Kraków, Poland
e-mail: golunska@agh.edu.pl

J. Kacprzyk
e-mail: Janusz.Kacprzyk@ibspan.waw.pl

© Springer International Publishing Switzerland 2016
G. De Tré et al. (eds.), *Challenging Problems and Solutions in Intelligent Systems*, Studies in Computational Intelligence 634,
DOI 10.1007/978-3-319-30165-5_11

changes of testimonies of the group that should result in reaching a consensus with respect to those testimonies. Virtually all consensus reaching processes proceed in a multistage setting, i.e., the agents change their preferences step by step until some consensus is reached [1, 3, 20, 21, 37]. In our works, we focus on the consensus reaching approaches in which there is a moderator, a "super-agent" who runs the process, as they are more promising in practice, i.e., a consensus reaching process is usually more effective and efficient in such a setting, and are the most often used in the literature [3, 13, 20, 24, 25, 31].

Basically, the comparison of different options or courses of action according to their utility or desirability in decision problems cannot be done by using a single criterion or by a single agent [1]. Thus, in this paper we assume a decision problem to be supported in the following setting. We assume that there is a set of options characterized by a set of attributes/criteria, and a group of agents (decision makers). The agents discuss the options under consideration and present their testimonies by providing values for the attributes/criteria of all options, as well as importance weights for the particular criteria [23]. We assume further that these values may initially belong to a set of linguistic terms in line with the linguistic direction presented in, e.g. [11, 13]. Then, based on such information the system automatically generates a preference structure that corresponds to, and results from the agents' testimonies provided, taking into account both multiple agents and multiple attributes/criteria.

In this paper, we concentrate on the derivation of a degree of (soft) consensus in the sense of [16–18] within the group of agents according to a generated preference structure. Furthermore, we consider the derivation of some suggestions/hints for a group discussion that could lead to a higher degree of consensus. Finally, we present an implementation of the system's session for a simple decision making problem with multiple criteria in which we first attain a consensus and then find a solution (a best accepted option).

2   Linguistic Representation of Data

We assume a multiple criteria decision making (MCDM) problem with the use of linguistic terms for the description of values of various entities. This should facilitate the expression of human judgments and assessments. The general multiple criteria decision problem as considered here consists of a group of agents (individuals, decision makers,...), denoted as $E = \{e_m\} = \{e_1, \ldots, e_M\}$ and a set of options which are potential solutions to the decision problem under consideration, denoted as $S = \{s_n\} = \{s_1, \ldots, s_N\}$. Moreover, each option is characterized by a set of evaluation criteria $C = \{c_k\} = \{c_1, \ldots, c_K\}$ and an importance weighting vector W^m for these criteria is provided by each agent e_m, $W^m = \{w_k^m\} = \{w_1^m, \ldots, w_K^m\}$, $w_k^m \in [0, 1]$, $m = 1, \ldots, M$, representing the importance of the criteria as in the opinion of this agent. For simplicity, we consider the same set of criteria for all decision makers and all options but the importance weights are assigned to the particular criteria independently by each agent. This should best serve the purpose of this paper which is meant to propose a new model.

Each agent assesses qualitatively how good a potential option (solution) $s_i (i = 1, \ldots, N)$ is with respect to each criterion $c_k (k = 1, \ldots, K)$. The criteria are classified as the subjective in the sense that they are evaluated by the humans and that is why these assessments may be naturally presented using linguistic values, i.e. terms in natural language which should yield a needed human consistency and ease of use. For instance, terms like Very good, Good, Bad, etc. [11, 13] may be employed. We assume that each agent is able to provide such an evaluation.

Let:

$$E_m : S \times C \to LL \tag{1}$$

be a mapping representing the evaluation of all options along all criteria as given by agent e_m, where $LL = \{$Excellent, Very good, Good, Fair, Bad, Very Bad$\}$. We assume that there is a linear order $\succ$ defined on the set LL of linguistic values (terms) such that $\forall l_1, l_2 \in LL$, $l_1 \succ l_2$ means that term l_1 denotes a better performance of an option than term l_2. In what follows, and in particular in the example given at the end of the paper, we assume the numerically expressed evaluations/assessments, i.e., $LL = [0, 1]$. However, the approach is in fact general and the linguistic ally expressed evaluations may be easily plugged-in.

A set of linguistic labels may be assumed also to express importance weights of the criteria. For example, the following set of linguistic terms may be used:

the most important $\succ$ very important $\succ$ important $\succ$ medium important $\succ$

less important $\succ$ weakly important

Formally these weights can be modelled as fuzzy sets I in the interval $[0, 1]$.

3 Derivation of a Corresponding Preference Structure

In the previous section we have distinguished such preference related data as weights of criteria and evaluations of each option with respect to all criteria. Notice that these data are directly provided by each agent. In this step, the system automatically generates for each agent a preference structure corresponding (equivalent) to the input data each decision maker has given. This will now be dealt with.

We adopt here the approach proposed in [38], with some simplifications. A fuzzy structure is a pair of fuzzy binary relations (P, I) each defined on the Cartesian product $S \times S$. The fuzzy binary relations in the pair correspond to the classic notions of the strong preference (P) and indifference (I). Usually, the properties required for these relations are the antisymmetry for P and the symmetry for I. The completeness of the fuzzy preference structure is stated as $\mu_P(., .) + \mu_{P^t}(., .) + \mu_I(., .) = 1$, where P^t is the transpose of P, i.e. $P^t(s_i, s_j) = P(s_j, s_i)$.

A fuzzy preference structure is automatically generated for an agent e_m by the following reasoning.

We define for each pair of options (s_i, s_j) four following indices (we denote here the order relation with ">" for simplicity and due to the further use of numerically expressed evaluations, as mentioned earlier, but in general this relation may refer to the order relation "≻" defined on LL being an ordered set, in particular an ordered set of linguistic terms such as the one mentioned at the end of the previous section.):

$$ALLC^m = \sum_{k \in 1..K} w_k^m \tag{2}$$

$$FFC^m : S \times S \to [1, K]; \quad FFC^m\left(s_i, s_j\right) = \sum_{k:\ E_m(s_i, c_k) > E_m(s_j, c_k)} w_k^m \tag{3}$$

$$FSC^m : S \times S \to [1, K]; \quad FSC^m\left(s_i, s_j\right) = \sum_{k:\ E_m(s_i, c_k) < E_m(s_j, c_k)} w_k^m \tag{4}$$

$$FBC^m : S \times S \to [1, K]; \quad FBC^m\left(s_i, s_j\right) = \sum_{k:\ E_m(s_i, c_k) = E_m(s_j, c_k)} w_k^m \tag{5}$$

The above formulas may be best analyzed when treated as referring to some fuzzy sets of criteria, with their membership degrees defined in terms of the importance weights w_k of the particular criteria. Thus, $ALLC$ is equal to the cardinality (in the sense of Zadeh, i.e. the ΣCount) of the fuzzy set of all criteria. All the three remaining indices count (in the same sense as $ALLC$) the criteria favoring the first option from the given pair of options (FFC), favoring the second option from the given pair of options (FSC), and indifferent with respect to both options (FBC).

Then, the above formulas are used to calculate the membership functions of the particular fuzzy preference relations constituting our fuzzy preference structure of the agent e_m:

$$\mu_P^m(s_i, s_j) = FFC^m(s_i, s_j)/ALLC^m \tag{6}$$

$$\mu_{Pt}^m(s_i, s_j) = FSC^m(s_i, s_j)/ALLC^m \tag{7}$$

$$\mu_I^m(s_i, s_j) = FBC^m(s_i, s_j)/ALLC^m \tag{8}$$

That is, the degree of strong preference, P, of option s_i over option s_j should be proportional to the fraction of the (weighted) number of criteria for which option s_i yields a better score than s_j. The same idea is applied for the computation of a membership degree of a pair options for the transposed strong preference relation. On the other hand, the degree of indifference of a pair of options is assumed to be proportional to the fraction of the (weighted) number of criteria for which the score for both options is equal.

4 Consensus Evaluation

Here we further develop the concept of a consensus reaching support system proposed in [8, 9]. Hence, the main goal here is to provide a framework for the selection of consensual solutions to the decision problem. We apply the idea of the soft degree of consensus [16–18] meant as "the agreement of most of the pairs of agents as to most pairs of the options". Thus, we employ in the definition of consensus some flexible linguistic quantifiers, Q_1 [39, 40].

In Zadeh's approach [40], a fuzzy linguistic quantifier Q_1 is assumed to be a fuzzy set in [0, 1]. For instance, $Q_1 = $ "*most*" may be defined as:

$$\mu_{\text{"most"}}(x) = \begin{cases} 1 & for & x > 0.8 \\ 2x - 0.6 & for & 0.3 \le x \le 0.8 \\ 0 & for & x < 0.3 \end{cases} \qquad (9)$$

what means, e.g., that a subset of E comprising at least 80 % agents definitely forms a majority corresponding to the notion of "most", a subset comprising less than 30 % of agents definitely does not form such a majority, while subsets of all intermediate cardinalities form such a majority to some extent, expressed by $\mu_{\text{"most"}}(x)$.

In the previous work we have concentrated on the evaluation of consensus with respect to the fuzzy preference relations of all agents via their pairwise comparison of options. In this section, we extend the previous approach to a richer form with a preference structure and other aggregation techniques, while preserving the basic idea of the soft consensus degree.

Basically, we follow the approach proposed in [16–18], i.e. in the first step, for each pair of options we calculate a degree of agreement among most of the agents due to the fuzzy preference relation from our fuzzy preference structure. Thus, for a given pair of options (s_i, s_j) we define three fuzzy properties of agents (effectively: three fuzzy sets of agents): $XP_{ij}, XP^t_{ij}, XI_{ij}$ which correspond to the three types a preference relations. The membership functions of the fuzzy sets representing these fuzzy properties are:

$$\mu_{XP_{ij}}(e_m) = \mu_P^m(s_i, s_j) \qquad (10)$$

$$\mu_{XP^t_{ij}}(e_m) = \mu_{P^t}^m(s_i, s_j) \qquad (11)$$

$$\mu_{XI_{ij}}(e_m) = \mu_I^m(s_i, s_j) \qquad (12)$$

Then, the degree of consensus of most agents as to the pair of options (s_i, s_j) is expressed as:

$$con_{Q_1}(E, s_i, s_j) = \max \left\{ Q_1 XP_{ij}(e_m), Q_1 XP^t_{ij}(e_m), Q_1 XI_{ij}(e_m) \right\} \qquad (13)$$

where Q_1 denotes a linguistic quantifier "most" meant in the sense of the Zadeh's approach; cf. (9), the index m varies from 1 to M, and the choice of the maximum operation is implied by the very essence of the problem.

Finally, we aggregate the results obtained above, for most (Q_2) pairs of options, finding the ultimate consensus degree:

$$con_{Q_1,Q_2}(E, S) = Q_2\{con_{Q_1}(E, s_i, s_j)\}_{(i,j):i<j} \tag{14}$$

Notice, that we take into account only those pairs of options (s_i, s_j) for which $i < j$ because the relation I is symmetric, while P and P^t are the transposes of each other.

5 Supporting Consensus Reaching Process via Linguistic Data Summaries

The consensus degree (14) is the basic indicator which may guide the consensus reaching process in the group. However, there should be an additional support to make consensus reaching process more efficient and faster. Since we apply the approach which is more promising in practice, namely the one in which a moderator plays an important role, we ought to facilitate the work of the moderator to get the group closer to the sufficient agreement. Usually, the moderator carries out two main tasks: checking the level of agreement and producing some advice for those agents that should change their minds. Thus, our task is to provide the moderator with some additional knowledge about the current state of the group, i.e. how far the group is from the consensus, which options are the most preferable, whose preferences are in the highest agreement and disagreement with the rest of the group, etc.

We propose a natural language based support in the form of linguistic summaries of data in the sense of Yager [39], what was first suggested in [22, 26] and extended in [19, 28]. Furthermore, as proposed in our recent papers [7, 8] we use here linguistic data summaries as a set of indicators measuring the current state of agreement concerning either the agents or options. These consensus measures assess how far the group is from consensus, determine those agents and options that are main obstacles in reaching consensus, and help to focus the discussion on issues which may resolve the conflict of opinions in the group, etc.

5.1 Additional Consensus Indicators

To effectively support the consensus reaching process in the setting of a multi-criteria evaluation of options, we had to adjust the consensus indicators used in our previous works [5–10] to the new type of preference related data. Due to some significant differences between the new type of input data and those implemented in the single criteria decision making problem in those previous works, in the first attempt we have adapted two additional consensus reaching related measures: *Option Consensus Degree*, and *Response To Omission*. Then we implement the third indicator *Criterion Consensus Degree* to simplify rules of recommendation given by the moderator in this specific multi-criteria decision making problem

Namely:

1. *Option consensus degree* [27, 30] for option s_i, $OCD(s_i) \in [0, 1]$, is the degree of truth of the statement (linguistic summary):

   "*Most* (Q_1) *pairs of agents agree in their preferences with respect to option* s_i".

The *OCD* index may take values from 0, implying that the preferences of Q_1 (e.g. most) agents differ fundamentally for a given option; to 1 when there is a widespread agreement among Q_1 (e.g. most) agents as to these preferences for the option in question.

The degree of truth of the statement above is now calculated in two steps:

$$v_i(e_{m_1}, e_{m_2}) = \frac{1}{N-1} \sum_{j=1:j \neq i}^{N} v_{ij}^{\alpha}(e_{m_1}, e_{m_2}) \tag{15}$$

$$OCD(s_i) = \mu_{Q_1} \left(\frac{2}{M(M-1)} \sum_{m_1=1}^{M-1} \sum_{m_2=m_1+1}^{M} \mu_{Q_2}(v_i(e_{m_1}, e_{m_2})) \right) \tag{16}$$

Here, $v_{ij}^{\alpha}(e_{m1}, e_{m2})$ denotes the degree of a *sufficient agreement* (at least to degree $\alpha \in [0, 1]$) between agents e_{m1} and e_{m2} as to their preferences between options (s_i, s_j). This agreement indicator for the numeric assessments $sc_{ik}^m = E_m(s_i, c_k) \in [0, 1]$ (cf. (1)) may be defined as follows:

$$v_{ij}^{\alpha}(e_{m_1}, e_{m_2}) = \begin{cases} 1 & \text{if } \forall_{k \in 1 \ldots K} \left| sc_{ik}^1 - sc_{ik}^2 \right| \leq 1 - \alpha \\ 0 & \text{otherwise} \end{cases} \tag{17}$$

2. *Criterion consensus degree*, for criterion c_k, $CCD(c_k) \in [0, 1]$, is the degree of truth of the statement:

   "*Most* (Q_1) *pairs of agents agree in their preferences with respect to criterion* c_k".

Likewise the *OCD* indicator, the *CCD* index might take values from 0, denoting that the preferences/evaluations of most agents differ substantially for a given criterion; to 1 when there is a substantial agreement among the agents as to this criterion.

The degree of truth of the statement above is now calculated analogously to formulas (15)–(17) with appropriate modifications:

$$u_k(e_{m_1}, \ e_{m_2}) = \frac{1}{N} \sum_{i=1}^{N} u_{ik}^{\alpha}(e_{m_1}, \ e_{m_2}) \qquad (18)$$

$$CCD\,(c_k) = \mu_{Q_1}\left(\frac{2}{M(M-1)} \sum_{i=1}^{N} \sum_{m_2=m_1+1}^{M} \mu_{Q_2}(u_k(e_{m_1}, e_{m_2}))\right) \qquad (18a)$$

Here, $u_{ik}^{\alpha}(e_{m1}, e_{m2})$ denotes the degree of a *sufficient agreement* (at least to degree $\alpha \in [0, 1]$) between agents e_{m1} and e_{m2} as to their assessments of option s_i with respect to the criterion c_k. This agreement indicator for the numeric assessments $sc_{ik}^{m} \in [0, 1]$ may be defined as follows:

$$u_{ik}^{\alpha}\left(e_{m_1}, e_{m_2}\right) = \begin{Bmatrix} 1 & if \left| sc_{ik}^{1} - sc_{ik}^{2} \right| \leq 1 - \alpha \\ 0 & otherwise \end{Bmatrix} \qquad (18b)$$

3. *Response to omission* of agent $e_m \in E$, $RTO(e_m) \in [-1, 1]$, is determined as a difference between the consensus degree computed for the whole group and the consensus degree for the group without taking into account agent e_m.

$$RTO\,(e_m) = con_{Q_1,Q_2}\,(E, S) - con_{Q_1,Q_2}(E - \{e_k\}, S) \qquad (19)$$

This index makes it possible to assess the contribution to consensus of a given agent. The range of values is from -1, for an absolutely negative influence, through 0 for lack of effect, to $+1$ for a definitely positive influence.

5.2   Rules of Recommendation

The use of the indicators *OCD*, *CCD* and *RTO* for the support of the consensus reaching process follows the general scheme proposed by Kacprzyk and Zadrożny [27, 30], and in the context of this paper can be summarized as follows. If the degree of agreement is not sufficient, our system, by providing some additional tools mentioned earlier, helps the moderator to control the decision making session according to some flexible strategy. With a full access to the values of the *OCD*,*CCD* and *RTO* indicators, the moderator sticks to the following scenario:

1. Chooses an agent with the lowest value of the *RTO* indicator, i.e., so called "outsider", an agent whose current opinion has a negative effect on the agreement in the group.
2. Selects an option with the lowest value of the *OCD* indicator, i.e., option that is the main obstacle to reaching consensus.

3. Chooses a criterion with the lowest value of the *CCD* indicator, i.e. a criterion which has a negative influence on the consensus degree.
4. Finally, to get the group closer to the sufficient agreement, suggests the agent who was selected in step 1 to change his/her opinion as to the option selected in step 2 according to the criterion chosen in step 3 What matters here is that the suggested change of evaluation aims at minimizing the distance between decision makers and leading the group closer to the consensus in the most effective and efficient way.

According to Kacprzyk et al. [30], we follow the scheme $\{sc_{ik}^1, \ldots, sc_{ik}^m\} \rightarrow sc_{ik}$, i.e., we assume a straightforward approach where the social opinion is given from the set of individual opinions by:

$$sc_{ik} = \frac{1}{M} \sum_{m=1}^{M} a_{ik}^m \qquad (20)$$

where

$$a_{ik}^m = \begin{cases} 1 & if \ sc_{ik}^m > 0.6 \\ 0 & otherwise \end{cases} \qquad (21)$$

Thus, the evaluation of a chosen option given by the agent who is a major obstacle to the reaching of the agreement most should be replaced by a new evaluation as follows: if $sc_{ik}^m < sc_{ik}$, then $sc_{ik}^m := \min(sc_{ik}^m + sc_{ik}, 1)$, otherwise $sc_{ik}^m := \max(sc_{ik}^m - sc_{ik}, 0)$.

For the further analysis of the proposed system, when the evaluation of a particular option with respect to a particular criterion of some agent is replaced by other one by a change of value (opinion) of 0.2 (or, e.g., "one linguistic term" in case the evaluations are expressed linguistically), it is assumed to constitute 1 point of cost. The motivation is simple: people usually do not want to change their opinions so that a change results in a "pain" or cost. Such a scenario should lead us effectively and efficiently to a satisfactory degree of consensus [7, 8]. In other words, we would like to get the largest increase in the degree of agreement as soon as possible and with the lowest cost of changes.

6 An Example of Implementation

6.1 Outline of the Problem

The decision problem concerns the simple selection of a proper candidate for the head of students' council at a university. A group of three students/electors needs to evaluate the quality of three candidates with respect to the following criteria: patience, experience, and communication skills. We assume that this set of criteria has been chosen taking into account to the personality traits required for a person who represents students and provides feedback to the university staff. We also assume that the agents (decision makers) know some facts about each candidate which allows for some justified evaluations. As mentioned, each decision maker may assign different weights to the particular criteria, depending on their own preferences. Notice that we assume, in general, the use of linguistic values from an ordered set to be used for options evaluation. Thus, in our example we use linguistic terms but, for simplicity, we interpret them as some real numbers from [0, 1], for instance 0.0, 0.2, 0.4, 0.6, 0.8, and 1.0. This is obviously a result of the granulation needed.

Then, the part of the source data may look as follows (options are identified by the first names of the candidates):

$S = \{Monika, Mateusz, Renata\}$
$C = \{Patience, Experience, Communication Skills\}$
$W^1 = \{important, less important, the most important\}$
$W^2 = \{less important, medium important, very important\}$
$W^3 = \{the most important, less important, very important\}$

and, obviously, the particular linguistic terms are represented as some real numbers as mentioned above.

Although the initial evaluation of particular options given by particular agents is expressed using the values from the set of linguistic terms, i.e. $LL = \{$Excellent, Very good, Good, Fair, Bad, Very Bad$\}$, in further calculations we will replace them by real numbers, as mentioned earlier, what leads to the following matrices:

$$\left[sc_{ik}^1\right]_{i=1..N,k=1..K} = \begin{pmatrix} 0.4 \ 0.6 \ 0.8 \\ 1.0 \ 0.2 \ 0.8 \\ 0.2 \ 0.8 \ 0.6 \end{pmatrix}$$

$$\left[sc_{ik}^2\right]_{i=1..N,k=1..K} = \begin{pmatrix} 0.4 \ 0.6 \ 0.0 \\ 0.0 \ 0.4 \ 0.8 \\ 0.2 \ 1.0 \ 0.4 \end{pmatrix}$$

$$\left[sc_{ik}^3\right]_{i=1..N,k=1..K} = \begin{pmatrix} 1 \ 1 \ 1 \\ 0 \ 0 \ 0 \\ 0 \ 0 \ 0 \end{pmatrix}$$

where $sc_{ik}^m = E_m(s_i, c_k)$; cf. (1).

The implementation proposed makes it possible to:

- set weights of the particular criteria by each agent,
- evaluate the quality of each option according to each criterion, and by each agent,
- compare opinions of the agents and automatically compute the consensus degree.

6.2 Results

Results of the application of the proposed consensus reaching support system for the example shown in Sect. 6.1 are as follows.

1. The first iteration of the system yields (the empty entries of matrices mean that the values do not matter):

$$ALLC^1 = 1.6 \quad ALLC^2 = 1.8 \quad ALLC^3 = 2.0$$

Here and later on in this section, if not otherwise specified, for each agent a particular indicator, i.e., FFC, FSC, FBC, is defined as a matrix for each pair of options (s_i, s_j):

$$FFC^1 = \begin{pmatrix} 0.2 & 1.4 \\ & 1.4 \end{pmatrix}, \quad FFC^2 = \begin{pmatrix} 0.6 & 0.2 \\ & 0.8 \end{pmatrix}, \quad FFC^3 = \begin{pmatrix} 2.0 & 2.0 \\ & 0 \end{pmatrix}$$

$$FSC^1 = \begin{pmatrix} 0.6 & 0.2 \\ & 0.2 \end{pmatrix}, \quad FSC^2 = \begin{pmatrix} 0.8 & 1.2 \\ & 0.6 \end{pmatrix}, \quad FSC^3 = \begin{pmatrix} 0 & 0 \\ & 0 \end{pmatrix}$$

$$FBC^1 = \begin{pmatrix} 0.8 & 0 \\ & 0 \end{pmatrix}, \quad FBC^2 = \begin{pmatrix} 0 & 0 \\ & 0 \end{pmatrix}, \quad FBC^3 = \begin{pmatrix} 0 & 0 \\ & 2.0 \end{pmatrix}$$

Likewise all properties given below are represented by matrices calculated for each pair of options (s_i, s_j); cf. formulas (13) and (14):

$$Q_1XP\,(e_m) = \begin{pmatrix} 0.52 & 0.67 \\ & 0.48 \end{pmatrix}, \quad Q_1XP^t\,(e_m) = \begin{pmatrix} 0.32 & 0.33 \\ & 0.18 \end{pmatrix}$$

$$Q_1XI\,(e_m) = \begin{pmatrix} 0.16 & 0 \\ & 0.33 \end{pmatrix}$$

$$con_{Q_1}\,(E, s_i, s_j) = \begin{pmatrix} 0.67 & 0.33 \\ & 0.33 \end{pmatrix}$$

where the linguistic quantifier Q_1 is a unitary linguistic quantifier, i.e., $\mu_{Q1}(x) = x$ while in what follows the quantifier Q_2 is represented by the membership function (9).

The degree of consensus among the group of agents equals:

$$con_{Q_1,Q_2}\,(E, S) = 0.51$$

The degree of consensus equals 0.51 what indicates that the group is far from the agreement. Therefore, according to the assumed scenario, the moderator suggests agent 2 (with the lowest value of the RTO indicator; cf. Table 3) to change his/her opinion about options "Monika" and "Mateusz" (the lowest value of the OCD indicator; cf. Table 1) with respect to the criterion "Communication skills" (the lowest value of the CCD indicator; cf. Table 2).

Table 1 Values of the OCD indicator for each option after the first iteration

Option, s	OCD (s)
"Monika"	0.0
"Mateusz"	0.0
"Renata"	0.14

Table 2 Values of the CCD indicator for each option after the first iteration

Criterion, c	CCD (c)
"Patience"	0.42
"Experience"	0.11
"Communication skills"	0.0

Table 3 Values of the RTO indicator for each agent after the first iteration

Agent, e	RTO (e)
3	−0.02
1	−0.07
2	−0.22

The evaluations of the agent e_2 that should be changed according to the reasoning given above are as follows:

$$
\begin{array}{c@{\quad}ccc}
 & c_1 & c_2 & c_3 \\
s_1 & - & - & \times \\
s_2 & - & - & \times \\
s_3 & - & - & -
\end{array}
$$

Thus; cf. (20) and (21) agent 2 is suggested to change his/her opinion in the following way: we use here, for illustrative purposes, and comprehensiveness, linguistic values instead of numeric ones:

(Monika, Communication skills): Very bad ➜ Good
(Mateusz, Communication skills): Very good ➜ Bad

2. The second iteration of the system yields (the bold entries indicate what has been changed):

$$
[sc^2_{ik}]_{i=1..N,k=1..K} = \begin{pmatrix} 0.4 & 0.6 & \mathbf{0.6} \\ 0.0 & 0.4 & \mathbf{0.2} \\ 0.2 & 1.0 & 0.4 \end{pmatrix}
$$

$$
FFC^2 = \begin{pmatrix} 1.4 & 1.0 \\ & 0.0 \end{pmatrix}, \quad FSC^2 = \begin{pmatrix} 0.0 & 0.4 \\ & 1.4 \end{pmatrix}, \quad FBC^2 = \begin{pmatrix} 0.0 & 0.0 \\ & 0.0 \end{pmatrix}
$$

$$
Q_1XP\,(e_m) = \begin{pmatrix} 0.71 & 0.86 \\ & 0.29 \end{pmatrix}, \quad Q_1XP^t\,(e_m) = \begin{pmatrix} 0.12 & 0.14 \\ & 0.38 \end{pmatrix}
$$

Table 4 Values of the *OCD* indicator for each option after the second iteration

Option, *s*	*OCD* (*s*)
"Renata"	0.15
"Monika"	0.11
"Mateusz"	0.0

Table 5 Values of the *CCD* indicator for each option after the first iteration

Criterion, *c*	*CCD* (*c*)
"Patience"	0.42
"Experience"	0.11
"Communication skills"	0.0

Table 6 Values of the *RTO* indicator for each agent after the second iteration

Agent, *e*	*RTO* (*e*)
3	0.03
2	−0.03
1	−0.27

$$Q_1 XI\,(e_m) = \begin{pmatrix} 0.17 & 0 \\ & 0.33 \end{pmatrix}$$

$$con_{Q_1}\,(E, s_i, s_j) = \begin{pmatrix} 0.86 & 0.38 \\ & 0.33 \end{pmatrix}$$

The degree of consensus among the group of agents equals:

$$con_{Q_1, Q_2}\,(E, S) = 0.7$$

After this iteration, the degree of consensus has grown from $0.51 \rightarrow 0.7$ but this level of agreement is still considered insufficient. According to the scenario, the moderator suggests agent 1 (with the lowest value of the *RTO* indicator; cf. Table 6) to change his/her opinion about option "Mateusz" (with the lowest value of the *OCD* indicator; cf. Table 4) with respect to the criteria "Communication skills" (the lowest value of the *CCD* indicator; cf. Table 5).

The evaluations that should be changed according to the reasoning given earlier are as follows:

$$\begin{array}{c c c c} & c_1 & c_2 & c_3 \\ s_1 & - & - & - \\ s_2 & - & - & \times \\ s_3 & - & - & - \end{array}$$

Thus; cf. (20) and (21), agent 1 is suggested to change his/her opinion in the following way:

(Mateusz, Communication skills): Very bad ➜ Fair

3. The third iteration of the system.

Matrices that have changed after modification of agent's 1 evaluation:

$$[sc^1_{ik}]_{i=1..N,k=1..K} = \begin{pmatrix} 0.4\ 0.6\ 0.8 \\ 1.0\ 0.2\ \mathbf{0.4} \\ 0.2\ 0.8\ 0.6 \end{pmatrix}$$

$$FFC^1 = \begin{pmatrix} 1.0\ 1.4 \\ 0.6 \end{pmatrix}, FSC^1 = \begin{pmatrix} 0.6\ 0.2 \\ 1.0 \end{pmatrix}, FBC^1 = \begin{pmatrix} 0\ 0 \\ 0 \end{pmatrix}$$

$$Q_1XP\,(e_m) = \begin{pmatrix} 0.86\ 0.86 \\ 0.13 \end{pmatrix}, Q_1XP^t\,(e_m) = \begin{pmatrix} 0.12\ 0.14 \\ 0.54 \end{pmatrix}$$

$$Q_1XI\,(e_m) = \begin{pmatrix} 0\ \ \ 0 \\ 0.33 \end{pmatrix}$$

$$con_{Q_1}\,(E, s_i, s_j) = \begin{pmatrix} 0.87\ 0.54 \\ 0.33 \end{pmatrix}$$

The degree of consensus among the group of agents equals:

$$con_{Q_1,Q_2}\,(E, S) = 0.92$$

After the third iteration we were able to obtain the sufficient degree of total agreement equal 0.92, and the session may be finished. Cost of all these changes, mentioned in the Sect. 5.2, equals 8 points, which constitute a satisfactory result in respect to the efficiency of the system proposed.

7 Conclusion

We presented an extension of a consensus reaching support system proposed in our previous work [5–10] meant to accommodate a multi-criteria evaluation of options and importance weights of all criteria given by each agent. Clearly, this implies some modifications of the concept, tools and techniques proposed in our previous papers. To improve the efficiency of the process we designed some additional suggestions/hints for the moderator in the form of linguistic summaries, adapted to the multi-criteria setting. We presented an application of the proposed techniques for a real world problem.

Our work was intended to attain the best increase in the consensus degree within some time limit. For that, we defined some tactic for the rules of moderator recommendations to guide the consensus reaching process towards the highest degree of agreement. However, our efficiency-oriented procedures [8] are only a point of departure for a further extension in the direction of multi-criteria evaluations. These may include novel agent-based computational models, for example, new direction in the analysis of multiagent systems related to human group and societies, notably an idea of fairness [5, 6, 10] or the concept of the ideal and anti-ideal point [7] which should make the consensus reaching process faster.

Acknowledgments This work is partially supported by the Foundation for Polish Science under the "International Ph.D. Projects in Intelligent Computing" financed from the Europe-an Union within the Innovative Economy Operational Programme 2007–2013 and European Regional Development Fund (D. Gołuńska), and partially by the National Science Centre under Grant No. UMO 2012/05/B/ST6/03068 (J. Kacprzyk).

References

1. Butle, C.T., Rothstein, A.: On Conflict and Consensus: A Handbook on Formal Consensus Decision Making. Food Not Bombs Publishing, Takoma Park (2006)
2. Carlsson, C., Fedrizzi, M., Fuller, R.: Group decision support systems. In: Carlsson, C., Fedrizzi, M., Fuller, R. (eds.) Fuzzy Logic in Management, pp. 57–125. Springer Science, Berlin (2004)
3. Fedrizzi, M., Kacprzyk, J., Zadrożny, S.: An interactive multi-user decision support system for consensus reaching process using fuzzy logic with linguistic quantifiers. Decis. Support Syst. **4**(3), 313–327 (1988)
4. Fedrizzi, M., Fedrizzi, M., Marques Pereira, R.A.: Consensus modelling in group decision making: a dynamical approach based on Zadeh's fuzzy preferences. In: Seising, R., Trillas, E., Moraga, C., Termini, S. (eds.) On Fuzziness (Homage to Lotfi A. Zadeh), pp. 165–170. Springer, Heidelberg (2013)
5. Gołuńska, D., Hołda, M.: The need of fairness in the group consensus reaching process in a fuzzy environment. Tech. Trans. Autom. Control **1–AC**, 29–38 (2013)
6. Gołuńska, D., Kacprzyk, J.: The conceptual framework of fairness in consensus reaching process under fuzziness. In: Proceedings of the 2013 Joint IFSA World Congress NAFIPS Annual Meeting, pp. 1285–1290. Edmonton, Canada, 24–28 June 2013
7. Gołuńska, D., Kacprzyk, J., Zadrożny, S.: A consensus reaching support system based on concepts of ideal and anti-ideal point. In: Proceedings of the 2014 North American Fuzzy Information Processing Society Conference (NAFIPS 2014), pp. 1–6 (2014)
8. Gołuńska, D., Kacprzyk, J., Zadrożny, S.: A model of efficiency-oriented group decision and consensus reaching support system in a fuzzy environment. In: Proceedings of the 15th International Conference on Information Processing and Management of Uncertainty in Knowledge-Based Systems, IPMU-2014, pp. 424–433 (2014)
9. Gołuńska, D., Kacprzyk, J., Zadrożny, S.: On efficiency-oriented support of consensus reaching in a group of agents in a fuzzy environment with a cost based preference updating approach. In: Proceedings of SSCI-2014. IEEE Press, Orlando (2014)
10. Gołuńska, D., Kacprzyk, J., Herrera-Viedma, E.: Modeling different advising attitudes in a consensus focused process of group decision making. In: Filev, D., et al. (ed.) Intelligent Systems'2014, Series: Advances in Intelligent Systems and Computing, vol. 322, pp. 279–288. Springer, Berlin (2015)
11. Herrera, F., Herrera-Viedma, E., Verdegay, J.L.: A rational consensus model in group decision making using linguistic assessments. Fuzzy Sets Syst. **88**(1), 31–49 (1997)
12. Herrera-Viedma, E., García-Lapresta, J.L., Kacprzyk, J., Fedrizzi, M., Nurmi, H., Zadrożny, S. (eds.): Studies in fuzziness and soft computing. Consensual Processes. Springer, Berlin (2011)
13. Herrera-Viedma, E., Cabrerizo, F.J., Kacprzyk, J., Pedrycz, W.: A review of soft consensus models in a fuzzy environment. Inf. Fusion **17**, 4–13 (2014)

14. Kacprzyk, J.: Group decision making with a fuzzy majority via linguistic quantifiers. Part I: a consensory like pooling. Cybern. Syst. Int. J. **16**, 119–129. Part II: a competitive like pooling. Cybern. Syst. Int. J. **16**, 131–144 (1985)
15. Kacprzyk, J.: Group decision making with a fuzzy linguistic majority. Fuzzy Sets Syst. **18**(2), 105–118 (1986)
16. Kacprzyk, J., Fedrizzi, M.: 'Soft' consensus measures for monitoring real consensus reaching processes under fuzzy preferences. Control Cybern. **15**(3–4), 309–323 (1986)
17. Kacprzyk, J., Fedrizzi, M.: A 'soft' measure of consensus in the setting of partial (fuzzy) preferences. Eur. J. Oper. Res. **34**, 315–325 (1988)
18. Kacprzyk, J., Fedrizzi, M.: A 'human-consistent' degree of consensus based on fuzzy logic with linguistic quantifiers. Math. Soc. Sci. **18**, 275–290 (1989)
19. Kacprzyk, J., Yager, R.R.: Linguistic summaries of data using fuzzy logic. Int. J. Gen. Syst. **30**, 33–154 (2001)
20. Kacprzyk, J., Zadrożny, S.: On the use of fuzzy majority for supporting consensus reaching under fuzziness. In: Proceedings of FUZZ-IEEE'97 - Sixth IEEE International Conference on Fuzzy Systems (Barcelona, Spain), vol. 3, pp. 1683–1988 (1997)
21. Kacprzyk, J., Zadrożny, S.: An internet-based group decision and consensus reaching support system. Management **7**(28), 4–10 (2003)
22. Kacprzyk, J., Zadrożny, S.: Supporting consensus reaching in a group via fuzzy linguistic data summaries. In: IFSA'2005 World Congress - Fuzzy Logic, Soft Computing and Computational Intelligence, pp. 1746–1751. Tsinghua University Press/Springer, Beijing (2005)
23. Kacprzyk, J., Zadrożny, S.: On a concept of a consensus reaching process support system based on the use of soft computing and Web techniques. In: Ruan, D., Montero, J., Lu, J., Martínez, L., D'hondt, P., Kerre, E.E. (eds.) Computational Intelligence in Decision and Control, pp. 859–864. World Scientific, Singapore (2008)
24. Kacprzyk, J., Zadrożny, S.: Towards a general and unified characterization of individual and collective choice functions under fuzzy and nonfuzzy preferences and majority via the ordered weighted average operators. Int. J. Intell. Syst. **24**(1), 4–26 (2009)
25. Kacprzyk, J., Zadrożny, S.: Soft computing and Web intelligence for supporting consensus reaching. Soft Comput. **14**(8), 833–846 (2010)
26. Kacprzyk, J., Zadrożny, S.: Supporting consensus reaching processes under fuzzy preferences and a fuzzy majority via linguistic summaries. In: Greco, S., Marques Pereira, R.A., Squillante, M., Yager, R.R. (eds.) Preferences and Decisions, vol. 257, pp. 261–279 (2010)
27. Kacprzyk, J., Zadrożny, S.: Computing with words is an implementable paradigm: fuzzy queries, linguistic data summaries and natural language generation. IEEE Trans. Fuzzy Syst. **18**(3), 461–472 (2010)
28. Kacprzyk, J., Yager, R.R., Zadrożny, S.: A fuzzy logic based approach to linguistic summaries of databases. Int. J. Appl. Math. Comput. Sci. **10**, 813–834 (2000)
29. Kacprzyk, J., Zadrożny, S., Wilbik, A.: Linguistic summarization of some static and dynamic features of consensus reaching. In: Reusch, B. (ed.) Computational Intelligence, Theory and Applications, pp. 19–28. Springer, Berlin (2006)
30. Kacprzyk, J., Zadrożny, S., Fedrizzi, M., Nurmi, H.: On group decision making, consensus reaching, voting and voting paradoxes under fuzzy preferences and a fuzzy majority: a survey and some perspectives. In: Bustince, H., Herrera, F., Montero, J. (eds.) Fuzzy Sets and Their Extensions: Representations, Aggregation and Models, pp. 263–295. Springer, Berlin (2008)
31. Kacprzyk, J., Zadrożny, S., Raś, Z.W.: How to support consensus reaching using action rules: a novel approach. Int. J. Uncertain. Fuzziness Knowl.-Based Syst. **18**(4), 451–470 (2010)
32. Loewer, B.: Special issue on consensus. Synthese **62**, 1–122
33. Loewer, B., Laddaga, R.: Destroying the consensus. Synthese **62**, 79–96 (1985)
34. Perez, I.J., Wikström, R., Mezei, J., Carlsson, C., Anaya, K., Herrera-Viedma, E.: Linguistic consensus models based on a fuzzy ontology. Procedia Comput. Sci. **17**, 498–505 (2013)
35. Spillman, B., Spillman, R., Bezdek, J.C.: A fuzzy analysis of consensus in small groups. In: Wang, P.P., Chang, S.K. (eds.) Fuzzy Automata and Decision Processes, pp. 331–356. North-Holland, Amsterdam (1980)

36. Szmidt, E., Kacprzyk, J.: A consensus-reaching process under intuitionistic fuzzy preference relations. Int. J. Intell. Syst. **18**(7), 837–852 (2003)
37. Turban, E., Aronson, J.E., Liang, T.P.: Decision Support Systems and Intelligent Systems, 6th edn. Prentice Hall, Upper Saddle River (2005)
38. Van de Walle, B., De Baets, B., Kerre, E.: A plea for the use of Lukasiewicz triplets in fuzzy preference structures. Part 1: general argumentation. Fuzzy Sets Syst. **97**, 349–359 (1998)
39. Yager, R.R.: A new approach to the summarization of data. Inf. Sci. **28**, 69–86 (1982)
40. Zadeh, L.A.: A computational approach to fuzzy quantifiers in natural languages. Comput. Math. Appl. **9**, 149–184 (1983)

Improving Spatial Estimates of Greenhouse Gas Emissions at a Fine Resolution: A Review of Approaches

Joanna Horabik-Pyzel and Zbigniew Nahorski

Abstract The paper presents a review of the methods which can be useful in quantification of greenhouse gas (GHG) emissions at a fine spatial resolution. The discussed approaches include: spatial disaggregation of GHG emissions based on proxy data and/or statistical modeling of spatial correlation, an estimation of fossil fuel emission changes from measuring rates (mixing ratios) of tracer (like $^{14}CO_2$) concentrations in the atmosphere, the atmospheric inversion methods, and flux tower observations.

Keywords CO_2 emission fluxes · Spatial resolution · Disaggregation · Emission inventory

1 Introduction

Quantification of CO_2 emissions at fine spatial (and time) scales is advantageous for many environmental, physical, and socio-economic analyzes; in principle, it can be easily integrated with other data in a gridded format. This is especially important for improved assessment of carbon cycle and climate change. Of particular importance are estimations of fossil fuel CO_2 fluxes, which are used to quantitatively estimate CO_2 sources and sinks, see e.g. [6]. A few institutions gather data on emissions from fossil fuels at national levels, like the US Department of Energy Carbon Dioxide Information Analysis Center (CDIAC) [7, 39]; the International Energy Agency (IEA), IEA [28]. IPCC gathers data from national GHG inventories within the Kyoto Protocol agreement and its continuation, IPCC [29]. British Petroleum company compiles energy statistics, BP [8] that, can be conveniently used for estimation of national CO_2 emissions. These datasets have been used for estimating global sources and sinks on a regional (e.g. continental) scale, [3, 11, 22, 48, 52, 54].

J. Horabik-Pyzel (✉) · Z. Nahorski
Systems Research Institute, Polish Academy of Sciences, Warsaw, Poland
e-mail: Joanna.Horabik@ibspan.waw.pl

Z. Nahorski
e-mail: Zbigniew.Nahorski@ibspan.waw.pl

© Springer International Publishing Switzerland 2016   237
G. De Tré et al. (eds.), *Challenging Problems and Solutions*
in Intelligent Systems, Studies in Computational Intelligence 634,
DOI 10.1007/978-3-319-30165-5_12

Much less data are available on emissions in the sub-national scales. Some countries publish data for provinces, but their scale is still too rough compared with other contemporary studies, like those presented in the sequel. The need for such fine resolutions comes from requirements of other applications, for example modeling of GHG dispersion at a local scale. Spatial disaggregation of emissions introduces additional uncertainty to a developed inventory. This is why the option of using additional information to reduce this uncertainty is of great interest.

A common approach to disaggregation of emissions is a usage of proxy data, most often areas of fine grids or population therein. However, independent estimates of carbon fluxes, like inverse modeling or eddy covariance, provide opportunity to incorporate additional knowledge and provide more comprehensive spatial quantification of carbon budget. Evidently, there is a mismatch between distinct approaches to estimation of CO_2 fluxes. In general, there are two kinds of estimates: it can be either accounting of emissions (bottom-up) or by measuring concentrations of CO_2 and inferring about original emission fluxes (top-down). As a result, merging such datasets is a challenging task due to incomplete accounting and uncertainties underlying each of the methods. Moreover, one should take into account various spatial scales and different scarcity of data. This paper outlines several relevant methods, and discusses advantages and limitations of using them for improving inventory emission estimates in fine scales. The review highlights the issues related to spatial dimension of the task.

The paper is organized as follows. Section 2 comments on traditional disaggregation with a usage of proxy data. Among them, independent emission assessments can be obtained with the methods that use nighttime lights observations. Section 3 presents a statistical method based on spatial correlation of emissions. Observations of $^{14}CO_2$ are discussed in Sect. 4, the atmospheric inversion method in Sect. 5, and flux tower observations in Sect. 6. Section 7 concludes with an outline of possibilities for integration of knowledge coming from different observations.

2   Disaggregation Based on Proxy Data

There exist few national inventories of high resolution, like the Vulcan inventory, [23], compiled for United States with the spatial resolution $10\,km \times 10\,km$ and the time resolution 1 month; this one has been considered as the most accurate in many publications, see e.g. [19, 42, 50]. Different disaggregation methods have been used to obtain high-resolution emissions. This is typically done using proxy data available in finer scales. The most straightforward approach to estimate data in a fine scale is to disaggregate national emissions proportionally to gridded population information, see e.g. [1, 43, 58]. These data are typically compiled with the resolution $1^0 \times 1^0$ (approximately $100\,km \times 100\,km$). The resolution admitted there cannot be much finer since various emissions, like those from power stations, are not correlated well with population distribution. Another easily applicable approach is to disaggregate emissions proportionally to area. This can provide satisfactory results provided that the emission sources are homogenous in space, which can be approximately true only in some cases. When more kinds of proxy variables are available, a regression model can be used, provided a method for estimating parameters of the regression function is invented. Such a method, in a more general formulation, is presented in the next section.

This way of disaggregation requires gathering very detailed information about emissions or activities generating emissions. In this context, the role of institutions gathering and publishing relevant data becomes crucial. For instance, a very high resolution of emission cadasters ($2\,km \times 2\,km$ grid) was obtained for Poland within the 7th FP GESAPU project [13]. It resulted from a detailed analysis of information from various sources, published by governmental and research agencies, as well as energy or industry plants (e.g. taking part in emission trading scheme); the analysis was followed by precise modeling. At present, this approach to disaggregation seems to provide the best results. Nevertheless, the relevant procedure, and particularly gathering data from numerous sources and publications, requires immense input of human work—especially for big regions.

Another proxy data used for disaggregation are the satellite observations of night-time lights, like those obtained by the Defense Meteorological Satellite Operational Line Scanner (DMSP-OLS) and released by the scientific community, see [27]. Early experiments, [14, 15], showed that the CO_2 emissions obtained are underestimated in comparison with the CIDIAC data for most countries. A few improvements have been then made. Rayner et al. [50] used a modified Kaya identity, in which emissions are modeled as a product of population density, per capita economic activity, energy intensity of economy, and carbon intensity of energy. This formula is used to predict emissions from several sectors, namely energy, manufacturing, transport (broken to land, sea, and air emissions), and other, with 0.25^0 resolution. Next, the predictions are constrained using statistics of national emissions, distribution of nightlights and population. Comparison with the Vulcan inventory showed better performance than the models based only on population or only on nightlight data.

Oda and Maksyutov [42] extracted emissions from point sources before disaggregating the non-point emissions proportionally to the nightlight distribution, and integrated them again to obtain $1\,km \times 1\,km$ emission data. The information on locations of high point sources and their emissions was taken from the Carbon Monitoring and Action database (CARMA, http://carma.org), and national emissions were calculated from the BP data on usage of fossil fuels.

Ghosh et al. [19] investigated two models. In the first one, the correlation between the nighttime lights and the Vulcan data is calculated and then used to calculate disaggregated emissions for other countries. This model did not provide satisfactory results. One of the reasons for discrepancies was underestimation of emissions in dark areas of the nightlight maps. Then, a second model was developed, in which both nighttime lights and population data were used. The CO_2 total emission data from the US Environmental Protection Agency (EPA) were used. Firstly, high point (electrical power plant) emissions were subtracted from the original data, and the difference (so-called non-utility CO_2 emissions) was disaggregated taking into account the nighttime lights and population distribution. For this, the nightlight map has been divided into lit and dark areas. Secondly, the dark areas were halved, and "mean" aerial emission was calculated dividing total non-utility CO_2 emission by the sum of the lit and halved dark areas. The "total emissions" from the lit areas were calculated multiplying the mean emission by the areas, while those for the dark areas were calculated by multiplying the halved mean emission by the areas. Then, the calculated non-utility CO_2 emissions in the lit areas were distributed proportionally to the nighttime lights, and those in the dark areas proportionally to the population. Finally, point emissions and non-utility emissions were integrated into the final grid.

The most troublesome aspect of the satellite observations of nighttime lights is connected with different habits of using lights in different regions. Usually, less developed regions use less nighttime lights compared to the usage of fossil fuel energy there. This can be observed not only on a country level, but also between regions of bigger countries, and even districts of big cities, which may become important in case of high resolution maps. In either case, the nightlight maps carry valuable information that can help in disaggregation of national emission totals.

Nighttime lights cannot, however, indicate the variability existing among sectors, like residential, commercial and industrial or transportation, at least within a usual observation resolution. Radiation from different sectors does not relate in the same way to the CO_2 emissions. With very high resolutions of disaggregated emission maps, this effect may introduce biases.

To find locations of point sources, the databases on world installations, like CARMA, are typically used. These databases do contain errors. For example, not all existing power plants may be listed or emission estimates may be only approximate. Also locations of point sources may be erroneous, see e.g. discussion in [42].

3 Inclusion of a Neighborhood Correlation

Spatial disaggregation of emission data can be also approached by means of the statistical inference methods that account for spatial correlation. In the context of national GHG inventories, spatial allocation is performed for activity data, and those are usually disaggregated from district or municipal level to smaller, regular grids. For some sectors or subsectors, activity data are highly correlated in space; this is for instance the case of agricultural datasets. The approach presented in [24] is to allocate the data to finer spatial scales conditional on covariate (proxy) information observable in a fine grid, and accounting for a neighborhood correlation. Spatial dependence is modeled with the conditional autoregressive (CAR) structure introduced into a linear model as a random effect.

First, the model is specified on a level of *fine* grid. Let Y_i denote a random variable associated with an unknown value of interest y_i defined at each cell i for $i = 1, \ldots, n$ of a fine grid (n denotes the overall number of cells in a fine grid). The random variables Y_i are assumed to follow the Gaussian distribution with the mean μ_i and variance σ_Y^2

$$Y_i|\mu_i \sim Gau\left(\mu_i, \sigma_Y^2\right)$$

Given the values μ_i and σ_Y^2, the random variables Y_i are assumed independent. The mean $\boldsymbol{\mu} = \{\mu_i\}_{i=1}^n$ represents the true process underlying emissions, and the (unknown) observations are related to this process through a measurement error with the variance σ_Y^2. The approach to modeling μ_i expresses an assumption that available covariates explain part of the spatial pattern, and the remaining part is captured through a spatial dependence. The CAR scheme follows an assumption of similar random effects in adjacent cells, and it is given through the specification of full conditional distribution functions of μ_i for $i = 1,...,n$

$$\mu_i|\boldsymbol{\mu}_i \sim Gau\left(\boldsymbol{x}_i^T\boldsymbol{\beta} + \rho\sum_{\substack{j=1\\j \neq i}}^{n} \frac{w_{ij}}{w_{i+}}\left(\mu_j - \boldsymbol{x}_j^T\boldsymbol{\beta}\right), \frac{\tau^2}{w_{i+}}\right)$$

where $\boldsymbol{\mu}_i$ denotes all elements in $\boldsymbol{\mu}$ but μ_i, w_{ij} are the adjacency weights (w_{ij} if j is a neighbour of i and 0 otherwise, also $w_{ii} = 0$); $w_{i+} = \sum_j w_{ij}$ is the number of neighbours of an area i; $\boldsymbol{x}_i^T\boldsymbol{\beta}$ is a regression component with proxy information available for area i and a respective vector of regression coefficients; τ^2 is a variance parameter. Thus, the mean of the conditional distribution $\mu_i|\boldsymbol{\mu}_{-i}$ consists of the regression part and the second summand, which is proportional to the average values of remainders $\mu_j - \boldsymbol{x}_j^T\boldsymbol{\beta}$ for neighbouring sites (i.e. when $w_{ij} = 1$). The proportion is calibrated with the parameter ρ, reflecting strength of a spatial association. Furthermore, the variance of the conditional distribution $\mu_i|\boldsymbol{\mu}_{-i}$ is inversely proportional to a number of neighbours w_{i+}.

The joint distribution of the process $\boldsymbol{\mu}$ is the following (for derivation see e.g. [5])

$$\boldsymbol{\mu} \sim Gau_n\left(\boldsymbol{X}\boldsymbol{\beta}, \tau^2\left(\boldsymbol{D} - \rho\boldsymbol{W}\right)^{-1}\right) \tag{1}$$

where $\boldsymbol{D}$ is an $n \times n$ diagonal matrix with w_{i+} on the diagonal; and $\boldsymbol{W}$ is an $n \times n$ matrix with adjacency weights w_{ij}. Equivalently, we can write (1) as

$$\boldsymbol{\mu} = \boldsymbol{X}\boldsymbol{\beta} + \boldsymbol{\varepsilon}, \quad \boldsymbol{\varepsilon} \sim Gau_n\left(\boldsymbol{0}, \boldsymbol{\Omega}\right) \tag{2}$$

with $\boldsymbol{\Omega} = \tau^2\left(\boldsymbol{D} - \rho\boldsymbol{W}\right)^{-1}$.

The model for a *coarse* grid of (aggregated) observed data is obtained by multiplication of (2) with the $N \times n$ *aggregation matrix* $\boldsymbol{C}$, where N is a number of observations in a coarse grid

$$\boldsymbol{C}\boldsymbol{\mu} = \boldsymbol{C}\boldsymbol{X}\boldsymbol{\beta} + \boldsymbol{C}\boldsymbol{\varepsilon}, \quad \boldsymbol{C}\boldsymbol{\varepsilon} \sim Gau_n\left(\boldsymbol{0}, \boldsymbol{C}\boldsymbol{\Omega}\boldsymbol{C}^{\mathrm{T}}\right) \tag{3}$$

The aggregation matrix C consists of 0's and 1's, indicating which cells have to be aligned together. The random variable $\lambda = C\mu$ is treated as the mean process for variables $Z = \{Z_i\}_{i=1}^N$ associated with observations $z = \{z_i\}_{i=1}^N$ of the aggregated model (in a coarse grid)

$$Z|\lambda \sim Gau_N\left(\lambda, \sigma_Z^2 I_N\right)$$

Also at this level, the underlying process λ is related to Z through a measurement error with variance σ_Z^2.

Model parameters β, σ_Z^2, τ^2 and ρ are estimated with the maximum likelihood method based on the joint unconditional distribution of observed random variables Z

$$Z \sim Gau_N\left(CX\beta, \sigma_Z^2 I_N + C\Omega C^T\right) \tag{4}$$

The log likelihood function associated with (4) is formulated, and the analytical derivation is limited to the regression coefficients β; further maximization of the profile log likelihood is performed numerically.

As to the prediction of missing values in a fine grid, the underlying mean process μ is of our primary interest. The predictors optimal in terms of the mean squared error are given by conditional expected value $E(\mu|z)$. The joint distribution of (μ, Z) is

$$\begin{bmatrix} \mu \\ Z \end{bmatrix} \sim Gau_{n+N}\left(\begin{bmatrix} X\beta \\ CX\beta \end{bmatrix}, \begin{bmatrix} \Omega & \Omega C^T \\ C\Omega & \sigma_Z^2 I_N + C\Omega C^T \end{bmatrix}\right) \tag{5}$$

The distribution (5) yields both the predictor $\widehat{E(\mu|z)}$ and its error $\widehat{Var(\mu|z)}$

$$\widehat{E(\mu|z)} = X\widehat{\beta} + \widehat{\Omega}C^T\left(\widehat{\sigma_Z^2} I_N + C\widehat{\Omega}C^T\right)^{-1}\left[z - CX\widehat{\beta}\right]$$

$$\widehat{Var(\mu|z)} = \widehat{\Omega} - \widehat{\Omega}C^T\left(\widehat{\sigma_Z^2} I_N + C\widehat{\Omega}C^T\right)^{-1}C\widehat{\Omega}$$

The standard errors of parameter estimators are calculated with the Fisher information matrix based on the log likelihood function, see [25]. Both the expected and observed Fisher information matrices are developed as two alternative approaches. To this end, respective elements of the second derivative matrix of the log likelihood function are calculated. All required derivatives are evaluated analytically, thus no numerical differentiation is needed. Due to numerical properties, the usage of the expected Fisher information matrix is recommended for accuracy assessment of parameters in the disaggregation model.

In comparison with the regression methods described in the previous section and based solely on proxy data, this approach can be regarded as a kind of generalization since in the regression methods the modeled values are assumed independent in space, i.e. $\mu_i = x_i^T\beta$, and variance σ_Y^2 is a constant value. The above model is particularly useful for areal data, highly correlated in space, and it is not applicable for point sources. In [24] it has been applied for forecasting ammonia (NH_3) emissions from fertilization, and clear improvement has been obtained for the CAR

Data – 5km   10km   15km

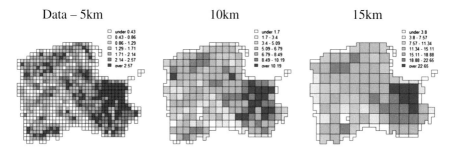

Fig. 1 Ammonia emissions: inventory data in 5 km grid, and aggregated values in 10 and 15 km grids

model over a regression. Figure 1 presents the map of original ammonia data in 5 km grid as well as the values to be disaggregated from 10 to 15 km grids. For the disaggregation from 10 km × 10 km to 5 km × 5 km grids, the mean square error (MSE) was 0.064 (CAR model) and 0.186 (regression). For the disaggregation from 15 km × 15 km to 5 km × 5 km grids, the MSE values were 0.136 and 0.189, respectively. Resulting maps of predicted values for the analysed settings were visually hardly distinguishable (not shown), but respective scatterplots of predicted values versus observations (for disaggregation from 10 km grid) clearly illustrate the difference, see Fig. 2. Introduction of spatial dependence evidently improved accuracy of prediction. In addition, the applied spatial CAR structure considerably limited a number of highly overestimated predictions i.e. the points below the straight line.

Currently, the main limitation of this approach is that it does not account for skewness of the data. The model assumes the Gaussian distribution of observations and usually this is quite far from reality, particularly for the values zero or close to zero in some cells. Overcoming this obstacle in the model proposed is not a straightforward task, since the procedure requires the usage of an aggregation matrix C, which in turn

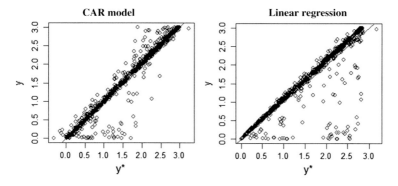

Fig. 2 Predicted (y^*) versus observed (y) values

calls for summation of random variables. The closed skew normal distribution [21], being closed under summation and conditioning, might be potentially considered here.

4 Estimation of Fossil Fuel Emission Changes from $^{14}CO_2$ Observations

Observations of atmospheric concentrations can be helpful in improving disaggregation accuracy. Emissions (or absorptions) of CO_2 fluxes cannot be measured directly in the atmosphere due to the fact that it is impossible to distinguish the sources from which they are released. However, fossil fuel emissions are characterized with a lack of the ^{14}C isotope. The ^{14}C isotope is produced by cosmic radiation in the upper atmosphere, and then it is transported down and absorbed by living organisms. The ^{14}C isotope decays in time of a few hundred years (its half-life equals approximately 5700 years), while the fossil fuels come from organisms which lived million to hundred million years ago. Intensive burning of the fossil fuels dilutes the atmospheric concentration of the ^{14}C isotope [56]. This way (a lack of) ^{14}C isotope may be used as a tracer of fossil fuel originated CO_2 emissions, and the rate of dilution can be used to assess local/regional/global emissions of fossil fuel CO_2.

The ^{14}C isotope has not been the only tracer of CO_2 emissions considered. Also, SF_6 and CO have been investigated [18, 34, 61], but ^{14}C has been found to be the most useful and directly available. Lopez et al. [38] used additional tracers of CO, NO_x, and $^{13}CO_2$, besides that of $^{14}CO_2$, to estimate relative fossil fuel (from liquid and gas combustion) and biospheric fossil fuel (from biofuels, human and plant respiration) CO_2 in Paris, and got good agreement.

Estimation of the fossil fuel CO_2 basically comes from two mass balance equations, for CO_2 and ^{14}C (or $^{14}CO_2$), which are presented in the concentration form (or, more often, in the mixing ratio form; the mixing ratio s is defined as $s = \rho_c / \rho_a$, where ρ_c is a CO_2 density and ρ_a is the air density)

$$CO_2^{obs} = CO_2^{bg} + CO_2^{ff} + CO_2^{bio} + CO_2^{other} \tag{6}$$

$$^{14}C^{obs} = {}^{14}C^{bg} + {}^{14}C^{ff} + {}^{14}C^{bio} + {}^{14}C^{other} \tag{7}$$

where the superscripts stand for, respectively, the observed (^{obs}) mixing ratio, background (^{bg}) mixing ratio—without the local fossil fuel emission, fossil fuel (^{ff}) mixing ratio, biosphere (photosynthesis and heterotrophic respiration) component (^{bio}), and other components, like those coming from burning of biomass, nuclear industry or ocean (^{other}). The ^{14}C isotope is typically measured as a relative difference between the (^{13}C corrected) sample and absolute rate [31, 55]

$$\Delta^{14}C = \frac{\left(\frac{^{14}C}{C}\right)_{obs} - \left(\frac{^{14}C}{C}\right)_{abs}}{\left(\frac{^{14}C}{C}\right)_{abs}} \tag{8}$$

where the absolute ($_{abs}$) value is the absolute radiocarbon standard (1.176×10^{-12} $\text{mol}^{14}C/\text{molC}$), related to oxalic acid activity. Equation (8) is usually expressed in per mill (‰) and written as

$$\Delta^{14}C = \left[\frac{\left(\frac{^{14}C}{C}\right)_{obs}}{\left(\frac{^{14}C}{C}\right)_{abs}} - 1\right] \times 1000[‰] \tag{9}$$

Equation (7) can be easily transformed to the $\Delta^{14}C$ form. Dividing both sides of (7) by $\left(\frac{^{14}C}{C}\right)_{abs}$ and subtracting (6) converted to the equation in C, we get for the left hand side

$$\frac{^{14}C^{obs}}{\left(\frac{^{14}C}{C}\right)_{abs}} - C^{obs} = \left[\frac{\frac{^{14}C^{obs}}{C^{obs}}}{\left(\frac{^{14}C}{C}\right)_{abs}} - 1\right] \times C^{obs} = \left[\frac{\left(\frac{^{14}C}{C}\right)_{obs}}{\left(\frac{^{14}C}{C}\right)_{abs}} - 1\right] \times C^{obs}$$

By similar transformation of the terms on the right hand side, back conversion of C after square brackets to CO_2 and multiplication of both sides by 1000, we obtain the final equation

$$\Delta^{14}C^{obs}CO_2^{obs} = \Delta^{14}C^{bg}CO_2^{bg} + \Delta^{14}C^{ff}CO_2^{ff} + \Delta^{14}C^{bio}CO_2^{bio} + \Delta^{14}C^{other}CO_2^{other} \tag{7'}$$

From (6), the concentration of one component can be calculated and inserted to (7'). The choice of the eliminated component depends in principle on possibility of measuring the values in the equations, and the case considered. For example, [44, 62] eliminate CO_2^{obs}, while [37] eliminate CO_2^{bio}. Having eliminated CO_2^{obs}, the equation for CO_2^{ff} is found as follows

$$CO_2^{ff} = \frac{(\Delta^{14}C^{obs} - \Delta^{14}C^{bg}) \times CO_2^{bg}}{\Delta^{14}C^{ff} - \Delta^{14}C^{obs}} + \frac{(\Delta^{14}C^{obs} - \Delta^{14}C^{bio}) \times CO_2^{bio}}{\Delta^{14}C^{ff} - \Delta^{14}C^{obs}}$$
$$+ \frac{(\Delta^{14}C^{obs} - \Delta^{14}C^{other}) \times CO_2^{other}}{\Delta^{14}C^{ff} - \Delta^{14}C^{obs}}$$

As the concentration (and mixing ratio) of ^{14}C in the fossil fuel CO_2 is equal to 0, then from (9) we have $\Delta^{14}C^{ff} = -1000$. It is often assumed that $CO_2^{other} = 0$, particularly when a site is far from other sources. Yet, for greater areas with small biosphere component, the influence of ocean may be important. Some authors have assumed $\Delta^{14}C^{bio} = \Delta^{14}C^{bg}$ [62] and others $\Delta^{14}C^{bio} = \Delta^{14}C^{obs}$ [32, 51]. Other assumptions may be appropriate for the area considered, as this methodology can be applied to the studies of different scales, ranging from the global ones to small-scale, such as flux tower measurements of CO_2.

Various authors discuss assumptions and assess underlying uncertainties. Turnbull et al. [62] present a systematic discussion and quantify uncertainties using modelling and the above equations. Uncertainty of the $\Delta^{14}C^{obs}$ measurements is very important; at best it amounts about 2‰, which gives an uncertainty in CO_2^{ff} of 0.7 ppm for a single measurement [62]. The uncertainty of the monthly mean of $\Delta^{14}C^{obs}$ may be about ± 1.5 ppm, and annual mean 0.3–0.45 ppm [37].

Another important aspect is the choice of location for measurements of background values. Since the measurements are usually taken for long time periods, the background values are taken from observations at high-altitude sites. In Europe, commonly used background observations come from the High Alpine Research station Jungfraujoch at 3450 m a.s.l. in the Swiss Alps. Other local sites considered in Europe are the Vermunt station in Austria (1800 m a.s.l.) and the Schauinsland in Germany (1205 m a.s.l.). In Poland, there is an observation site at Kasprowy Wierch (1989 m a.s.l.) in the High Tatra Mountains, which can be used as a regional reference station [32]. Turnbull et al. [62] estimate differences of 1–3 ‰ due to the choice of a background site.

Discussion of other kinds of uncertainties, particularly in the global modelling, can be found in [62]. They estimate that the total uncertainty in the calculation of CO_2^{ff} is about 1 ppm for a single observation.

The resolution of CO_2^{ff} determination depends, first of all, on a spatial distribution of $\Delta^{14}C^{ff}$ measurements. The $\Delta^{14}C^{ff}$ measuring observation stations are rather scarce. For instance, there are only 10 measurement sites in Europe [44]. As a consequence, only rough spatial resolution can result from them. Levin & Rödenbeck [37] attempted to answer the question of detecting reduction of fossil fuel CO_2 emissions from measurements done at two sites in Germany: at the earlier mentioned Schauinsland station, and in the Suburbs of Heidelberg in upper Rhine valley. They used the t-test statistic for comparison of 5-year mean $^{14}CO_2$ concentrations in the first and last years of the Kyoto Protocol treaty. To test the difference between the means of 5 year base and 5 year commitment periods, they assumed their statistical independence and the Gaussian distributions, and then applied a statistical t-test. With the 95 % confidence level, they found that the changes of emissions would have to be larger than 36 % to be detectable (with 5 % significance) at Schauinsland, and 10 % to be detectable in Heildelberg, while the real changes were much smaller. Rayner et al. [50] used measurements from 194 stations and a transport model to conclude from the Monte Carlo analysis that this way they can get a reduction of uncertainty on pixel (0.25^0 resolution) level fluxes only about 15 %, but a reduction of the uncertainty on national emission level of 70 %.

Much better spatial resolution can be obtained using measurements in plant materials, like corn leaves [26], rice [53], grape wine ethanol [10, 44], grass [46, 51], tree leaves [36], and tree rings [2, 35, 57]. Most of them allow only for annual estimation, so measurements have to be done for many years to get longer time series. Only wine ethanol and tree rings enable historical records. This way Palstra et al. [44] was able to measure ^{14}C in 165 different wines from 32 different regions in 9 different European countries. The measurements were compared with those obtained from a regional atmospheric transport model, predicting fossil fuel CO_2 with the resolution 55 km × 55 km, with a good compatibility. Riley et al. [51] used measurements from winter annual grasses collected at 128 sites across California, USA, to model transport of fossil fuel CO_2 by using a regional transport model with the resolution 36 km × 36 km.

Above examples show that, at present, the ^{14}C measurements are not readily useful for high resolution disaggregation of emissions. They can provide some constraints on regional values, which can be somehow helpful in checking conformity of the disaggregation results, or as additional source of information. The measurements might be also useful for time disaggregation.

5   Atmospheric Inversion for Estimating CO_2 Fluxes

The main idea of the inversion methods is to use the atmosphere measurements to estimate CO_2 fluxes. For this, the Bayes estimator is generally applied [16, 59]. We begin with the basic linear model, which has been used in the IIASA RAINS project

$$y_{obs} = \mathbf{H}x + \psi \qquad (10)$$

where y_{obs} is an m-vector of the measured atmospheric concentrations (mixing ratios) in the receptor points, in space and time, above the background value, x is an n-vector of fluxes (emissions) from sources in the region considered, and $\mathbf{H}$ is the matrix that relates emissions in sources to the measurements. The elements of the $m \times n$ matrix $\mathbf{H}$ are computed using a transport model. It is assumed that they are constant in the considered time period, which may be a rough approximation. ψ is an m-vector of uncertainties of the relation (10); it is modeled as a random variable with the Gaussian distribution

$$p(\psi) = \left[(2\pi)^m \det \mathbf{C}_y\right]^{-1} \exp\left\{-\frac{1}{2}\psi^T \mathbf{C}_y^{-1} \psi\right\} \qquad (11)$$

The real fluxes are unknown but it is assumed that uncertain information on fluxes x_{prior} is given, so that

$$x = x_{prior} + \vartheta \qquad (12)$$

where again, the uncertainty is modeled as a random vector with the Gaussian distribution, independent on $p(\psi)$,

$$p(\vartheta) = \left[(2\pi)^m \det \mathbf{C}_x\right]^{-1} \exp\left\{-\frac{1}{2}\vartheta^T \mathbf{C}_x^{-1} \vartheta\right\} \qquad (13)$$

We are looking for the conditional probability $p(x|y_{obs})$. From the Bayes theory we have

$$p\left(x|y_{obs}\right) = \frac{p\left(y_{obs}|x\right) p(x)}{p(y_{obs})} \qquad (14)$$

As the Jacobians in the transformations (10) and (12) equal 1, then

$$p\left(y_{obs}|x\right) = \left[(2\pi)^m \det \mathbf{C}_y\right]^{-1} \exp\left\{-\frac{1}{2}\left(y_{obs} - \mathbf{H}x\right)^T \mathbf{C}_y^{-1}\left(y_{obs} - \mathbf{H}x\right)\right\} \qquad (15)$$

$$p\left(x\right) = \left[(2\pi)^m \det \mathbf{C}_x\right]^{-1} \exp\left\{-\frac{1}{2}\left(x - x_{prior}\right)^T \mathbf{C}_x^{-1}\left(x - x_{prior}\right)\right\} \qquad (16)$$

Thus, the conditional probability $p\left(x|y_{obs}\right)$ is proportional to

$$p\left(x|y_{obs}\right) \sim \exp\left\{-\frac{1}{2}\left[\left(y_{obs} - \mathbf{H}x\right)^T \mathbf{C}_y^{-1}\left(y_{obs} - \mathbf{H}x\right) + \left(x - x_{prior}\right)^T \mathbf{C}_x^{-1}\left(x - x_{prior}\right)\right]\right\} \qquad (17)$$

Now, assuming that it is unique, the value $\widehat{x}$ which maximizes the above conditional probability is taken as the estimator. Namely, it is the value which minimizes the following cost function

$$J = \left(y_{obs} - \mathbf{H}x\right)^T \mathbf{C}_y^{-1}\left(y_{obs} - \mathbf{H}x\right) + \left(x - x_{prior}\right)^T \mathbf{C}_x^{-1}\left(x - x_{prior}\right) \qquad (18)$$

The cost function has an appealing interpretation. The final estimate, i.e. a prediction of atmospheric concentrations, should be as close as possible to the prior estimate (the second summand) and agree as closely as possible with the measurements (the first summand). The weights of both subcriteria depend inversely on error estimates of the priors and measurements.

For a nonlinear dependence of y_{obs} on x, the cost function has to be minimized numerically. For the linear case, as the one above, the solution can be found analytically. Since the matrices $\mathbf{C}_y$ and $\mathbf{C}_x$ are symmetric, the derivative of J with respect to x is

$$\frac{1}{2}\frac{dJ}{dx} = -\mathbf{H}^T \mathbf{C}_y^{-1}\left(y_{obs} - \mathbf{H}x\right) + \mathbf{C}_x^{-1}\left(x - x_{prior}\right) \qquad (19)$$

Supposing that the below inverted matrix is nonsingular, the derivative is zero for

$$\widehat{x} = \left(\mathbf{H}^T \mathbf{C}_y^{-1}\mathbf{H} + \mathbf{C}_x^{-1}\right)^{-1}\left(\mathbf{H}^T \mathbf{C}_y^{-1} y_{obs} + \mathbf{C}_x^{-1} x_{prior}\right)$$
$$= \left(\mathbf{H}^T \mathbf{C}_y^{-1}\mathbf{H} + \mathbf{C}_x^{-1}\right)^{-1}\left[\mathbf{H}^T \mathbf{C}_y^{-1}\left(y_{obs} - \mathbf{H}x_{prior}\right) + \left(\mathbf{H}^T \mathbf{C}_y^{-1}\mathbf{H} + \mathbf{C}_x^{-1}\right) x_{prior}\right]$$

which finally gives the Bayes estimator of the fluxes

$$\widehat{x} = x_{\text{prior}} + \left(\mathbf{H}^T \mathbf{C}_y^{-1} \mathbf{H} + \mathbf{C}_x^{-1}\right)^{-1} \mathbf{H}^T \mathbf{C}_y^{-1} \left(y_{\text{obs}} - \mathbf{H} x_{\text{prior}}\right) \qquad (20)$$

If the matrix $\mathbf{H}^T \mathbf{C}_y^{-1} \mathbf{H} + \mathbf{C}_x^{-1}$ is singular, then a singular value decomposition (SVD) can be used.

It can be demonstrated that having inserted $\widehat{x}$ to (17), one gets the Gaussian distribution. Then, as the expression under the exponent in the Gaussian distribution is quadratic, we can find the inverse of the covariance matrix $\widehat{\mathbf{C}}_x$ of the estimator from the second derivative of J. Differentiating (19) gives

$$\frac{1}{2}\frac{d^2 J}{dx^2} = \mathbf{H}^T \mathbf{C}_y^{-1} \mathbf{H} + \mathbf{C}_x^{-1}$$

thus

$$\widehat{\mathbf{C}}_x = \left(\mathbf{H}^T \mathbf{C}_y^{-1} \mathbf{H} + \mathbf{C}_x^{-1}\right)^{-1} = \mathbf{C}_x - \mathbf{C}_x \mathbf{H}^T \left(\mathbf{H} \mathbf{C}_x \mathbf{H}^T + \mathbf{C}_y\right)^{-1} \mathbf{H} \mathbf{C}_x \qquad (21)$$

This matrix allows us to estimate statistical uncertainty of the Bayesian estimator. The most right expression in (21) is more convenient for numerical calculations since only $m \times m$ matrix has to be inverted, while in the middle expression it is necessary to invert $n \times n$ matrices; usually, there is $n \gg m$.

Let us look at interpretation of the expressions (20) and (21). The estimate $\widehat{x}$ is the sum of the prior estimate plus a correction, which depends on the deviation of observations from their predicted values. This correction improves the initial estimate of fluxes (e.g. obtained from disaggregation of the inventory estimates). The expression (21) informs us that the errors of the improved estimates (the values on the diagonal of $\widehat{\mathbf{C}}_x$) are not bigger (and very likely smaller) than the errors of the á priori estimate.

To use the above expressions, one has to know estimates of the covariance matrices $\widehat{\mathbf{C}}_x$ and $\mathbf{C}_y$. This issue is discussed in numerous papers (e.g. [33, 45, 52]). Various methods of finding appropriate values have been proposed; very often diagonal matrices have been used. Exponential decay of covariance values, both in space and/or time, has been found to match the reality better. Michalak et al. [41] develop a maximum likelihood method for estimating the covariance parameters. The likelihood function is formulated and the Cramér–Rao bound is derived. However, the likelihood function is nonlinear, and it has to be minimized numerically.

The idea to use the likelihood function approach has been also used in the so-called geostatistical inverse modelling [20]. In this setting, instead of using prior information, emissions are modelled as linear combinations of trends

$$x = \mathbf{X} b + \xi \qquad (22)$$

where $\mathbf{X}$ is a pre-specified matrix defining the trends, b is a vector of parameters, and ξ is a vector of errors. The parameters and model structure are estimated using observations y_{obs} in the relation

$$y_{\text{obs}} = \mathbf{H}\mathbf{X}b + \zeta \tag{23}$$

where ζ is a linear function of ξ and ψ, see (10). Looking for a linear estimator

$$\hat{z} = \Lambda y_{\text{obs}} \tag{24}$$

and requiring its unbiasedness, the following equation is obtained

$$(\Lambda \mathbf{H} X - X)\,b = 0$$

which is satisfied when

$$\Lambda \mathbf{H} X - X = 0 \tag{25}$$

Assuming the Gaussian distributions, maximization of the likelihood function reduces to minimization of the loss function

$$J_1 = \left(y_{\text{obs}} - \mathbf{H}x\right)^T \mathbf{C}_y^{-1} \left(y_{\text{obs}} - \mathbf{H}x\right) + (x - \mathbf{X}b)^T \mathbf{C}_x^{-1} (x - \mathbf{X}b) \tag{26}$$

over x and b. Following the geostatistical approach, this minimization is subject to a vector equality constraint (25) arising from the requirement of unbiased prediction of the model. Its solution enables finding the vectors $\hat{x}$ and $\hat{b}$ which minimize the problem, as well as the estimates of their covariance matrices. Details are given in [20] and are not discussed here.

More advanced modelling of the fluxes has been proposed in the so-called assimilation data method proposed in Kaminski et al. [30], and then used e.g. in [49]. In this method, a more thorough model of emissions from the biosphere is included. The models depend on unknown parameters which are estimated using the Bayesian approach. Instead of a simple linear dependence of concentrations (mixing ratios) in the receptor points on fluxes, they use an atmospheric transport model. Then, the following cost function is optimized

$$J_2 = \left[y_{\text{obs}} - M(p)\right]^T \mathbf{C}_y^{-1} \left[y_{\text{obs}} - M(p)\right] + \left(p - p_{\text{prior}}\right)^T \mathbf{C}_p^{-1} \left(p - (x - x_{\text{prior}})^T\right.$$
$$\left. \times \mathbf{C}_x^{-1} \left(x - x_{\text{prior}}\right)\right) \tag{27}$$

where $M(\cdot)$ is the model, p is the vector of parameters, and x_{prior} is the vector of prior parameter estimates. To minimize this cost function, it is necessary to use a numerical nonlinear optimization method.

The above expressions have been used mostly in flux inversion studies. Ciais et al. [12] provide various comments on practical applications of this sort of methods. Peylin et al. [45] use them for estimating monthly European CO_2 fluxes and report 60 % reduction of errors. Rivier et al. [52] apply them for estimating monthly fluxes of CO_2 from the biosphere and ocean for the global and European scale. The Bayesian estimate errors are reduced therein by 76% for the western and southern Europe, and by 56% for the central Europe. Lauvaux et al. [33] give inversion results for a 300 km × 300 km region in the South-West of France near Bordeaux with the 8 km × 8 km resolution of CO_2 fluxes, reporting about 50% error reduction. Continuous measurements were taken in two towers, and two aircrafts measuring CO_2 were used. Thompson et al. [60] estimated the N_2O fluxes in the western and central Europe. With only one in-situ measurement point used for inversion, they obtained between 30 and 60% error reduction for Germany.

The idea of atmospheric inversion methods is very general, and it can be used for improving estimates given any additional information in a suitable form. Atmospheric measurements are rather rare in space, so it may be difficult to obtain significant improvement for a very fine spatial grid. Some expectations could be associated with assimilation data method having suitable parameterization. Nevertheless, atmospheric inversion methods are nowadays the most important approaches used to constrain estimates of emission fluxes from the biosphere.

The Bayesian methodology proposed in the atmospheric inversion may be applied for combining very high resolution space estimates of the fossil fuel with information from the nighttime lights. These two sources give independent information, which can complement each other. In this case, the cost function (18) takes a simpler form

$$J = \left(y_{\text{obs}} - x\right)^T C_y^{-1} \left(y_{\text{obs}} - x\right) + \left(x - x_{\text{prior}}\right)^T C_x^{-1} \left(x - x_{\text{prior}}\right) \tag{28}$$

where y_{obs} is the vector of the emission estimates from the nighttime observations and x_{prior} is the vector of estimates from disaggregated inventories. The solution of the minimization problem is

$$\widehat{x} = x_{\text{prior}} + \left(C_y^{-1} + C_x^{-1}\right)^{-1} C_y^{-1} \left(y_{\text{obs}} - x_{\text{prior}}\right) = x_{\text{prior}} + C_x \left(C_y + C_x\right)^{-1} \left(y_{\text{obs}} - x_{\text{prior}}\right) \tag{29}$$

and the estimate of the improved estimate covariance matrix takes the form

$$\widehat{C}_x = \left(C_y^{-1} + C_x^{-1}\right)^{-1} = C_x - C_x \left(C_x + C_y\right)^{-1} C_x \tag{30}$$

Particularly simple computations are obtained for diagonal covariance matrices C_x and C_y. In this case the above formulae read

$$\widehat{x}_i = x_{i,\text{prior}} + \frac{c_{ii,x}}{c_{ii,x} + c_{ii,y}} \left(y_{i,\text{obs}} - x_{i,\text{prior}}\right), \qquad i = 1, \dots, n \tag{31}$$

$$\widehat{c}_{ii,x} = \frac{1}{\frac{1}{c_{ii,x}} + \frac{1}{c_{ii,y}}} = \frac{c_{ii,x} c_{ii,y}}{c_{ii,x} + c_{ii,y}}, \qquad i = 1, \dots, n \tag{32}$$

It is readily seen that $\widehat{c}_{ii,x} \le c_{ii,x}$ and $\widehat{c}_{ii,x} \le c_{ii,y}$. Rayner et al. [50] assess roughly the uncertainty of their disaggregation method as 50%, and of earlier nightlight observation methods as approaching 100 % for high emissions and greater than 100 % for small values. These values give about 30 % reduction of the covariance values. However, the elements of the covariance matrices may considerably vary in space.

6 Flux Tower Observations

Flux towers offer possibility of direct measurements of emission source and sink fluxes, usually those coming from the biosphere. The measurements are taken above the plant canopies, and use the so-called eddy covariance method. The basic idea of the eddy covariance is as follows, see also [9].

Let F_z be the vertical flux of a gas. In a turbulent flow, the flux can be represented as

$$F_z = \overline{v \rho_c} = \overline{\rho_a v s} \tag{33}$$

where v is the vertical wind velocity, ρ_c is the gas (CO_2) density, ρ_a is the air density, and $s = \rho_c / \rho_a$ is the earlier introduced mixing ratio. The bar above the variables means averaging over all flows coming from different turbulences (different eddies). Introducing the mean values of each variable (denoted by bar) and its deviation from the means (denoted with Δ before the variable), the above expression can be written as

$$F_z = \overline{(\bar{\rho}_a + \Delta\rho_a)(\bar{v} + \Delta v)(\bar{s} + \Delta s)}$$
$$= \overline{\bar{\rho}_a \bar{v} \bar{s} + \bar{\rho}_a \bar{v} \Delta s + \bar{\rho}_a \Delta v \bar{s} + \bar{\rho}_a \Delta v \Delta s + \Delta\rho_a \bar{v} \bar{s} + \Delta\rho_a \bar{v} \Delta s + \Delta\rho_a \Delta v \bar{s} + \Delta\rho_a \Delta v \Delta s}$$

But the mean values of deviations equal zero, so we obtain

$$F_z = \overline{\bar{\rho}_s \bar{v} \bar{s} + \bar{\rho}_a \Delta v \Delta s + \Delta\rho_s \bar{v} \Delta s + \Delta\rho_a \Delta v \bar{s} + \Delta\rho_a \Delta v \Delta s}$$

Now, under the assumption that the deviations of the air density are negligible (equal 0), the expression reduces to

$$F_z = \overline{\bar{\rho}_a \bar{v} \bar{s} + \bar{\rho}_a \Delta v \Delta s}$$

Finally, assuming zero average wind velocity $\bar{v}$ we get

$$F_z = \overline{\bar{\rho}_a \Delta v \Delta s} = \bar{\rho}_a \overline{\Delta v \Delta s} \tag{34}$$

This expression is often further simplified by reducing $\bar{\rho}_a$ with the denominator in the definition of s to obtain

$$F_z = \overline{\Delta v \Delta \rho_c} = \overline{(v - \bar{v})(\rho_c - \bar{\rho}_c)} \tag{35}$$

In probability theory, the above expression is the covariance of v and s, from which the name of the method has been coined.

It is further assumed that the stochastic flow processes are ergodic, and then the averaging over the flows is turned over to averaging in time. In practical applications the means are calculated by averaging the values v and ρ_c for half an hour, and covariances are calculated from very frequent measurements of v and ρ_c (with the frequency 10–20 Hz).

Many of the assumptions taken in deriving the formula (34), may be not satisfied in practical measurements. Foken & Wichura [17] discuss the errors connected with them.

The flux tower observations could be a perfect way to provide very high resolution emission fluxes from the biosphere both in space and time provided that a net of flux towers is dense enough. Unfortunately, flux towers are rather scarce. Even over the large area of USA and Canada, only 36 flux tower observations are reported [47]. In Poland there is only one experimental flux tower. Their use can be therefore considered in the future, when more flux towers are constructed. At present, they are used mainly for an assessment of biosphere emission models, see [4] or [47].

7 Conclusions

As mentioned, additional observations can be used for improving disaggregation results, but this is effective rather for larger grids. Out of the presented methods, only a single one provides the results comparable with those obtained for a very fine disaggregation, like that of $2\,\text{km} \times 2\,\text{km}$ for Poland in the GESAPU project [13]—namely, the one based on nighttime lights. For example, ODIAC inventory provides data with the resolution $1\,\text{km} \times 1\,\text{km}$. However, nighttime lights allow only for estimation of energy emissions. To reliably integrate both datasets, the estimates of uncertainties are needed, and precise estimates are missing. Another problem to be solved in such a case is a mismatch of the grids, compare Fig. 3.

Atmospheric tracer measurements have been used for estimation of fossil fuel emissions. Dilution of $^{14}CO_2$ in the atmosphere due to fossil fuel combustion allows for constraining emissions, particularly from bigger areas and/or bigger sources. The approach allows also to get a reduction of uncertainty using a transport model; see e.g. Rayner et al. [50], where the area grid of 0.25^0 resolution was used. However, applying their method for a much higher resolution would bring problems with intensive computation expenses and much smaller improvement. Another issue is how to properly define the areas from which fossil fuel emissions arise. Additionally, the estimates of fluxes obtained from the atmospheric tracer measurements are much more uncertain than those obtained from inventories. On the other hand, the measurements can be regarded as more "objective", while inventories are much more subject to human error or expert biases.

Atmospheric inversion for estimating fluxes is another method that can help to improve inventory results. It has been used mainly for estimation of CO_2 fluxes from

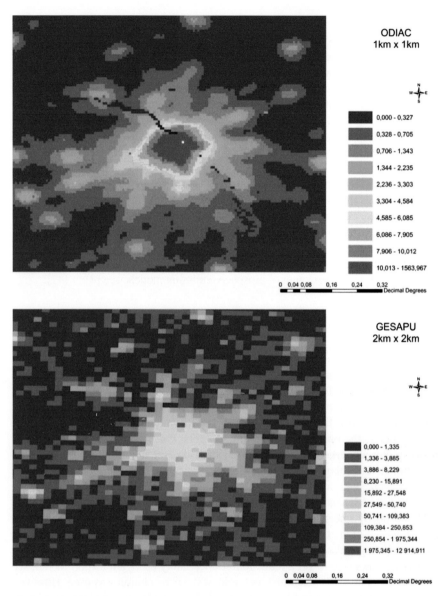

Fig. 3 Fossil fuel CO_2 emissions in Warsaw agglomeration and its surroundings [Gg CO_2]—assessments from two projects in two different grids

biospheric sources and oceans, assuming the fossil fuel emissions as exact, or at least much less uncertain. Acknowledging relatively low uncertainty of biospheric inventory estimates, the inversion methods may introduce useful additional information. They can help to improve inventory assessments and to reduce underlying uncertainty. As before, the problems stem from scarce measurement sites, and uncertainty as to the area from which emissions arise. Due to those reasons, the method was applied to the tasks of rough spatial resolution. Nevertheless, Lauvaux et al. [33] have been able to get the results with the $8\,km \times 8\,km$ resolution, although for a relatively small region ($300\,km \times 300\,km$) and with intensive measurement efforts, including aircrafts. The inverse methods are quite general, and can be used for other greenhouse gases as well. For example, it was used in [60] for estimation of N_2O fluxes. Following this idea, also other gridded information can be incorporated into inventory estimates.

Probably the most promising results for comparison and assessment of inventory emissions for the biosphere fluxes can be obtained from flux tower observations. Unfortunately, at present existing nets of flux towers are scarce. The measurements from a flux tower can be representative for a single cell or, at most, for a few cells of very high resolution. With this respect, flux tower measurements can be more useful for calibration of biosphere emission models applied for inventorying, rather than to significantly improve annual inventories given in a form of high resolution grids. Given very high uncertainty of the biosphere emissions, any additional precise information can introduce valuable knowledge.

Independent sources of spatial data often have mismatched grids; this problem is practically unavoidable when two maps overlay. This incompatibility introduces errors that can form a substantial part of uncertainty resulting from a common processing of maps. As of now, solutions to this problem are still under development. Usually, the approach is simply to partition emissions proportionally to area or to some other proxy variables, like population. Typically, a lot of additional information is available, which can be used for more efficient, and potentially more accurate, allocation of emissions. When this knowledge is given in a non-numerical form, it may be difficult, and in a strict sense even impossible, to apply probability terms and statistical methods. In this context, intelligent computation and fuzzy logic approaches are being developed, see e.g. [63].

Acknowledgments This study was partly conducted within the 7th FP Marie Curie Actions project *Geoinformation technologies, spatio-temporal approaches, and full carbon account for improving accuracy of GHG inventories*, Grant Agreement No. PIRSES-GA-2009-247645. Joanna Horabik-Pyzel was supported by the Foundation for Polish Science under International PhD Projects in Intelligent Computing; project financed from The European Union within the Innovative Economy Operational Programme 2007–2013 and European Regional Development Fund.

References

1. Andres, R.J., Marland, G., Fung, I., Matthews, E.: A $1^0 \times 1^0$ distribution of carbon dioxide emissions from fossil fuel consumption and cement manufacture, 1950–1990. Glob. Biogeochem. Cycles **10**, 419–429 (1996)
2. Babst, F., Alexander, M.R., Szejner, P., Bouriaud, O., Klesse, S., Roden, J., Ciais, P., Poulter, B., Frank, D., Moore, D.J.P., Trouet, V.: A tree-ring perspective on the terrestrial carbon cycle. Oecologia **176**(2), 307–322 (2014). doi:10.1007/s00442-014-3031-6

3. Baker, D.F., Law, R.M., Gurney, K.R., Rayner, P., Peylin, P., Denning, A.S., Bousquet, P., Bruhwiler, L., Vhen, Y.H., Ciais, P., Fung, I.Y., Heimann, M., John, J., Maki, T., Maksyutov, S., Masarie, K., Prather, M., Pak, B., Taguchi, S., Zhu, Z.: TransCom 3 inversion intercomparison: Impact of transport model errors on the interannual variability of regional CO_2 fluxes, 1988–2003. Glob. Biogeochem. Cycles **20**(1), GB10002 (2006). doi:10.1029/2004GB002439. 1988–2003

4. Baldocchi, D., Meyers, T.: On using eco-physiological, micrometeorological and biochemical theory to evaluate carbon dioxide, water vapour and trace gas fluxes over vegetation: a perspective. Agric. For. Meteorol. **90**, 1–25 (1998)

5. Banerjee, S., Carlin, B.P., Gelfand, A.E.: Hierarchical Modeling and Analysis for Spatial Data. Chapman & Hall/CRC, Boca Raton (2004)

6. Battle, M., Bender, M.L., Tans, P.P., White, J.W.C., Ellis, J.T., Conway, T., Francey, R.J.: Global carbon sinks and their variability inferred from atmospheric O_2 and $\delta^{13}C$. Science **287**, 2467–2470 (2000)

7. Boden, T.A., Marland, G., Andres, R.J.: Global, regional, and national fossil-fuel CO_2 emissions. Department of Energy, Oak Ridge, Tennessee, USA, Carbon Dioxide Information Analysis Center, Oak Ridge National Laboratory, U.S (2010). doi:10.3334/CDIAC/00001_V2010

8. BP (2014) Statistical Review of World Energy. London. http://www.bp.com/en/global/corporate/energy-economics/statistical-review-of-world-energy.html

9. Burba, G., Anderson, D.: Introduction to the Eddy Covariance method: general guidelines and conventional workflow. LI-COR Biosciences (2007). http://www.instrumentalia.com.ar/pdf/Invernadero.pdf

10. Burchuladze, A.A., Chudy, M., Eristave, I.V., Pagava, S.V., Povinec, P., Sivo, A., Togonidze, G.I.: Antropogenic ^{14}C variations in atmospheric CO_2 and wines. Radiocarbon **31**(3), 771–776 (1989)

11. Canadell, J.G., Le Quéré, C., Raupach, M.R., Field, C.B., Buitenhuis, E.T., Ciais, P., Conway, T.J., Gillett, N.P., Houghton, R.A., Marland, G.: Contributions to accelerating atmospheric CO_2 growth from economic activity, carbon intensity, and efficiency of natural sinks. Proc. Natl. Acad. Sci. **104**(47), 18866–18870 (2007). doi:10.1073/pnas.0702737104

12. Ciais, P., Rayner, P., Chevallier, F., Bousquet, P., Logan, M., Peylin, P., Ramonet, M.: Atmospheric inversion for estimating CO_2 fluxes: methods and perspectives. Clim. Change **103**(1–2), 69–92 (2010)

13. CORDIS (2015) Geoinformation technologies, spatio-temporal approaches, and full carbon account for improving accuracy of GHG inventories (GESAPU). http://cordis.europa.eu/project/rcn/97282_en.html, http://ec.europa.eu/research/infocentre/article_en.cfm?id=/research/headlines/news/article_14_06_30_en.html?infocentre&item=Industrial%20research&artid=32296&caller=other

14. Doll, C.N.H., Muller, J.P., Elvidge, C.D.: Nighttime imagery as a tool for global mapping of socioeconomic parameters and greenhouse gas emissions. Ambio **29**, 157–162 (2000)

15. Elvidge, C.D., Baugh, K.E., Kihn, E.A., Koehl, H.W., Davis, E.R., Davis, C.W.: Relations between satellite observed visible near-infrared emissions, population, economic activity and power consumption. Remote Sens. **18**, 1373–1379 (1997)

16. Enting, I.G.: Inverse Problems in Atmospheric Constituent Transport. Cambridge University Press, New York (2002)

17. Foken, T., Wichura, B.: Tools for quality assessment of surface-based flux measurements. Agric. For. Meteorol. **78**, 83–105 (1996)

18. Gamnitzer, U., Karstens, U., Kromer, B., Neubert, R.E.M., Meijer, H.A.J., Schroeder, H., Levin, I.: Carbon monoxide: a quantitative tracer for fossil fuel CO_2? J. Geophys. Res. **111**, D22302 (2006). doi:10.1029/2005JD006966

19. Ghosh, T., Elvidge, C.D., Sutton, P.C., Baugh, K.E., Ziskin, D., Tuttle, B.T.: Creating a global grid of distributed fossil fuel CO_2 emissions from nighttime satellite imagery. Energies **3**, 1895–1913 (2010). doi:10.3390/en312

20. Gourdji, S.M., Mueller, K.L., Schaefer, K., Michalak, A.: Global monthly averaged CO_2 fluxes recovered using a geostatical inverse modelling approach: 2. Results including auxiliary environmental data. J. Geophys. Res. **113**, D21115 (2008). doi:10.1029/JD009733

21. Gonzalez-Farias, G., Dominiguez-Molina, A., Gupta, A.K.: Additive properties of skew normal random vectors. J. Stat. Plan. Inference **126**, 512–534 (2004)
22. Gurney, K.R., Rachel, L.M., Denning, A.S., Rayner, P.J., Baker, D., Bousquet, P., Bruhwiler, L., Chen, Y.H., Ciais, P., Fan, S., Fung, I.Y., Gloor, M., Heimann, M., Higuchi, K., John, J., Maki, T., Maksyutov, S., Masarie, K., Peylin, P., Prather, M., Pak, B.C., Randerson, J., Sarmiento, J., Taguchi, S., Takahashi, T., Yuen, C.W.: Towards robust regional estimates of CO_2 sources and sinks using atmospheric transport models. Nature **415**, 626–630 (2002)
23. Gurney, K.R., Mendoza, D.L., Zhou, Y., Fisher, M.L., Miller, C.C., Geethakumar, S., da la Rue du Can, S., : High resolution fossil fuel combustion CO_2 emission fluxes for the United States. Environ. Sci. Technol. **43**(14), 5535–5541 (2009). doi:10.1021/es900806c
24. Horabik, J., Nahorski, Z.: Improving resolution of a spatial air pollution inventory with a statistical inference approach. Clim. Change **124**, 575–589 (2014a)
25. Horabik J., Nahorski Z.: The Cramér-Rao lower bound for the estimated parameters in a spatial disaggregation model for areal data. In: Filev, D., Jabłkowski, J., Kacprzyk, J., Popchev, I., Rutkowski, L., Sgurev, V., Sotirova, E., Szynkarczyk, P., Zadrożny S.: Intelligent Systems 2014, Springer International Publishing, Berlin, pp. 661–668 (2014b)
26. Hsueh, D.Y., Krakauer, N.Y., Randerson, J.T., Xu, X., Trunbore, S.E., Southon, J.R.: Regional patterns of radiocarbon and fossil fuel-derived CO_2 in surface air across North America. Geophys. Res. Lett. **34**, L02816 (2007). doi:10.1029/2006GL027032
27. Huang, Q., Yang, X., Gao, B., Yang, Y., Zhao, Y.: Application of DMSP/OLS nighttime light images: a meta-analysis and a systematic literature review. Remote Sens. **6**, 6844–6866 (2014). doi:10.3390/rs6086844
28. CO_2 Emissions From Fuel Combustion: 1971–2005. International Energy Agency, Paris (2007)
29. IPCC Revised IPCCC 1996 guidelines for national greenhouse gas inventories. Technical Report IPCC/OECD/IEA, Paris (1996). http://www.ipcc-nggip.iges.or.jp/public/gl/invsl.html
30. Kaminski, T., Knorr, W., Rayner, P.J., Heimann, M.: Assimilating atmospheric data into a terrestrial biosphere model: A case study of the seasonal cycle. Global Biogeochem. Cycles **16**(4), 1066 (2002). doi:10.1029/2001GB001463
31. Karlen, I., Olsson, I.U., Kilburg, P., Kilici, S.: Absolute determination of the activity of two ^{14}C dating standards. Arkiv Geophysik **4**, 465–471 (1968)
32. Kuc, T., Rozanski, K., Zimnoch, M., Necki, J., Chmura, L., Jelen, D.: Two decades of regular observations of $^{14}CO_2$ and $^{13}CO_2$ content in atmosphere carbon dioxide in Central Europe: Long-term changes of regional anthropogenic fossil CO_2 emissions. Radiocarbon **49**(2), 807–816 (2007)
33. Lauvaux, T., Uliasz, M., Sarrat, C., Chevallier, F., Bousquet, P., Lac, C., Davis, K.J., Ciais, P., Denning, A.S., Rayner, P.J.: Mesoscale inversion: first results from the CERES campaign with synthetic data. Atmos. Chem. Phys. **8**, 3459–3471 (2008)
34. Levin, I., Karstens, U.: Inferring high-resolution fossil fuel CO_2 records at continental sites from combined $^{14}CO_2$ and CO observations. Tellus. Ser. B **59**, 245–250 (2007)
35. Levin, I., Kromer, B.: Twenty years of atmospheric $^{14}CO_2$ observations at Schauinsland station Germany. Radiocarbon **39**(2), 205–218 (1997)
36. Levin, I., Münnich, K.O., Weiss, W.: The effect of anthropogenic CO_2 and ^{14}C sources on the distribution of ^{14}C sources on the distribution of ^{14}C in the atmosphere. Radiocarbon **22**, 379–391 (1980)
37. Levin, I., Rödenbeck, C.: Can the envisaged reductions of fossil fuel CO_2 emissions be detected by atmospheric observations? Naturwissenshaften **95**, 203–208 (2008). doi:10.1007/s00114-007-0313-4
38. Lopez, M., Schmidt, M., Delmotte, M., Colomb, A., Gros, V., Janssen, C., Lehman, S.J., Mondelain, D., Perrussel, O., Ramonet, M., Xueref-Remy, I., Bousquet, P.: CO, NOx, $^{13}CO_2$ as tracers for fossil fuel CO_2: results from a pilot study in Paris during winter 2010. Atmos. Chem. Phys. **13**, 7343–7358 (2013). doi:10.5194/acp-13-7343-2013
39. Marland, G., Boden, T.A., Andres, R.J. Global, regional, and national fossil fuel CO_2 emissions. In: Trends: A Compendium of Data on Global Change, Carbon Dioxide Information Analysis Center. Oak Ridge National Laboratory, Oak Ridge, Tennessee (2008)

40. Meijer, H.A.J., Smid, H.M., Perez, E., Keizer, M.G.: Isotopic characterisation of anthropogenic CO_2 emissions using isotopic and radiocarbon analysis. Phys. Chem. Earth **21**(5–6), 483–487 (1996)
41. Michalak, A., Hirsch, A., Bruhwiler, L., Gurney, K.R., Peters, W., Tans, P.P.: Maximum likelihood estimation of covariance parameters for Bayesian atmospheric trace gas surface flux inversion. J. Geophys. Res. **110**, D24107 (2005). doi:10.1029/2005JD005970
42. Oda, T., Maksyutov, S.: A very high-resolution (1 km × 1 km) global fossil fuel CO_2 emission inventory derived using a point source database and satellite observations of nighttime lights. Atmos. Chem. Phys. **11**, 543–556 (2011). doi:10.5194/acp-11-543-2011
43. Olivier, J.G.J., Aardenne, J.A.V., Dentener, F.J., Pagliari, V., Ganzeveld, L.N., Peters, J.A.H.W.: Recent trends in global greenhouse gas emissions: Regional trends 1970–2000 and spatial distribution of key sources in 2000. Environ. Sci. **2**(2–3), 81–99 (2005). doi:10.1080/15693430500400345
44. Palstra, S.W., Karstens, U., Streurman, H.J., Meijer, H.A.J.: Wine ^{14}C as a tracer for fossil fuel CO_2 emissions in Europe: measurements and model comparison. J. Geophys. Res. **113**, D21305 (2008). doi:10.1029/2008JD010282
45. Peylin, P., Rayner, P.J., Bousquet, P., Carouge, C., Hourdin, F., Heinrich, P., Ciais, P., AEROCARB contributors, : Daily CO_2 flux estimates over Europe from continuous atmospheric measurements: 1, inverse methodology. Atmos. Chem. Phys. **5**, 3173–3186 (2005)
46. Quarta, G., D'Elia, M., Rizzo, G.A., Calganile, L.: Radiocarbon dilution effects induced by industrial settlements in southern Italy. Nucl. Instrum. Methods Phys. Res. Sec. B **240**, 458–462 (2005)
47. Raczka, B.M., Davis, K.J., Hutzinger, D., Neilson, R.P., Poulter, B., Richardson, A.D., Xiao, J., Baker, I., Ciais, P., Keenan, T.F., Law, B., Post, W.M., Ricciuto, D., Schaefer, K., Tian, H., Tomelleri, E., Verbeeck, H., Viovy, N.: Evaluation of continental carbon cycle simulations with North American flux tower observations. Ecol. Monogr. **83**(4), 531–556 (2013)
48. Raupach, M.R., Marland, G., Ciais, P., Le Quéré, C., Canadell, J.G., Klepper, G., Field, C.B.: Global and regional drivers of accelerating CO_2 emissions. Proc. Natl. Acad. Sci. **104**(24), 10288–10293 (2007). doi:10.1073/pnas.0700609104
49. Rayner, P.J., Scholze, M., Knorr, W., Kaminski, T., Giering, R., Widmann, H.: Two decades of terrestrial carbon fluxes from a carbon cycle data assimilation system (CCDAS). Glob. Biogeochem. Cycles **19**, GB2026 (2005). doi:10.1029/2004GB002254
50. Rayner, P.J., Raupach, M.R., Paget, M., Peylin, P., Koffi, E.: A new global gridded dataset if CO_2 emissions from fossil fuel combustion: methodology and evaluation. J. Geophys. Res. **115**, D19306 (2010). doi:10.1029/2009JD013439
51. Riley, W.G., Hsueh, D.Y., Randerson, J.T., Fischer, M.L., Hatch, J., Pataki, D.E., Wang, W., Goulden, M.L.: Where do fossil fuel carbon dioxide emissions from California go? an analysis based on radiocarbon observations and an atmospheric transport model. J. Geophys. Res. **113**, G04002 (2008). doi:10.1029/2007JG000625
52. Rivier, L., Peylin, P., Ciais, P., Gloor, M., Rödenbeck, C., Geels, C., Karstens, U., Bousquet, P., Brandt, J., Heimann, M., Aerocarb experimentalists, : European CO_2 fluxes from atmospheric inversions using regional and global transport models. Clim. Change **103**(1–2), 93–115 (2010)
53. Shibata, S., Kawano, E., Nakabayashi, T.: Atmospheric $[^{14}C]CO_2$ variations in Japan during 1982–1999 based on ^{14}C measurements in rice grains. Appl. Radiat. Isot. **63**, 285–290 (2005)
54. Stephens, B.B., Gurney, K.R., Tans, P.P., Sweeney, C., Peters, W., Bruhwiler, L., Cias, P., Ramonet, M., Bousquet, P., Nakazawa, T., Aoki, S., Innoue, T.M.G., Vinnichenko, N., Lloyd, J., Jordan, A., Heimann, M., Shibistova, O., Langefelds, R.L., Steele, L.P., Francey, R.J., Denning, A.S.: Weak northern and strong tropical land carbon uptake from vertical profiles of atmospheric CO_2. Science **316**, 1732–1735 (2007)
55. Stuiver, M., Polach, H.A.: Discussion. Reporting of ^{14}C data. Radiocarbon **19**(3), 355–363 (1977)
56. Suess, H.E.: Radiocarbon concentration in modern wood. Science **122**, 415–417 (1955). doi:10.1126/science.122.3166.415-a

57. Tans, P.P., se Jong A.F.M., Mook W.G., : Natural atmospheric ^{14}C variation and the Suess effect. Nature **208**, 826–828 (1979)
58. Tans, P.P., Fung, I.Y., Takahashi, T.: Observational constraints on the global atmospheric CO_2 budget. Science **247**, 1431–1438 (1990)
59. Tarantola, A.: Inverse Problem Theory and Methods for Model Parameter Estimation. SIAM, Philadelphia (2005)
60. Thompson, R.L., Gerbig, C., Rödenbeck, C.: A Bayesian inversion estimate of N_2O emissions for western and central Europe and the assessment of aggregation error. Atmos. Chem. Phys. **11**, 3443–3458 (2011). doi:10.5194/acp-11-3443-2011
61. Turnbull, J.C., Miller, J.B., Lehman, S.J., Tans, P.P., Sparks, R.J.: Comparison of $^{14}CO_2$, CO and SF_6 as tracers for recently added fossil fuel CO_2 in the atmosphere and implications for biological CO_2 exchange. Geophys. Res. Lett. **33**, L01817 (2006). doi:10.1029/2005GL024213
62. Turnbull, J., Rayner, P., Miller, J., Naegler, T., Ciais, P., Cozic, A.: On the use of $^{14}CO_2$ as a tracer for fossil fuel CO_2: quantifying uncertainties using an atmospheric transport model. J. Geophys. Res. **14**, D22302 (2009). doi:10.1029/2009JD012308
63. Verstraete, J.: Solving the map overlay problem with a fuzzy approach. Clim. Change **124**(3), 591–604 (2014)
64. Zondervan, A., Meijer, H.A.J.: Isotopic characterization of CO_2 sources during regional pollution events using isotopic and radiocarbon analysis. Tellus Ser. B **48**, 601–612 (1996)

A New Approach to the Multiaspect Text Categorization by Using the Support Vector Machines

Sławomir Zadrożny, Janusz Kacprzyk and Marek Gajewski

Abstract In our earlier work we introduced the concept of the *multiaspect text categorization* (MTC) task which has its roots in relevant practical problems of managing collections of documents at many, if not all, commercial companies and, above all, public institutions. Specifically, it is a well defined general problem which boils down to the classification of textual documents at two levels: first, to a general category, and—second—to a specific sequence of documents within such a category. While the former task may be dealt with the use of some standard text categorization techniques, the latter one is more challenging due to, first of all, a limited number of training documents. On the other hand, it is assumed that there is some natural logic, for instance, resulting from rules and regulations, behind the succession of documents within the sequences which can be exploited to make a decision as to the assignment of a new document to a proper sequence. We have studied the MCT problem in a number of papers and proposed some solutions to it. Here we propose a new solution which is based on the use of the support vector machines (SVMs) which are known as a very effective technique to solve various classification tasks. We consider the application of SVMs in a specific context, determined by the characteristics of the MTC problem, and by a specific data set used for the experimentation. The use of the SVMs has implied a new, more sophisticated representation of the documents and their sequences which has made it possible to obtain promising results in computational experiments. Moreover, the proposed approach is flexible and may be considerably modified and extended to cover many possible problem versions.

S. Zadrożny (✉) · J. Kacprzyk · M. Gajewski
Systems Research Institute, Polish Academy of Sciences, ul. Newelska 6, 01-447
Warszawa, Poland
e-mail: Slawomir.Zadrozny@ibspan.waw.pl

J. Kacprzyk
e-mail: Janusz.Kacprzyk@ibspan.waw.pl

M. Gajewski
e-mail: Marek.Gajewski@ibspan.waw.pl

© Springer International Publishing Switzerland 2016   261
G. De Tré et al. (eds.), *Challenging Problems and Solutions
in Intelligent Systems*, Studies in Computational Intelligence 634,
DOI 10.1007/978-3-319-30165-5_13

1 Introduction

Manual management of large document collections is a tedious and error-prone task, in most non-trivial cases even impossible due to the sheer size of those collections. Luckily enough, techniques proposed and implemented in the framework of *information retrieval* (IR), which are getting more and more effective and efficient, have made automatically solvable many problems in this area, and their number is constantly increasing. In particular, *text categorization* (TC), c.f. e.g., [17], provides us with tools to automatically classify documents depending on their topics, genre, intended audience, authorship, etc. There are, however, still some complex problems which call for some dedicated solutions since general solutions may be too complicated or even intractable.

Recently, we have defined [22–24] a new, "unorthodox" problem of text categorization, referred to as the *multiaspect text categorization*. It is rooted in some relevant practical problem of document management in a company that is faced by its employees. It basically consists in the need to classify a document, which arrives and has to be handled, according to two interrelated categorization schemes. The initial part of this classification well fits the standard text categorization setting but the second is less liable to classic approaches. Namely, it boils down to the assignment of the document, in scenarios usually occurring in practice, to short sequences of documents, referred to as *cases* and related to some topic, purpose, etc. Thus, on the one hand, we have potentially some extra information which may be used to properly classify documents and which is related to some assumed logic, for instance resulting from some regulations or procedures, that implies the succession of documents within the cases. On the other hand, the training datasets which are needed, for automatic classification algorithms that are in the problem considered usually to be based on the supervised learning, are of a limited size, for obvious reasons. Moreover, in the MTC problem, by the very essence, the number of cases is not fixed in advance, i.e., a document may initiate a new case within a category rather than to be just assigned to one of the existing cases. The list of the categories is, on the other hand, assumed to be fixed. Thus, classic tools and techniques may be difficult to be applied to this kind of conceptually different problem.

In our previous work we proposed a number of methods to deal with the MTC problem, c.f., e.g., [21–24]. In this paper we propose a new solution technique based on the use of the support vector machines (SVMs) which are known for excellent performance in many traditional text categorization problems. For the assignment of a category to a document we use the same technique as in [23], i.e., the k-nearest neighbors classifier. Hence, in this paper we focus on the second stage, i.e. the classification of a document to a case. The direct application of an SVM to a multiclass problem in which each class corresponds to a case (within a given category), and the data to be classified are represented in the space of keywords does not seem to work as in our experiments we have obtained poor results. Thus, we propose a new, more sophisticated representation of the data to be classified. Basically, the idea boils down to the classification of a document in question separately combined with each case candidate. Such a pair is further represented in terms of several measures of similarity between the document and the case. Only the data represented in such a way are processed by the SVM in the usual way.

The paper is organized as follows. In Sect. 2 we briefly remind the formal definition of the multiaspect text categorization problem (MTC). In Sect. 3 we discuss some possible ways to use the support vector machines to solve the problem and we present in detail our new approach. Finally, we discuss some preliminary results of the computational experiments and concluding remarks.

2 Multiaspect Text Categorization: A Brief Problem Statement

We consider a collection of documents D:

$$D = \{d_1, \ldots, d_n\} \tag{1}$$

which belong to some predefined *categories* from the set C:

$$C = \{c_1, \ldots, c_m\} \tag{2}$$

A multiclass single label classification task is considered at the level of categories, i.e., each document $d \in D$ belongs to exactly one class $c \in C$ and the cardinality of C is, in general, greater than 2.

Documents are further organized within particular categories. Each document $d \in D$ belongs additionally to exactly one *case* σ, i.e., a sequence of documents:

$$\sigma_k =< d_{k_1}, \ldots, d_{k_l} > \tag{3}$$

and the set of all cases is denoted as Σ:

$$\Sigma = \{\sigma_1, \ldots, \sigma_p\} \tag{4}$$

Thus, again, at the level of cases we are dealing with the multiclass single label classification problem but this time the number of classes may be growing and, as we noticed earlier, the number of training documents for particular classes (i.e., cases) may be small.

Let us denote as d_* a new document which has to be classified, both to a proper category $c \in C$ and to a case $\sigma \in \Sigma$ within category c. Different strategies may be employed during this classification. First, the document may be classified directly to the case as this implies also the classification to a category. Second, the document may be first classified to a category and then it may be easier to classify it to a proper case, provided the correct category has been assigned. This is the approach adopted in this paper as well as in our previous work [23]. Finally, classifications at both levels may be carried out in parallel and support each other; c.f. our approach proposed in [24].

The classification of a document d_* to a category $c \in C$ seems to be an easier task as the set of categories C is fixed in advance and each category is assumed to be represented by a sufficient number of documents in the collection D. Thus, the standard text categorization techniques may be employed; c.f., e.g., [17].

We have introduced at the conceptual level the problem of the multiaspect text categorization, as meant here, in [21, 22], however not using this name explicitly. It may be considered as a kind of the hierarchical text categorization (c.f., e.g., [5]), however it possesses, as mentioned earlier, a specific characteristic feature due to the combination of the following inherent properties: lack of a fixed list of categories, and a relatively small number of training documents at the lower level of classification. Moreover, it is assumed that the documents follow some logical succession within a case which may be specific for a category to which the case belongs. This may clearly be related to what the particular case represents, deals or is concerned with, etc. This may help in classifying a document to a case if this logic is discovered. However it may also make such a classification more difficult as different cases within the same category may look equally attractive as a proper match for the document as all they follow the same logic as meant above. Obviously, the Topic Detection and Tracking (TDT) [1] problem shares some properties of our MTC problem. The basic task considered in the TDT is the grouping together of news stories concerning the same events/topics. The set of events is not fixed and thus, as in case of the MTC, the classifier has to decide if a new story belongs to an existing topic or starts a new one. Several subproblems are distinguished within the TDT. For example, the problem of deciding if a story starts a new topic or not is referred to as the *first story detection*. We study the relationship between our MTC and the known TDT in more depth in [9, 23]. The reader can also find there a brief review of better known, and more effective and efficient methods to solve the TDT related problems, some of which may be also applicable to some extent to our MTC problem.

3   Solution by Using the Support Vector Machines

In the new approach proposed in this paper, we follow the general concept of a two-level solution technique adopted by us earlier in [23]. In fact, we leave the first step of the method proposed in [23] intact and we focus on the second step, i.e. the classification of a document to a case. Our motivation is the fact that, in general, this is the more difficult of the two steps, and – from an analytical and practical points of view—a more effective and efficient, and more flexible solution, easier adjustable to a specific application—is desired.

The main idea is to employ the support vector machines (SVMs) which are known as a very effective and efficient classification tool. We start, as previously, with the vector space model based standard representation of documents. Namely, the

$tf \times IDF$ weighting scheme [16] is employed. This is accompanied by standard text operations such as stemming or stopword elimination as well as the dimension reduction (we assume that the documents in question are rather short and thus the original document-term matrix is very sparse). Details are described in the next section, reporting the results of computational experiments.

The collection of documents D, introduced in (1), is structured as described earlier, i.e., each document belongs to a category and within a category to a case. Moreover, there are cases which are completed/closed and other which are still on-going.

In the first step, as in [23], we employ the k-nearest neighbors classifier (k-NN) to classify a new document d_* to a category c^*. Thus, no special representation for the categories is adopted—all training documents from D participate in determining the category for a document. The k documents closest to d_*, in the sense of a selected distance measure, are used to decide the category to which d_* is to be assigned. Namely, the category to which the majority from among the k closest documents belong to, is chosen for d_*. Formally, using the notation introduced in (1)–(4), the category c^* assigned to the document d_* is defined as follows:

$$c^* = \arg\max_{c_i} |\{d \in D : \text{Category}(d) = c_i \wedge d \in \mathbf{NN}_k(d_*)\}| \qquad (5)$$

where $\text{Category}(d)$ denotes the category $c \in C$ assigned to a training document d and $\mathbf{NN}_k(d_*)$ denotes the set of k documents $d \in D$ which are the closest to d_*.

The second step, which is the main subject of this paper, consists in the choice for document d_* of one of the on-going cases belonging to a category that has been selected earlier. In this paper we do not consider the counterpart of the first story detection problem.

3.1 Support Vector Machines

The support vector machine (SVM) is a popular supervised learning technique, known to be effective and efficient in many application domains, notably for the classification, regression as well as novelty detection tasks [12, 19]. Here, we will employ them for the classification only. It is assumed that the objects to be classified are represented by vectors in some space endowed with the inner product. The most basic case is that of the binary classification, i.e., where there are only two classes of objects considered and thus each training vector is assigned to one of these classes. Then, the learning process of the SVM boils down to the solution of an optimization problem which yields the parameters of a hyperplane that satisfies two conditions: separates vectors belonging to two classes, and is located as far as possible from the closest training vector (i.e., there is as wide as possible *margin* around the hyperplane not containing any training vectors). If such an optimal hyperplane exists, it is then used to classify new vectors depending on the side of the hyperplane they are located in. In order to decide that, it is enough to compute and combine the inner products of the vector to be classified and the *support vectors* of the computed hyperplane, i.e., those training vectors that are the closest to this hyperplane. Thus, both the learning step and the classification step may be executed effectively and efficiently.

A hyperplane separating training vectors belonging to two classes may, however, not exist. This problem may be addressed in the two ways. Firstly, the hyperplane may be allowed to imperfectly separate the vectors of two classes. However, a better hyperplane is the one for which more training vectors lie on its correct side. Secondly, the vectors may be mapped to a higher dimensional space hoping that there it will be easier to find a separating hyperplane. The mapping is usually not carried out explicitly because what is really needed is the possibility to compute the inner product between the vectors in this higher dimensional space. And for some mappings, the inner product in the higher dimensional space may be computed in the original, lower dimensional space, and is expressed in this space by a so-called *kernel function*.

The concept of the support vector machines may be fairly easy extended to the case of multiple classes. The simplest solution consists in building in the learning step an ensemble of binary SVM classifiers according to either the "one-against-one" or "one-against-all" paradigms and then the execution of all of them in the classification step. Another approach is to form and solve the optimization problem which yields multiple hyperplanes separating vectors of particular classes.

In our works we have considered first a "quick-and-dirty" application of the SVM to solve the problem of the assignment of a document to a case. Namely, in this attempt each case is treated as a separate class and each document is represented as a vector in the space of the keywords. Then, an SVM is constructed in a standard way using one of the possible extensions to the multiclass problem. Our experiments have shown that this approach may be inadequate. Presumably, the reasons are the following. Firstly, there are not enough training documents representing particular classes (i.e., cases) and, secondly, the logic behind the grouping of documents into cases may be quite subtle and is expressed by the order of the documents within cases what may be difficult to discover by an SVM using such a straightforward representation of documents and cases.

Next, we decided to experiment with a different representation of training data which eventually has proven to be effective and efficient enough. It will be described in detail, with its further transformation, in the next section, together with the presentation of the new overall approach we have proposed here. In the beginning, however, we have used it in a different setting which will be briefly presented in what follows.

The original training data set consisting of a set of on-going and completed cases $\Sigma_c = \{\sigma_1, \ldots, \sigma_q\}$, $\Sigma_c \subseteq \Sigma$ belonging to a category $c \in C$ (c.f. Sect. 2) is transformed to a set of pairs: (prefix of case, document) which will be denoted in what follows as (σ, d) since a prefix of a case may be treated as a case on its own. The resulting transformed training data set is meant for a binary (two-class) classification problem. The positive class comprises such pairs (σ_i, d_j) in which the document d_j actually occurs as the continuation of σ_i, i.e., there exists σ_i^+ such that

$$\sigma_i^+ = < d_{i_1}, \ldots, d_{i_f}, d_{i_{f+1}}, \ldots, d_{i_l} > \tag{6}$$

and σ_i is its prefix, i.e.,

$$\sigma_i = < d_{i_1}, \ldots, d_{i_f} > \tag{7}$$

and the document d_j appears in σ_i^+ as document $d_{i_{f+1}}$.

The negative class, on the other hand, comprises such pairs (σ, d) that document d is not a continuation of case σ while σ is the prefix of a case belonging to the training data set and d also belongs to the training data set. Then, the idea is to train the SVM to distinguish the positive examples from the negative ones. Then, during the classification step, the document d to be classified, and first recognized as belonging to the category c, is added to each on-going case belonging to the category c, and all these pairs are classified as positive or negative by the SVM that has been trained earlier (we will provide details and a formal presentation of the whole process in the next section). However, in order to employ the SVM technique we have to turn the pairs (σ, d) into vectors of features. The simplest solution would be to use the concatenation of the vectors representing the particular documents of the case σ and the document d as they are assumed to be represented in the same space of keywords. However, vectors resulting from this concatenation would be of different lengths as the cases occurring therein may be composed of a different number of documents. Thus, we decided to precompute the kernel function which yields the inner product for any two training pairs of (σ, d_j). Having such a kernel function we can train the SVM and then use it for the classification as sketched earlier. We will omit here the details as this approach has not yielded satisfactory results in our preliminary computational experiments. However, it has provided a starting point to our final proposal which is described in the next section, and which has proven to be successful.

3.2   A New Solution Proposed

As mentioned in the previous section, to solve the problem of classification of a document to a case, we use the SVM and a special representation of training data by the pairs (case, document). Briefly stating, the SVM is trained to distinguish the positive examples from the negative examples which is described in Sect. 3.1; c.f. (6) and (7).

We will now formally present the steps executed to classify a document to one of the on-going cases, assuming that it has been earlier classified to a category $c \in C$.

3.2.1   Generating Positive and Negative Examples

The idea of the training data construction is presented in Sect. 3.1—c.f. formulas (6) and (7) and the explanation following them. Here we will describe this construction in a more precise and complete way.

A positive example is constructed for each document, except for the first one, in each of the on-going and completed cases belonging to a given category $c \in C$ in the following way:

$$\sigma_i = <d_{i_1}, \ldots, d_{i_l}> \longmapsto \{(<d_{i_1}, \ldots, d_{i_{k-1}}>, d_{i_k})\} \quad k = 2, 3, \ldots, l \quad \forall \sigma \in \Sigma_c \tag{8}$$

where $\Sigma_c \subseteq \Sigma$ denotes the set of all completed and on-going cases belonging to the category c. Thus, the set of all generated positive examples, denoted E_{pos}, may be expressed as:

$$E_{pos} = \bigcup_{\sigma_i \in \Sigma_c} g_{pos}(\sigma_i)$$

where g_{pos} denotes the mechanism used to generate sets of positive examples for a given case, as described by (8).

The negative examples are generated in a similar way, with the exception that the document in the pair is randomly selected from among all documents belonging to other cases, that is:

$$\sigma_i = <d_{i_1}, \ldots, d_{i_l}> \longmapsto \{(<d_{i_1}, \ldots, d_{i_{k-1}}>, d_{j_m})\} \quad k = 2, 3, \ldots, l \quad \forall \sigma \in \Sigma_c \tag{9}$$

where $\Sigma_c \subseteq \Sigma$, as previously, denotes the set of all completed and on-going cases belonging to the category c and d_{j_m} denotes a randomly selected document belonging to a randomly selected case $\sigma_j \in \Sigma_c$ and $\sigma_j \neq \sigma_i$. The document d_{j_m} is randomly selected each time a negative example is generated. Thus, the set of all generated negative examples, denoted E_{neg}, may be expressed as:

$$E_{neg} = \bigcup_{\sigma_i \in \Sigma_c} g_{neg}(\sigma_i)$$

where g_{neg} denotes the mechanism used to generate sets of negative examples for a given case, as described by (9).

3.2.2 Transforming the Training Data

The sets of positive and negative examples together form the training data set for the binary classifier. However, as argued earlier, it cannot be directly applied for the SVM training. Thus, this set is transformed in the following way. Each pair (case, document) is transformed into an n-dimensional vector in which one of the dimensions corresponds to the binary class of the example: positive or negative. The remaining dimensions correspond to some measures of matching between the document and the case. The very idea of such a transformation is inspired by the feature functions used in the conditional random fields [13] or by the use of so-called re-writing rules in an approach to sentence-rewriting learning proposed in [4]. In both cases, the original data are replaced by their transformations using sets of mappings.

In the latter case this is specifically employed in the context of the kernel function application.

We propose to use the following measures of similarity between a document and (the prefix of) a case:

1. Jaccard index between the document and a representation of the case, both treated as fuzzy sets of keywords,
2. Jaccard index between the document and the last document of the case, both treated as fuzzy sets of keywords,
3. a more sophisticated version of the first measure, defined using the Sugeno integral (c.f., e.g., [10])

The detailed form and the motivation for the use of the particular measures is the following. Let us consider a pair consisting of a case and a document, denoted as (σ, d_*). Let $d_* =< d_*^1, \ldots, d_*^K >$ and $\sigma =< d_1, \ldots, d_k >$ where $d_i =< d_i^1, \ldots, d_i^K >$ and d^j is the $tf \times IDF$ weight of the jth keyword in the representation of a given document, and K is the total number of keywords used for the representation of the whole collection of documents. Then, the Jaccard index for the pair (σ, d^*) is defined as follows. First, the representation of the case σ is defined as

$$d_\sigma =< \max_{i=1,\ldots,k} d_i^1, \max_{i=1,\ldots,k} d_i^2, \ldots, \max_{i=1,\ldots,k} d_i^K > \tag{10}$$

and then:

$$Jaccard1(\sigma, d_*) = \frac{\sum_{j=1}^{K} \min(d_\sigma^j, d_*^j)}{\sum_{j=1}^{K} \max(d_\sigma^j, d_*^j)} \tag{11}$$

Thus, the classic definition of the Jaccard coefficient for two crisp sets A and B, given by $Jaccard(A, B) = \frac{|A \cap B|}{|A \cup B|}$ is here employed but sets A and B are assumed fuzzy and the the so-called $\Sigma Count$ [20] is adopted as the cardinality of a fuzzy set. This measure indicates the similarity of a document to the case if most of the keywords representing d_* are also present in at least one of the documents comprising the case. At the same time, those pairs (σ, d_*) are favored where there are not too many keywords with high weights in d_σ which are absent in d_*. If this measure is high we can be strongly convinced about the similarity of the case and the document.

The second measure, denoted as $Jaccard2$ is defined in a similar way as $Jaccard1$ but this time the similarity is measured between document d_* and the last document of the case. Thus, formally:

$$Jaccard2(\sigma, d_*) = \frac{\sum_{j=1}^{K} \min(d_k^j, d_*^j)}{\sum_{j=1}^{K} \max(d_k^j, d_*^j)}$$

where $\sigma =< d_1, \ldots, d_k >$.

The motivation for using this measure are our earlier results [23] showing that the similarity of neighboring documents within a case may be expected to be high.

The third measure employed may be seen as a modification of the first one (11). One can expect that it is enough to declare the matching between the document and the case if *most* of the *important* keywords present in the representation of the document are also present in the representation of the case. It may be stated in a slightly different way as the requirement that there should exist a set of keywords containing most of the keywords important for the representation of the document such that all of them are also important for the representation of the case. The importance of a keyword for the representation of a document or case is expressed by its weight in this document or in the representation of the case. The degree to which a set of keywords contains most of the keywords which are important for the representation of a document is equated with the truth of the appropriate linguistic proposition [11, 20]. Thus, for each set of keywords one obtains the degree to which it contains most of the keywords important for the representation of the document and the degree to which they are also important for the representation of the case. All these degrees are proposed to be aggregated using the Sugeno integral [10]. Formally, the measure sketched above may be presented as follows.

Let us denote as $T = \{t_1, \ldots, t_K\}$ the set of all keywords (terms) used to represent the whole collection of documents. The importance of a keyword t_j for the representation of the document d_* is equated with the weight of t_j in d_*, i.e., it is equal to d_*^j. Let us define the *importance measure* of a (crisp) set of keywords $A \subseteq T$, denoted as μ_{d_*}, as the truth degree of the linguistic proposition "A contains *most* of the keywords important for the representation of the document d_*":

$$\mu_{d_*}(A) = \frac{\sum_{t_j \in A} d_*^j}{\sum_{t_j \in T} d_*^j} \tag{12}$$

Formula (12) is a special case of the general formula for the truth value of the proposition "*Most x*'s satisfying A satisfy also B" [11, 20]:

$$\text{truth}(QA's \text{ are } B) = \frac{\sum_{x \in U} \min(B(x), A(x))}{\sum_{x \in U} A(x)} \tag{13}$$

where A and B are fuzzy sets defined in the universe U and representing gradual properties denoted by the same symbols; $A(x)$, $B(x)$ denote the values of their membership functions. The set A in (12) is a crisp set of keywords ($A(x) \in \{0, 1\}$) while B is a fuzzy set representing all keywords which are important to represent the document d_*, i.e., $B(t_j) = d_*^j$. Let us note that μ_{d_*} is a *fuzzy measure* on T [10], i.e., if $A_1 \subseteq A_2$ then $\mu_{d_*}(A_1) \leq \mu_{d_*}(A_2)$.

Now we will define a function:

$$o_{\sigma, d_*} : T \longrightarrow [0, 1] \tag{14}$$

which expresses the degree to which the keyword present in the representation of the document d_* is also present in (the representation of) the case σ. The semantics of such a function is the following: if the keyword t_j is not important at all for the representation of the document d_*, i.e., its weight is $d_*^j = 0$, then it does not matter at all if it is present in the representation of the case, i.e., $o(t_j) = 1$, no matter what the weight of t_j in the representation of the case is. On the other hand, if t_j's weight d_*^j is high, then we require t_j to also have a high weight in the representation of the case σ. Such a requirement may be conveniently expressed using a fuzzy implication operator [6, 8], i.e.:

$$o_{\sigma,d_*}(t_j) = d_*^j \rightarrow d_\sigma^j \qquad (15)$$

where d_σ is the representation of the case defined by (10) and "$\rightarrow$" is an operator representing a fuzzy implication, i.e., $\rightarrow: [0, 1] \times [0, 1] \longrightarrow [0, 1]$, "$\rightarrow$" is monotonic in both arguments and, using the infix notation, $\forall y\ 0 \rightarrow y = 1$, $\forall x\ x \rightarrow 1 = 1$, and $1 \rightarrow 0 = 0$. We tested a number of well known implication operators and have finally chosen the Gougen operator [8] defined as

$$x \rightarrow y = \begin{cases} 1 & \text{for } x \le y \\ \frac{y}{x} & \text{otherwise} \end{cases} \qquad (16)$$

So, we have defined how a keyword t_j contributes to the matching between the case σ and the document d_*, which is expressed by $o_{\sigma,d_*}(t_j)$ in (15), and how a set of keywords $A \subseteq T$ covers the set of all keywords important for the representation of the document d_*, which is expressed by $\mu_{d_*}(A)$ in (12). Now, using the Sugeno integral we aggregate these matching degrees to obtain the degree to which there exists a subset of keywords which gathers most of the keywords important for the representation of the document d_* and which are also present in the representation of the case σ. Let us first briefly remind the definition of the *discrete Sugeno integral*, using the notation suitable for the context considered here.

Definition 1 ([10]) Let μ be a fuzzy measure defined on $T = \{t_1, \ldots, t_K\}$. The discrete Sugeno integral of a function $o : T \longrightarrow [0, 1]$, $o(t_j) = h_j$ with respect to a measure μ is the function $S_\mu : [0, 1]^n \longrightarrow [0, 1]$ such that

$$S_\mu(o) = S_\mu(h_1, \ldots, h_n) = \max_{j=1,\ldots,n} \min(h_{\delta(j)}, \mu(T_j)) \qquad (17)$$

where δ is such a permutation of the set $\{1, \ldots, n\}$ that $h_{\sigma(j)}$ is jth lowest element among all the elements h_j, and $T_j = \{t_{\delta(j)}, \ldots, t_{\delta(n)}\}$.

Thus, we define this similarity measure between a case σ and a document d_* as

$$Sugeno(\sigma, d_*) = S_\mu(o_{\sigma,d_*}) \qquad (18)$$

where μ is defined as in (12) and o_{σ,d_*} is defined as in (15).

Summarizing, a pair (σ, d_*) is in fact finally represented in the proposed approach as a triple (extended with the binary class assignment for the training data but we will omit it here for clarity of the notation):

$$(\sigma, d_*) \longmapsto (Jaccard1(\sigma, d_*), Jaccard2(\sigma, d_*), Sugeno(\sigma, d_*)) \qquad (19)$$

and thus the SVM may be directly used to classify such data.

3.2.3 Classification

In order to classify a new document d_* we construct an SVM using the training data set obtained as described in Sects. 3.2.1–3.2.2. As mentioned earlier, such an SVM is constructed for each category separately. While classifying the document it is assumed that its category $c \in C$ is known and the corresponding SVM is used. The classification of a document d_* proceeds as follows

1. the test data set is constructed which comprises a pair (σ, d_*) for each on-going case σ belonging to the category c,
2. each pair is transformed to a triple (19), as described in Sect. 3.2.2,
3. the triples are classified using the SVM and the result is expressed for each triple as a pair of the probabilities: that it is a positive example and that it is a negative example, respectively (which sum up to one, of course),
4. the document d_* is classified to the case corresponding to that triple for which the probability of being a positive example is the highest.

In step 3., basically, the binary classification is carried out. Namely, it is checked on which side of the hyperplane, in a possibly high-dimensional space, the given triples are located. However, it may happen that either none of them or more than one will be classified as positive. In both cases it is not clear to which case the document d_* should be classified. The former case may happen when the document is actually starting a new case, i.e., is a first story (c.f. Sect. 2). However, here we assume that in both cases the document d_* should be classified to an on-going case. Thus, the SVM is assumed to yield the probability of a triple to be positive, basically using Platt's approach [14].

4   Computational Experiments

4.1   The Data Set

No benchmark datasets has been constructed so far for the multiaspect text categorization problem considered in this paper. Thus, in our work we are using a collection of scientific articles (ACL ARC) [3] which are preprocessed to obtain a set of cases (c.f. also [22–24]). The ACL Anthology Reference Corpus (ACL ARC) [3] consists of selected scientific papers on computational linguistics. The papers are available in the XML format with the `Section` elements comprising explicitly distinguished sections. We transform the whole collection of papers in such a way that each paper is turned into a case, documents of which are sections of the paper in exactly the same order as they appear within the paper. We are using a subset of the ACL ARC comprising 113 papers and thus we get 113 cases gathering 1453 documents. First, however, we group the papers in 7 categories using the standard k-means algorithm. This number of categories have been chosen after a series of experiments so as to obtain possibly high number of categories but comprising a reasonable number of cases. This way the categories are also assigned to the particular cases as each case is created from a paper.

Thus, the obtained collection of documents ordered within the cases is split into the training and testing data sets. This is done by randomly choosing some percentage of cases as the on-going cases (in the following we report the results for 50 % of cases chosen as on-going). Then, each of the cases selected as on-going is split into two parts: the documents preceding the split point are the training data while the rest are kept for testing. The whole training data set comprises also all cases which have not been selected as on-going (and thus are treated as completed) and their documents.

All the steps of data preprocessing and documents classification are implemented using the R platform [15] with the help of several packages, including the `tm` package [7] to preprocess the collection of text documents, `FNN` package [2] to cluster the papers and the `kernlab` package [12] to perform classification using its implementation of the SVM (`ksvm` function). The code preparing the data collection and implementing the categorization algorithm takes the form of a few R scripts.

The document collection is prepared in the following steps, as described in our previous papers (c.f. [23]):

1. creation of the corpus of all selected papers using the `tm::Corpus` function; for this purpose the whole papers are reconstructed from text files representing their sections;
2. the corpus created is normalized, i.e., punctuation, numbers and multiple white spaces are removed, stemming is applied, the case is changed to the lower case, stopwords and words shorter than 3 characters are eliminated;
3. a document-term matrix is constructed for the normalized corpus using the `tm::DocumentTermMatrix` function and $tf \times IDF$ terms weighting scheme; sparse keywords, i.e. appearing in less than 10 % of papers are removed from the document-term matrix; finally, the vectors representing papers are normalized in such a way that each coordinate is divided by the norm of the vector;
4. the documents, i.e., the whole papers of the ACL ARC collection, are then clustered using the k-means algorithm implemented in R via the `stats::kmeans` function; this way 7 clusters are obtained which define the categories of particular documents in our collection;

5. another corpus is created, this time comprising all documents, i.e., sections of the papers of the ACL ARC collection; it is normalized as previously, i.e., via the punctuation, numbers and multiple white spaces removal, stemming, changing all characters to the lower case, stopwords and words shorter than 3 characters elimination;
6. a document-term matrix is constructed for the above corpus using $tf \times IDF$ terms weighting scheme; sparse keywords, i.e. appearing in less than 10 % of documents are removed from the document-term matrix resulting in 125 keywords left in the representation of the documents; the vectors representing documents are normalized as in step 3;
7. the set of documents is divided into the training and testing parts as described earlier and that completes the data collection preparation.

4.2 Classification and Results

The approach proposed in Sect. 3 has been tested on the data set described earlier. Several runs of the algorithm have been executed using each time a data set randomly split into the testing and training parts.

For each category of documents a separate SVM has been constructed. The SVM employed has been trained using the following parameters, selected after an extensive testing. The type of the kernel is the `laplacedot`, i.e., the kernel function has the form: $k(x, y) = exp(-\sigma \|x - y\|)$. The parameter σ is automatically estimated by the package. The parameter C, expressing the cost of the soft margin violation, is set to 10. The SVM is executed in the `C-svc` mode, i.e., for classification, with the parameter yielding class probabilities as a part of the output.

For comparison, we have employed the k-NN algorithm to classify documents also to the cases (k-NN is used also to assign a category to a document, as described earlier; c.f. (5)). Therefore, each case is treated as a distinct class and the document d_* is assigned to the case to which there belongs the majority of 10 closest documents to d_* documents in the sense of the Euclidean distance.

In Table 1 we show the summary of the obtained results. The first column shows the fraction of properly assigned categories (this is given for completeness although in this paper we are mostly interested in classifying documents to cases). The second

Table 1 The averaged results of 100 runs of the proposed algorithm, compared to the direct application of the k-NN classifier (first row shows the mean value over the all runs of the fraction of documents properly assigned to its case; second row shows the standard deviation of the fractions)

Category assignment accuracy	Case assignment accuracy (our approach)	Case assignment accuracy (k-NN)
0.5386	0.4846	0.4661
0.0373	0.0684	0.0590

column shows the fraction of the documents properly assigned to their cases using our approach described in Sect. 3. Finally, the third column shows the fraction of the documents properly assigned to their cases using the k-NN technique. All fractions mentioned are averaged over all 100 runs. The second row shows the standard deviation of these fractions over all 100 runs.

It should be noted that only the documents at the cut-off point randomly chosen during the data set split are classified (56 in total in each run). Moreover, while classifying a document to the case it is assumed that it has been properly classified to its current category (via the first application of the k-NN technique).

The results of the experiments show a promising potential of the proposed method. They are comparable to those obtained using the k-NN classifier. However, our approach has much more parameters to set which may be, on the one hand, a bit cumbersome but, on the other hand, makes it possible to adjust the method to the characteristics of a given collection of documents. The proposed technique also provides a framework for finding a good solution for the difficult first story detection problem. For example, the use of the SVM in the mode of the novelty detection [18] may be conceived. Our first attempts in this direction have not succeeded so far but the further work is under way.

5 Concluding Remarks

We have proposed a new approach to the problem of the multiaspect text categorization (MTC) in which textual documents have to be classified along different schemes, referred to as a set of categories and a set of cases. We focused on the solution of the latter classification as it is more difficult to carry out due to, among others, a limited number of training documents. From the technical point of view, the new solution is based on the use of the support vector machines. However, the novelty of the approach consists mainly in the proposal of a new representation of the data to be classified. A two-level representation is proposed which starts with the pairs of cases and documents which exemplify positive and negative matching, and then transforming them to low-dimensional representation with the use of some case/document similarity measures.

We present the results of some computational tests which are encouraging. Further research will concentrate on studying the properties of the proposed representation, selection of parameters of the method and extending the approach to also address the problem of the first story detection.

Acknowledgments This work is supported by the National Science Centre (contract no. UMO-2011/01/B/ST6/06908).

References

1. Allan, J. (ed.): Topic Detection and Tracking: Event-Based Information. Kluwer Academic Publishers, Boston (2002)
2. Beygelzimer, A., Kakadet, S., Langford, J., Arya, S, Mount, D., Li, S.: FNN: fast nearest neighbor search algorithms and applications (2013). http://CRAN.R-project.org/package=FNN. R package version 1.1
3. Bird, S., Dale, R., Dorr, B., Gibson, B., Joseph, M., Kan, M.Y., Lee, D., Powley, B., Radev, D., Tan, Y.: The ACL anthology reference corpus: a reference dataset for bibliographic research in computational linguistics. In: Proceedings of Language Resources and Evaluation Conference (LREC 08), pp. 1755–1759. Marrakesh, Morocco
4. Bu, F., Li, H., Zhu, X.: String re-writing kernel. In: The 50th Annual Meeting of the Association for Computational Linguistics, Proceedings of the Conference, 8–14 July 2012, Jeju Island, Korea - Volume 1: Long Papers, pp. 449–458. The Association for Computer Linguistics (2012)
5. Ceci, M., Malerba, D.: Classifying web documents in a hierarchy of categories: a comprehensive study. J. Intell. Inf. Syst. **28**(1), 37–78 (2007)
6. Dubois, D., Prade, H.: Weighted minimum and maximum operations in fuzzy set theory. Inf. Sci. **39**, 205–210 (1986)
7. Feinerer, I., Hornik, K., Meyer, D.: Text mining infrastructure. R. J. Stat. Softw. **25**(5), 1–54 (2008)
8. Fodor, J., Roubens, M.: Fuzzy Preference Modelling and Multicriteria Decision Support. Series D: System Theory, Knowledge Engineering and Problem Solving. Kluwer Academic Publishers, Boston (1994)
9. Gajewski, M., Kacprzyk, J., Zadrożny, S.: Topic detection and tracking: a focused survey and a new variant. Informatyka Stosowana (to appear)
10. Grabisch, M.: Fuzzy integral as a flexible and interpretable tool of aggregation. In: Bouchon-Meunier, B. (ed.) Aggregation and Fusion of Imperfect Information. Studies in Fuzziness and Soft Computing, pp. 51–72. Physica-Verlag, Heidelberg (1998)
11. Kacprzyk, J., Zadrożny, S.: Power of linguistic data summaries and their protoforms. In: Kahraman, C. (ed.) Computational Intelligence Systems in Industrial Engineering. Atlantis Computational Intelligence Systems, vol. 6, pp. 71–90. Atlantis Press, Amsterdam (2012)
12. Karatzoglou, A., Smola, A., Hornik, K., Zeileis, A.: Kernlab - an S4 package for kernel methods. R. J. Stat. Softw. **11**(9), 1–20 (2004). http://www.jstatsoft.org/v11/i09/
13. Lafferty, J.D., McCallum, A., Pereira, F.C.N.: Conditional random fields: probabilistic models for segmenting and labeling sequence data. In: Brodley, C.E., Danyluk, A.P. (eds.) Proceedings of the Eighteenth International Conference on Machine Learning (ICML 2001), Williams College, Williamstown, 28 June–1 July 2001, pp. 282–289. Morgan Kaufmann (2001)
14. Platt, J.C.: Probabilistic outputs for support vector machines and comparisons to regularized likelihood methods. Advances in Large Margin Classifiers, pp. 61–74. MIT Press, Cambridge (1999)
15. R Core Team: R: A language and environment for statistical computing. R Foundation for Statistical Computing. Vienna, Austria (2014). http://www.R-project.org
16. Salton, G., Buckley, C.: Term-weighting approaches in automatic text retrieval. Inf. Process. Manag. **24**, 513–523 (1988)
17. Sebastiani, F.: Machine learning in automated text categorization. ACM Comput. Surv. **34**(1), 147 (2002)
18. Tax, D.M.J., Duin, R.P.W.: Support vector domain description. Pattern Recognit. Lett. **20**(11–13), 1191–1199 (1999)
19. Vapnik, V.: Statistical Learning Theory. Wiley, New York (1998)
20. Zadeh, L.: A computational approach to fuzzy quantifiers in natural languages. Comput. Math. Appl. **9**, 149–184 (1983)
21. Zadrożny, S., Kacprzyk, J., Gajewski, M., Wysocki, M.: A novel text classification problem and two approaches to its solution. In: Proceedings of the International Congress on Control and Information Processing (ICCIP'13). Cracow University of Technology (2013)

22. Zadrożny, S., Kacprzyk, J., Gajewski, M., Wysocki, M.: A novel text classification problem and its solution. Tech. Trans. Autom. Control **4–AC**, 7–16 (2013)
23. Zadrożny, S., Kacprzyk, J., Gajewski, M.: A new two-stage approach to the multiaspect text categorization. In: IEEE Symposium on Computational Intelligence for Human-Like Intelligence, CIHLI 2015, Cape Town, South Africa, 8–10 December 2015. IEEE (to appear)
24. Zadrożny, S., Kacprzyk, J., Gajewski, M.: A novel approach to sequence-of-documents focused text categorization using the concept of a degree of fuzzy set subsethood. In: Proceedings of the Annual Conference of the North American Fuzzy Information processing Society NAFIPS'2015 and 5th World Conference on Soft Computing 2015, Redmond, 17–19 August 2015

Part V
Intelligent Text and Data Retrieval: Towards a Better Representation of Users Intensions

Content Data Based Schema Matching

Marcin Szymczak, Antoon Bronselaer, Sławomir Zadrożny
and Guy De Tré

Abstract A novel automatic method for detecting corresponding attributes in schemas based on content data is studied. More specifically, our proposed method for the detection of coreferent attributes in schemas is based on a statistical and lexical comparison of content data and detected coreferent tuples across multiple datasets, which increase the possibility of correct schema matching. We will show that knowledge of even a small number of coreferent tuples is sufficient to establish correct matching between corresponding attributes of heterogeneous schemas. The behaviour of the novel schema matching technique has been evaluated on several real life datasets, giving a valuable insight in the influence of the different parameters of our approach on the results obtained.

1 Introduction

The existence of coreferent content data (coreferent tuples, duplicates) which describe the same entity but in a different way across multiple, related databases significantly lowers data quality and should be avoided. However, a small number of coreferent tuples can be useful in the data integration process, which involves importing data from one source to another. Namely, coreferent tuples may be helpful

---

M. Szymczak (✉) · S. Zadrożny
Systems Research Institute, Polish Academy of Sciences, Newelska 6,
01-447 Warsaw, Poland
e-mail: szymczak@ibspan.waw.pl

S. Zadrożny
e-mail: zadrozny@ibspan.waw.pl

M. Szymczak · A. Bronselaer · G. De Tré
Department of Telecommunications and Information Processing, University Ghent,
St-Pietersnieuwstraat 41, 9000 Ghent, Belgium
e-mail: antoon.bronselaer@ugent.be

G. De Tré
e-mail: guy.detre@ugent.be

© Springer International Publishing Switzerland 2016   281
G. De Tré et al. (eds.), *Challenging Problems and Solutions
in Intelligent Systems*, Studies in Computational Intelligence 634,
DOI 10.1007/978-3-319-30165-5_14

Table 1 Example of objects extracted from the source dataset S

Key	Name	Lon.	Lat.	Category	Address
1	Belfry & Cloth Hall	3.724911	51.053653	Tourist attract	Sint-Baafsplein, 9000 Ghent
2	Saint Bavo	3.797826	50.984194	Church	Sint-Baafsplein, 9000 Ghent
3	Cafe-Restaurant De Ster	4.050876	51.281777	Restaurant	Grotestraat 91, 7471 BL Goor
4	Het Kouterhof	3.665122	51.034331	Lodging	Stoopkensstraat 24, 3320 Hoegaarden
5	Borluut B&B	3.657992	51.018882	Lodging	Kleine Gentstraat 69, 9051 St-Denijs-Westrem
6	Gravensteen Hotel	3.719741	51.056485	Hotel	Jan Breydelstraat 35,9000,Ghent
7	Carlton Hotel	3.713951	51.036280	Lodging	Chartreuseweg 20, 8200 Brugge
8	Vlaamse Opera	3.722336	51.049746	Theater	Schouwburgstraat 3, 9000 Ghent

in establishing a true matching between the corresponding attributes of heterogeneous database schemas. This is known as the *schema matching* problem, which is the first step in data integration and is investigated in this paper.

1.1 Problem Illustration

As a motivating example let us consider the schema matching scenario of two datasets of points of interest (POIs) in which corresponding attributes in the source dataset S in Table 1 and the target dataset T in Table 2 have to be aligned as in Fig. 1.[1] The attributes "Key", "Name", "Lat", "Lon" and "Category" in Table 1 have to be matched to the attributes "ID", "POI", "Geo1", "Geo2" and "Type" in Table 2, respectively. It is obvious that matching techniques which are based on the attributes' names are not capable to establish all of these matchings. Semantic matching of corresponding attributes has to be established as coreferent attributes may have different names. Moreover, an attribute "Address" in the left part of Fig. 1 and in Table 1 is decomposed into a number of (sub)attributes: "Street", "City", "ZipCode" in the right part of Fig. 1 and in Table 2. Thus, a one-to-many matching between these attributes is required; namely, a concatenation function has to be applied to solve the *attribute*

---

[1]The order of datasets does not matter, i.e., there exists schema matching between corresponding attributes from the source dataset and the target dataset, and vice versa.

Table 2 Example of objects extracted from the target dataset T

ID	POI	Geo1	Geo2	Type	Street	City	ZipCode
1	Belfort en Lakenhalle	51.054898	3.721675	Bell tower	Emile Braunplein	Gent	9000 BE
2	Sint-Bavokerk	51.054898	3.721675	Church	Sint-Baafsplein	Gent	9000 BE
3	Cafe Theatre	51.049830	3.722015	Restaurant	Schouwburgstraat 5-7	Gent	9000 BE
4	Het Kouterhof	51.034379	3.665140	Hotel	Stoopkensstraat 24	Hoegaarden	3320 BE
5	Borluut Bed Breakfast	51.018938	3.657975	Hotel	Kleine Gentstraat 69	St-Denijs-Westrem	9051 BE
6	Hotel Gravensteen	51.056465	3.719741	Hotel	Jan Breydelstratt 35	Gent	9000 BE

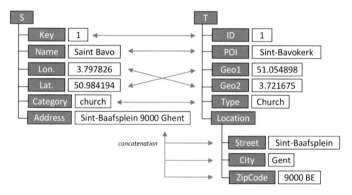

Fig. 1 Example of schema matching. *Arrows* represent matchings of coreferent schema attributes

granularity problem. In general also the *coverage* problem of matched different attributes exists, i.e., coreferent attributes do not necessarily completely have to represent the same information; for example, the attribute "Address" in Fig. 1 and in Table 1 does not contain information about the country. Moreover, due to errors, inaccuracies and lack of standard, coreferent data are not bound to be equal, i.e., the Belfry in Ghent has a different category in the considered tables. It should be clear that detected coreferent tuples do not guarantee perfect schema matching, i.e., the attributes "Name" and "Type" may contain similar values, e.g., *cafe* or *theatre*, which may mislead the matching system. Therefore, all of this makes the finding of coreferent data in schemas using content data a challenging task.

Examples of coreferent tuples are the objects described in the first, second, fourth, fifth and sixth rows in Tables 1 and 2, respectively. They have slightly different names, similar geographic coordinates and different categories and addresses, but they are still describing coreferent objects. These detected coreferent tuple pairs in the considered datasets are used to derive schema matching, known as *horizontal matching*. The same or similar attribute values among coreferent tuple pairs imply coreference of the corresponding attributes of the schemas.

However, detecting coreferent tuple pairs without having knowledge about the correspondences between the attributes of heterogeneous schemas (known as schema alignment) is time-consuming and error prone. It requires the comparison of the values of each attribute from one schema with the values of each attribute from the other schema. Thus, one of the main challenges in the efficient detection of coreferent tuple pairs is the reduction of the set of attributes involved in the comparison to those that may correspond to each other. For this purpose our content data-based approach statistically and lexically compares the attributes' effective domains and selects potentially corresponding (coreferent) attributes which are called *candidate* attributes. This method is known as vertical matching. It significantly decreases the number of comparisons and increases the quality of coreferent tuple pairs detection. Candidate attributes give the first tips of the coreference among attributes, which is confirmed or rejected by detected coreferent tuples or even may be the basis to establish schema matching in case of a lack of coreferent tuples. However, it should be clear that vertical matching is necessary but not sufficient for efficient attribute coreference identification. Thus, our approach is a combination of vertical and horizontal schema matching methods used to establish the matching of corresponding attributes.

Many problems have to be addressed while devising such a schema matching algorithm. To sum up, the most important among them are the following:

- How can content data be useful in schema matching?
- How can one-to-one and one-to-many semantic matching be established between corresponding attributes?
- How can attribute granularity and the coverage matching problems occurrence be recognized?

1.2 Contributions

The objective of this paper is to propose a novel automatic semantical matching method of corresponding (coreferent) attributes in schemas based on data and metadata. More specifically, the detection of coreferent attributes in schemas is based on statistical and lexical analysis of content data and detected coreferent tuples across pairs of datasets, which increases the confidence in schemas matching. In other words, our method is a combination of vertical and horizontal schema matching techniques that applies possibilistic truth values (PTVs) and a kind of cardinality of a set of PTV to express the uncertainty about the matchings. Apart from this, our approach copes with the attribute granularity problem and the information coverage problem.

We will show that even a small number of coreferent tuples is sufficient to establish a correct matching between corresponding attributes of heterogeneous schemas. Such methods can then later be used to improve the coreference detection of data described by schema which are considered as metadata of content data.

1.3 Outline

The remainder of this paper is organized as follows. In Sect. 2, an extensive overview of work related to the topic of this paper is provided. Next, in Sect. 3, some preliminary concepts are introduced that serve as a theoretical foundation of this paper. In Sect. 4.2, an overview of our novel content data-based schema matching algorithm is presented. In Sects. 5 and 6, the details of the algorithm are studied. In Sect. 7 an experimental study of the proposed methods and techniques is reported. Finally, Sect. 8 summarizes the most important contributions of this paper.

2 Related Work

Schema matching can be established by using different methods. Some methods use only data (e.g., duplicates [1, 10]), others use only metadata (e.g., schema information [17, 25–27], knowledge base [24]), whereas other methods use both data and metadata, e.g., [8, 9]. In this paper the content data-based schema matching approach is proposed which uses coferent tuples. There is a large body of work on schema matching which uses content data [21]; for example, LSD [10] extracts information from a training set and consists of a learning and classification phase. More specifically, given a user-supplied mapping between schema elements, the learning step looks at content data to train the classifier, thereby discovering characteristic content data patterns and matching rules. Next, these patterns and rules can be applied to match other schema elements. Moreover, the approach in [18] captures valuable knowledge about the domain of the attribute. This approach uses regular expressions as a formalism to characterize a set of attribute values. Having these expressions, the corresponding attributes are detected by matching the regular expression of one attribute with the value of another attribute using the match function. In many cases it is still not clear which attributes correspond to each other. Thus, regular expressions are a valuable and useful tool but should be supported by other techniques. As opposed to most instance-based solutions which use summary information (e.g., average value) for attribute classification, we derive schema matching from detected coferent tuples in the datasets. One schema matching approach using duplicates is ILA [19], which is a domain-independent program that learns the meaning of external information by explaining it in terms of internal categories. ILA considers a pair of objects as duplicates if both objects contain at least one attribute value in common and relies on a high extensional overlap (a number of coferent tuples). In our opinion these assumptions are unrealistic.

IMap [8] is based on both schema and instance information as well as on a domain ontology and uses past matchings. The duplicates are identified by the user and only exact matches of attribute values are considered by a matcher. IMap copes with various attribute granularities, as in Chua et al. [6] and Lu et al. [16], but the focus is on numerical and differently scaled data (as opposed to our approach which focuses on textual data). Statistical analysis is employed to data in duplicates which are assumed to be identified by a common ID attribute. This means that at least one attribute is already aligned. The approach of Chua et al. [6] classifies attributes into domain classes (e.g., categorical) and forms attribute groups (sets of attributes from the same relation which may correspond one to another) based on predefined rules. Then, the correspondence scores of pairs of attribute groups are calculated. Finally, attributes are matched based on these scores. The approach of Lu et al. [16], one the other hand, uses correlation analysis techniques (supervised by the user) to identify attributes which are potentially semantically related; secondly, they apply regression analysis to generate the relevant conversion function that allows the attribute values of one database to be transformed into attribute values of the other database.

DUMAS [1], just as in our approach, drops several of the assumptions that were made in the above works: coreferent tuples are automatically detected using unaligned schemas; a few coreferent tuples being sufficient to establish schema matching (low extensional overlap). Moreover, it does not use any external source of information, such as an ontology; it is a content data-based approach. In contrast to our approach, DUMAS does not apply possibility theory and does not combine vertical and horizontal schema matching methods to detect coreferent tuples and establish schema matching.

3 Preliminaries

Before we present the details of elaborated algorithms we will first introduce some relevant basic concepts. We start with the multiset definition. Then, we more formally define the problem of coreference detection, which is the main problem addressed in this paper, and the cardinality of a set of pairs of coreferent objects.

3.1 Necessity Measure

In possibility theory, the certainty concerning the statement that the value of X is in A, denoted $Necessity(X$ is $A)$, is expressed by *the necessity measure* $N_X(A)$, defined with respect to a possibility measure $\Pi_X(A)$ as follows:

$$Necessity(X \text{ is } A) \triangleq N_X(A) \triangleq 1 - \Pi_X(\overline{A}) \qquad (1)$$

with $\overline{A}$ denoting the complement of the fuzzy set A [30].

3.2 Multisets

Within the context of this work, the framework of set theory (which is the basis of the relational model) will not suffice to present our approach. Instead, the more general framework of multisets (also called bags) will be used where necessary and the definitions by Yager [29] are adopted here. A multiset A over a universe U is defined by a function $A : U \rightarrow \mathbb{N}$. For each $u \in U$, $A(u)$ denotes the multiplicity (i.e., the number of occurrences) of u in A. The set of all multisets drawn from a universe U is denoted $\mathcal{M}(U)$. Yager has defined some basic operations on multisets. The j-cut of a multiset A is a regular set, denoted as A_j and is given by $A_j = \{u | u \in U \wedge A(u) \geq j\}$. Counterparts of classical set intersection, union operations and of the notion of subsethood are defined as follows:

$$\forall u \in U : (A \cup B)(u) = \max(A(u), B(u)) \tag{2}$$
$$\forall u \in U : (A \cap B)(u) = \min(A(u), B(u)) \tag{3}$$
$$A \subset B \equiv \forall u \in U : A(u) < B(u) \tag{4}$$
$$A \subseteq B \equiv \forall u \in U : A(u) \leq B(u). \tag{5}$$

The theory of multisets provides also an addition operator and a subtraction operator:

$$\forall u \in U : (A \oplus B)(u) = A(u) + B(u) \tag{6}$$
$$\forall u \in U : (A \ominus B)(u) = \max(A(u) - B(u), 0). \tag{7}$$

The cardinality of a multiset A is calculated as the sum of all multiplicities:

$$|A| = \sum_{u \in U} A(u). \tag{8}$$

Finally, it is said that an element u belongs to the multiset A, denoted as $u \in A$, if $A(u) \geq 1$.

3.3 Object Coreference Detection

Considering a more abstract view of entity representation, denoting the universe of the ith feature of an object by U_i, we can model the universe O of objects by:

$$O = U_1 \times \cdots \times U_n. \tag{9}$$

Two objects $o_1 \in O$ and $o_2 \in O$ are said to be *coreferent* (denoted $o_1 \leftrightarrow o_2$) if and only if they describe the same real world entity.

Two elementary operators play an important role in establishing the coreference of objects: a *comparison operator* working at the level of object features (or metadata features, e.g., tags, paths) and an *aggregation operator* combining the comparison scores obtained for particular features.

Definition 1 (*Comparison operator*) A comparison operator on the universe O is defined by a function $\mathcal{C}$:

$$\mathcal{C} : O^2 \to \mathbb{L} \tag{10}$$

where $(\mathbb{L}, \leq)$ is a totally ordered and bounded lattice.

A comparison operator $\mathcal{C}$ compares (a feature of) two objects o_1 and o_2 and expresses the result of this comparison as a *matching degree*. This matching degree may be interpreted as expressing how certain it is that both objects are coreferent and belongs to a totally ordered and bounded lattice $\mathbb{L}$. In the case of probabilistic methods, $\mathbb{L}$ can be instantiated with the unit interval $[0, 1]$ in order to express the result of comparison as a probability of coreference of the objects. Other practical examples of $\mathbb{L}$ are the set of truth values $\mathbb{B} = \{T, F\}$, where T and F denote full certainty of the match and mismatch, respectively or the set of possibilistic truth values (PTVs) [2].

In our approach we use PTVs to express the confidence (certainty) in the validity of the mappings produced by an algorithm. Hereby, a PTV is a normalized possibility distribution [30] defined over the set of Boolean values $\mathbb{B}$ [20]:

$$\text{PTV} \longmapsto \pi : \mathbb{B} \to [0, 1]$$

A PTV expresses the uncertainty about the Boolean value of a *proposition p*. We will often use the notation $\mu(T)$ and $\mu(F)$ instead of $\pi(T)$ and $\pi(F)$ assuming that the (un)certainty as to the truth of a proposition is expressed as "certainly true", "true or false" etc., represented by appropriate fuzzy sets in $\mathbb{B}$; e.g., respectively, $\mu(T) = 1$ and $\mu(F) = 0$, and $\mu(T) = \mu(F) = 1$, for the previous examples. In the context considered here, the propositions p of interest are of the form:

$$p \equiv o_1 \text{ and } o_2 \text{ are coreferent}$$

where o_1 and o_2 are two objects.

Let P denote a set of all propositions under consideration. Then each $p \in P$ can be associated with a PTV denoted $\tilde{p} = \{(T, \mu_{\tilde{p}}(T)), (F, \mu_{\tilde{p}}(F))\}$, where $\mu_{\tilde{p}}(T)$ represents the possibility that p is true and $\mu_{\tilde{p}}(F)$ denotes the possibility that p is false. In what follows, PTVs are often noted in couple notation as $(\mu_{\tilde{p}}(T), \mu_{\tilde{p}}(F))$. It is assumed that each PTV is normalized, which means that $\max(\mu_{\tilde{p}}(T), \mu_{\tilde{p}}(F)) = 1$. The domain of all possibilistic truth values is denoted $\mathcal{F}(\mathbb{B})$, i.e., is the fuzzy power set of (normalised) fuzzy sets over $\mathbb{B}$.

Let us define the order relation $\geq$ on the set $\mathcal{F}(\mathbb{B})$ by:

$$\tilde{p} \geq \tilde{q} \Longleftrightarrow \text{if}((\mu_{\tilde{p}}(F) \leq \mu_{\tilde{q}}(F)) \text{ and } (\mu_{\tilde{p}}(T) = \mu_{\tilde{q}}(T) = 1)) \text{ or } (\mu_{\tilde{q}}(T) \leq \mu_{\tilde{p}}(T)) \tag{11}$$

Moreover, two thresholds ($threshold_T$ and $threshold_F$) are employed to decide on object coreference. If $\mu_{\tilde{p}}(F)$ is lower than the $threshold_F$, then coreference is declared. If $\mu_{\tilde{p}}(T)$ is lower than the $threshold_T$, then a lack of coreference is declared. Finally, if both of the thresholds are exceeded then the coreference status is declared as being *unknown*.

Comparison of complex objects is usually a two-stage process. First, parts of objects, notably values of their features, are compared using a comparison operator. Thus, we extend our definition of the comparison operator (10) so as to make it applicable also to scalar feature values:

$$\mathcal{C}_i : U_i^2 \to \mathbb{L} \tag{12}$$

In this way a separate comparison operator $\mathcal{C}_i$ can be defined for each feature. Then, the results of those comparisons are aggregated to obtain an overall matching degree reflecting the coreference of the whole objects being compared. Therefore, another elementary operator, an *aggregation operator*, is needed.

Definition 2 (*Aggregation operator*) An aggregation operator on $\mathbb{L}$ is defined by a function $\mathcal{A}$:

$$\mathcal{A} : \mathbb{L}^n \to \mathbb{L} \tag{13}$$

where $(\mathbb{L}, \leq)$ is a totally ordered and bounded lattice.

For more information on aggregation operators the reader is referred to [5]. We assume an aggregation operator $\mathcal{A}$ to be idempotent:

$$\forall l \in \mathbb{L} : \mathcal{A}(l, l, \ldots, l) = l \tag{14}$$

Besides that, we assume that $\mathcal{A}$ is monotone in the following sense:

$$\forall (\boldsymbol{l}, \boldsymbol{l'}) \in \mathbb{L}^n \times \mathbb{L}^n : \boldsymbol{l} \leq \boldsymbol{l'} \Rightarrow \mathcal{A}(\boldsymbol{l}) \leq \mathcal{A}(\boldsymbol{l'}) \tag{15}$$

where the relation $\leq$ is generalized from $\mathbb{L}$ to vectors from $\mathbb{L}^n$ in a point wise way.

Based on the definition of these two elementary operators, a comparison of two objects can be generally written as:

$$\mathcal{C}(o_1, o_2) = \overset{n}{\underset{i=1}{\mathcal{A}}} (\mathcal{C}_i (u_{i1}, u_{i2})) \tag{16}$$

where u_{i1} and u_{i2} denote the value of the ith feature of o_1 and o_2, respectively.

In our approach $\mathbb{L}$ is the space of all PTVs endowed with the relation given in (11). Aggregation of PTVs may be carried out using the *Sugeno integral for possibilistic truth values* as defined in [3]; cf. also [23] for the original, general definition of the Sugeno integral. This integral uses two *fuzzy measures* (γ^T and γ^F) which are defined below. Let us first remind briefly the definition of a *fuzzy measure*.

Definition 3 (*Fuzzy measure*) A *fuzzy measure* on a finite universe U is a set function $\gamma : \mathcal{P}(U) \to [0, 1]$ that satisfies the following properties:

$$\gamma(\emptyset) = 0 \tag{17}$$

$$\gamma(U) = 1 \tag{18}$$

$$A \subseteq B \Rightarrow \gamma(A) \leq \gamma(B) \tag{19}$$

Then, in the context considered here, the measure $\gamma^T(A)$ (resp. $\gamma^F(A)$) provides the assessment of certainty that two complex objects are (not) coreferent, given that the set of (metadata) features A are (not) coreferent. As required by the definition of fuzzy measures, γ^T and γ^F are monotonic and satisfy the boundary conditions of a fuzzy measure.

Definition 4 (*Sugeno integral for PTVs* [3]) Given a set of propositions $P = \{p_1, \ldots, p_n\}$ and a corresponding set of PTVs $\tilde{P} = \{\tilde{p}_1, \ldots, \tilde{p}_n\}$, let γ^T and γ^F be two fuzzy measures defined on P which satisfy the condition:

$$\forall Q \subseteq P : \min(\gamma^T(Q), \gamma^F(\bar{Q})) = 0 \tag{20}$$

where $\bar{Q}$ denotes the complement of Q.

Then the Sugeno integral of $\tilde{P}$ with respect to γ^T and γ^F is defined by:

$$S_{\gamma^{T,F}}(\tilde{P}) : \mathcal{F}(\mathbb{B})^n \to \mathcal{F}(\mathbb{B}) : \tilde{P} \mapsto \tilde{p}, \text{ where} \tag{21}$$

$$\mu_{\tilde{p}}(T) = 1 - \bigvee_{i=1}^{n} N_{\tilde{p}_{(i)}}\left(F\right) \wedge \gamma^F\left(P_{(i)^F}\right) \tag{22}$$

and

$$\mu_{\tilde{p}}(F) = 1 - \bigvee_{i=1}^{n} N_{\tilde{p}_{(i)}}\left(T\right) \wedge \gamma^T\left(P_{(i)^T}\right) \tag{23}$$

where $(\cdot)^T$ (respectively $(\cdot)^F$) is a permutation that orders the elements of $\tilde{P}$ non-increasingly (non-decreasingly), while $P_{(i)^F}$ and $P_{(i)^T}$ are sets of propositions p_j with, respectively, i largest values $\mu_{\tilde{p}_j}(F)$ and i largest values $\mu_{\tilde{p}_j}(T)$.

Remark. The motivation to use PTVs and the Sugeno integral is the following. We would like to show that taking into account similarity and dissimilarity of objects in each step separately may be advantageous. In fact, De Cooman [7] has shown in his formal analysis of PTVs that it is essential that possibilities for true and false can be measured separately. To this aim, the aggregated PTVs indicate both the coreference and the lack of coreference of paths/schemas. The choice for the Sugeno integral is motivated by the ability of the related fuzzy measures to model complex preferences in the regular case, making the Sugeno integral a very powerful and flexible aggregation operator [3]. The research on the aggregation of *bipolar* information (here: for and against the coreference) is not that developed in the literature and the Sugeno integral is a prominent example of an aggregation operator adopted for this setting. Besides that, the experimental results confirm that this is a promising choice. An alternative aggregator can be, e.g., an Ordered Weighted Conjunction (OWC) [2] which is in fact a special case of the Sugeno integral.

3.4 Cardinality of a Set of Pairs of Coreferent Objects

In this paper we will often use (multi)sets of Boolean propositions, certainty of truth of which will be expressed with a PTV associated with each proposition. We will then use the concept of a kind of the cardinality of such a (multi)set which counts those propositions fully certain to be true (i.e., with a PTV $(1, 0)$ assigned) as 1, does not count at all propositions fully certain to be false (i.e., with a PTV $(0, 1)$ assigned), and counts the remaining propositions to some degree belonging to $[0, 1]$ and depending on how their PTVs are close to $(1, 0)$ or $(0, 1)$. In fact, this cardinality is similar to a fuzzy cardinality of fuzzy sets and will be expressed as a possibility distribution on the set of integers. We will denote this cardinality as $\pi_{\mathbb{N}}$ (provided that from the context it will be clear which set of propositions it concerns). We will call it also sometimes as a *fuzzy integer* due to the fact that its possibility distribution is assumed to be a convex function, in the same sense as membership functions of fuzzy numbers are assumed, i.e., every α-cut of this function (interpreted as a membership function of a fuzzy set) is an interval, i.e., contains all integers between the lowest and highest integers belonging to this α-cut.

In [12], a method is proposed to construct such a possibility distribution (fuzzy integer) for a (multi)set of propositions associated with PTVs $\tilde{P}$. In fact, this method may be treated as constructing a possibility distribution which expresses the possibility that an integer k represents the number of true propositions in P.

Definition 5 (*Cardinality of a set of PTV qualified propositions*) Let P be a multiset of independent Boolean propositions and let $\tilde{P}$ be the multiset of corresponding possibilistic truth values, i.e., $\forall p \in P : \tilde{p}$ is the PTV associated with p and expressing the (un)certainty as to the truth of p and let $\tilde{p}_{(i)}$ denote the ith largest possibilistic truth value with respect to the order relation defined by Eq. 11. The quantity of true propositions in P is given by the following possibility distribution on the set of all integers (fuzzy integer):

$$\pi_{\mathbb{N}}(k) = \begin{cases} \mu_{\tilde{p}_{(1)}}(F), & k = 0 \\ \mu_{\tilde{p}_{(k)}}(T), & k = |P| \\ \min\left(\mu_{\tilde{p}_{(k)}}(T), \mu_{\tilde{p}_{(k+1)}}(F)\right), & \text{else.} \end{cases} \tag{24}$$

This definition states that $\pi_{\mathbb{N}}(k)$ is the minimum of the possibility that at least k propositions are true and the possibility that at least $|P| - k$ propositions are false.

Let us define an order relation $\prec_{sup}$ on the set of such possibility distributions (fuzzy integers).

Definition 6 (*Sup-order of fuzzy integers*) For two fuzzy integers, $\tilde{n}$ and $\tilde{m}$, the order relation $\prec_{sup}$ is defined as:

$$\tilde{n} \prec_{sup} \tilde{m} \Leftrightarrow \sup \tilde{n}_\alpha < \sup \tilde{m}_\alpha \tag{25}$$

Hereby, $\tilde{n}_\alpha$ is the α-cut of $\tilde{n}$, which is treated here as a fuzzy set, and α is chosen such that:

$$\alpha = \sup\{x | \sup \tilde{n}_x \neq \sup \tilde{m}_x\} \tag{26}$$

4 Content Data-Based Schema Matching

Before we continue to describe our method for schema matching, first of all we should define the problem more formally.

4.1 Problem Definition

Within the scope of this paper it is assumed that entities from the real world are described as objects (tuples) which are characterised by a number of *attributes* (features). A *schema* $\mathcal{R}$ of a given dataset, which consists of tuples, is identified by a set of attributes A. For each attribute $a \in A$, let $dom(a)$ denote the domain of a (the set of possible values for attribute a) and let $dom'(a)$ denote the subset of $dom(a)$ comprising the values of a that are actually present in the (tuples of the) dataset.

Two datasets are considered. The source dataset over the schema $\mathcal{R}_S$ with the set of attributes $A_S = \{a_1^S, \ldots, a_n^S\}$ is denoted as S, while the target dataset over the schema $\mathcal{R}_T$ with the set of attributes $A_T = \{a_1^T, \ldots, a_m^T\}$ is denoted as T. The one-to-many schema matching is defined as follows.

Definition 7 (*One-to-many schema matching*) A relation M is a schema matching if:

$$M \subseteq 2^{A_S} \times 2^{A_T} \times \tilde{M} \qquad (27)$$

where $M = \{m_i\} = \{(A'_S, A'_T, \tilde{m})\}$, $A'_S \subseteq A_S$, $A'_T \subseteq A_T$ and A'_S or A'_T is a singleton set, $\tilde{M}$ is the set of PTVs and $\tilde{m} \in \tilde{M}$ expresses the certainty degree to which A'_S matches A'_T.

Some additional properties may be associated with each matching $m \in M$. For example, the *local matching cardinality*, denoted $card^l_m$, is the number of matched attributes in m, i.e., $card^l_m = |A'_S| + |A'_T|$ (e.g., $card^l_m$ is equal to 2 for a one-to-one matching). Moreover, particular matchings may be classified to a *type*. The following matching types are distinguished:

- *full matching* (coverage level 1): corresponding attributes have the same meaning and cover completely the same concept, e.g., "Name" and "POI" or "Type" and "Category" in Fig. 1;
- *inclusion matching* (coverage level 0.5): corresponding attributes have partially the same meaning and do not cover completely the same concept, e.g., "Address" in the source S in Fig. 1 represents the address of a POI which consists of a street, house number, city and zip code, and this is a part of the concatenation of the attributes "Street", "City" and "ZipCode" in the target T in Fig. 1, which consists of the same information as address from the source but is extended by a country code. "Street" in the target T in Fig. 1 represents only a part of the address from the source. Thus, two sub-types of matching are considered: the *source is a part of the target* and the *target is a part of the source*, respectively;
- *has a common part matching* (coverage level 0.3): corresponding attributes have partially the same meaning, do not cover completely the same concept and are not an inclusion matching. E.g., the matching between "Address" in the source S and "ZipCode" in the target T in Fig. 1. "Address" represents the address of a POI which consists of a street name, house number, city and zip code without the country code; while "ZipCode" in the target T in Fig. 1 represents the zip code and country code of a POI, thus only the zip code is a common part.
- *unknown* (coverage level 0): if attributes do not match.
- *concatenation*. This is a special case of attribute matching which combines two or more attributes. Combining matching types might result in another matching type. For instance, a combination of two inclusion matchings may give a full matching (of attributes) or an inclusion matching. E.g., let assume inclusion matchings between attributes from the source S and the target T in Fig. 1: "Address" and "Street"; "Address" and "City"; "Address" and "ZipCode". Concatenation of these matchings gives a full matching.

The matching $m \in M$ can be interpreted as a one-to-one matching of corresponding attributes if the cardinalities of A'_T and A'_S are equal to 1 or as a one-to-many matching of corresponding attributes if the cardinalities of A'_T or A'_S are greater than 1.

Furthermore, in the context of a one-to-many schema matching M, we consider a set D of coreferent tuple pairs which is defined as follows.

Definition 8 (*A set D of coreferent tuple pairs*) A set D of coreferent tuple pairs consists of 4-tuples $d = (t^S, t^T, M_V, \tilde{d})$ where t^S and t^T are coreferent tuples from the source and target datasets, respectively, M_V is a set of attributes matchings for which there are coreferent values in both particular tuples t^S and t^T, and $\tilde{d}$ is a PTV representing the (un)certainty that two tuples $t^S \in S$ and $t^T \in T$ are coreferent.

4.2 Algorithm

The novel content data-based schema matching Algorithm 1 creates matchings between corresponding attributes A_S and A_T of the source dataset S and the target dataset T, respectively, using content data. Therefore, the inputs for the algorithm are the source and target datasets (S and T, respectively), and a set of parameters (P_V and P_H for each phase) which are used to establish a schema matching. The objective of our algorithm is to establish as many valid one-to-one or one-to-many schema matchings M for coreferent attributes as possible.

Algorithm 1 SCHEMAMATCHINGALGORITHM

Require: Dataset S, Dataset T, Parameters P_V, Parameters P_H
Ensure: Schema Matching M
1: $M_V \leftarrow$ getVerticalMatchings($\{dom'(a^S)\}_{a^S \in A_S}$, $\{dom'(a^T)\}_{a^T \in A_T}$, P_V)
2: $M \leftarrow$ getHorizontalMatchings(S,T,M_V,P_H)

The Algorithm 1 is composed of two main phases. First, vertical schema matchings are established by the method getVerticalMatchings which compares the domains of particular attributes (line 1 in Algorithm 1, which is further discussed in Sect. 5). Second, the established vertical matchings M_V are used to detect coreferent tuple pairs in the heterogeneous data sources which, in turn, constitute a basis to generate horizontal schema matchings M by using the method getHorizontalMatchings (line 2 in Algorithm 1, which is further discussed in Sect. 6). These steps are described in detail in the following sections.

5 Phase I: Vertical Matching

The first phase of our novel schema matching approach is the generation of one-to-one and one-to-many vertical matchings between corresponding attributes. These matchings are established by Algorithm 2 based on statistical analysis and lexical comparison of attribute domains. Thus the input for the algorithm are the subsets of the domains consisting of these values that actually occur in tuples of respective datasets, $\{dom'(a^S)\}_{a^S \in A_S}$ and $\{dom'(a^T)\}_{a^T \in A_T}$, and also a set P_V of parameters which define the thresholds and submatcher settings and is detailed further on. This phase consists of three steps.

In the first step, "Statistical analysis of content data", the subsets of the attribute domains are statistically compared by the *statistical matcher*, and if particular subsets are coreferent, then the matching between their corresponding attributes is established (lines 2–18 in Algorithm 2); otherwise the attribute domains are lexically compared in the second step called "Overlapping" by the *lexical matcher* (lines 19–25 in Algorithm 2). More specifically, each pair of attributes is processed sequentially by the following techniques. First, the results of the statistical analysis of the subsets of the attribute domains (such as the analysing the average length, average values, called attribute properties) are compared, which is a relatively computationally non-expensive statistical technique. Second, only if coreference between two attributes is not declared then the intersection of the subsets of their domains, which are represented by multisets of terms, is calculated based on the equality relation, i.e., two terms are added to the intersection if they are equal. Thus, two attributes are considered as coreferent if a cardinality of the intersection exceeds threshold. Third, if coreference between the attributes is still not declared, then the subsets of their domains are calculated analogously to the second technique but based on the low-level string comparison technique [2] instead of the equality relation. This is the most computationally expensive method of the three, but it is also the most valuable because non-equal but coreferent terms can be detected. The established matchings are added to the set $M_V^{1:1}$ of the one-to-one schema matchings. Next, in the third step, called "Generalization", from the established one-to-one schema matchings in $M_V^{1:1}$, a one-to-many schema matching ($\in M_V^{1:n}$) is generated (line 28 in Algorithm 2). Finally, the vertical schema matching M_V is composed of the one-to-one schema matchings $M_V^{1:1}$ and the one-to-many schema matchings $M_V^{1:n}$ (line 29 in Algorithm 2). These steps are described in detail in the following subsections.

5.1   Step 1: Statistical Analysis of Content Data

In the first step the attribute domains of each schema are statistically analysed separately using predefined *Data Analysers* $P_V.AN$ (lines 2–8 in Algorithm 2). This returns a set of *properties* for each attribute which are considered as a basis for some heuristics for determining the coreference of attributes. There is a large body of work of such properties and heuristics [13–15, 22]. Thus, we give only some examples of such properties and also give an example of an application. These aspects are subject to further research and outside the scope of the work. The proposed examples of such heuristics are the following:

Algorithm 2 VERTICALMATCHINGALGORITHM

Require: $\{dom'(a^S)\}_{a^S \in A_S}$, $\{dom'(a^T)\}_{a^T \in A_T}$, Parameters P_V
Ensure: Schema Matching M_V
 1: Schema Matching $M_V^{1:1} \leftarrow$ null
 2: Properties $P_S[]$, $P_T[]$
 3: **for all** $a^S \in A_S$ **do**
 4:     $P_S[a^S] \leftarrow$ getProperties($dom'(a^S)$,$P_V.AN$)
 5: **end for**
 6: **for all** $a^T \in A_T$ **do**
 7:     $P_T[a^T] \leftarrow$ getProperties($dom'(a^T)$,$P_V.AN$)
 8: **end for**
 9: **for all** $a^S \in A_S$ **do**
10:     **for all** $a^T \in A_T$ **do**
11:         Matching $m \leftarrow$ null
12:         **if** compareStats($P_S[a^S]$,$P_T[a^T]$) $> P_V.thrStats$ **then**
13:             $m.A'_S \leftarrow a^S$
14:             $m.A'_T \leftarrow a^T$
15:             $m.\pi_{\mathbb{N}} \leftarrow \pi_{\mathbb{N}}^1$
16:             $M_V^{1:1} \leftarrow M_V^{1:1} \cup m$
17:             **continue**
18:         **end if**
19:         $m.\pi_{\mathbb{N}} \leftarrow$ compareDom($dom'(a^S)$,$dom'(a^T)$,P_V)
20:         FuzzyInteger $\pi_{(dom^S,dom^T)}^{thr} \leftarrow$ getThr($P_V.thrOverlap$)
21:         **if** $\pi_{(dom^S,dom^T)}^{thr} \prec_{sup} m.\pi_{\mathbb{N}}$ **then**
22:             $m.A'_S \leftarrow m.A'_S \cup a^S$
23:             $m.A'_T \leftarrow m.A'_T \cup a^T$
24:             $M_V^{1:1} \leftarrow M_V^{1:1} \cup m$
25:         **end if**
26:     **end for**
27: **end for**
28: Schema Matching $M_V^{1:n} \leftarrow$ getGeneralization($M_V^{1:1}$)
29: $M_V \leftarrow M_V^{1:1} \cup M_V^{1:n}$

- average, minimum and maximum length as a number of characters in a value without white spaces (numbers are considered as character strings, e.g., telephone numbers, bank accounts, etc.);
- average, minimum and maximum number of tokens for alphabetic and alphanumerical data types. Each value is tokenised, which results in a set of substrings which are called tokens. In most cases, the tokens are separate words. In our approach, the tokenisation of a value is equivalent to subdividing the value in a multiset of tokens and deleting all white spaces in a value;
- average, minimum and maximum value for numerical data types;
- data type: numerical (values contain only numbers), alphabetic (values contain only letters and special characters), alphanumerical (values contain any characters);

Next, attributes a^S and a^T that have similar properties are considered as potentially coreferent (candidate attributes, line 12 in Algorithm 2) and the established matching m between them is added to the set of matchings $M_V^{1:1}$ (lines 13–16 in Algorithm 2, similarity for all properties is assumed here). The statistical criteria are very strict, thus this matching is assigned a full certainty which is expressed by the fuzzy integer $\pi_{\mathbb{N}}^1$ that $\forall x \in \mathbb{N} \, \pi_{\mathbb{N}}^1(x) = 1$. The basis to decide if properties are similar is the similarity function. We use a simple function that calculates the similarity of properties for particular attributes as a normalised difference of property values. The returned values are within the unit interval [0, 1], where 1 means strong similarity and 0 means a complete lack of similarity. Properties with a similarity above threshold $P_V.thrStats$ are considered to be similar. The similarity function is defined by Eq. (28) and is applied for all properties, except for data type property which is considered similar only if compared data types are the same.

$$\text{simProp}(a^S, a^T) = 1 - \frac{|\text{propVal}(a^S) - \text{propVal}(a^T)|}{|\text{propVal}(a^S)| + |\text{propVal}(a^T)|} \qquad (28)$$

Hereby $a^S \in A_S$ and $a^T \in A_T$, and *propVal* is a method which gets the value of a particular property, e.g., the maximum length of the values for an attribute a^S (or a^T).

Remark. Information from the schema, e.g., maximum value, etc., is not considered because it might be too general and may mislead the matching algorithm. For instance, let assume a database of students with an attribute "Age" of type INTEGER in the range of -2^{31} to $2^{31} - 1$. This statistical analysis of the values of the attribute "Age" which are actually present in the database may return a range of 20–29. This information can be more useful than data type restriction which is defined in the database schema.

Example Let us consider the attributes from the source dataset in Table 1 and the target dataset in Table 2. The calculated properties of the subsets of the attribute domains from these datasets are presented in Tables 3 and 4, respectively. Next, the similarities between these properties are calculated by Eq. 28, e.g., for the attributes $a^S = $ "Lon" and $a^T = $ "Geo2" we obtain:

$$
\begin{aligned}
\text{MinLength:} \quad & \text{simProp}(a^S, a^T) = 1 - \frac{|8-8|}{|8|+|8|} = 1 \\
\text{MaxLength:} \quad & \text{simProp}(a^S, a^T) = 1 - \frac{|8-8|}{|8|+|8|} = 1 \\
\text{AvgLength:} \quad & \text{simProp}(a^S, a^T) = 1 - \frac{|8-8|}{|8|+|8|} = 1 \\
\text{MinValue:} \quad & \text{simProp}(a^S, a^T) = 1 - \frac{|3.657992-3.657975|}{|3.657992|+|3.657975|} = 0.999998 \\
\text{MaxValue:} \quad & \text{simProp}(a^S, a^T) = 1 - \frac{|4.050876-3.722015|}{|4.050876|+|3.722015|} = 0.957691 \\
\text{AvgValue:} \quad & \text{simProp}(a^S, a^T) = 1 - \frac{|3.756594-3.701370|}{|3.756594|+|3.701370|} = 0.992595
\end{aligned}
\qquad (29)
$$

Assuming that the threshold $P_V.thrStats$ is equal to 0.8, these attributes are considered as being coreferent because the similarity of all properties of these attribute domains

Table 3 Properties of attribute domains from the source dataset in Table 1

Property	Name	Lon.	Lat.	Category	Address
Min length	10	8	9	5	27
Max length	23	8	9	15	44
Avg length	14.86	8	9	8	31.36
Min #tokens	2	–	–	1	3
Max #tokens	4	–	–	2	5
Avg #tokens	2.38	–	–	1.13	3.86
Min value	–	3.657992	50.984194	–	–
Max value	–	4.050876	51.281777	–	–
Avg value	–	3.756594	51.064419	–	–
Data type	str	num	num	str	str-num

Num means numerical datatype, *str* means alphabetic data type, and *str-num* means alphanumeric datatype

Table 4 Properties of attribute domains from the target dataset Table 2

Property	POI	Geo1	Geo2	Type	Street	City	ZipCode
Min length	12	9	8	5	15	4	7
Max length	21	9	8	10	20	17	7
Avg length	16.17	9	8	8.83	18.17	7.17	7
Min #tokens	1	–	–	1	1	1	2
Max #tokens	3	–	–	2	3	1	2
Avg #tokens	2.16	–	–	1.17	2.16	1	2
Min value	–	51.018938	3.657975	–	–	–	–
Max value	–	51.056465	3.722015	–	–	–	–
Avg value	–	51.044901	3.701370	–	–	–	–
Data type	str	num	num	str	str-num	str	str-num

Num means numerical datatype, *str* means alphabetic data type, and *str-num* means alphanumeric datatype

exceeds 0.8. The same holds for the attribute pair "Lat" and "Geo1". Thus, these two attribute pairs determine two one-to-one matchings which are added to the set $M_V^{1:1}$ of one-to-one matchings. However, the other attribute pairs are not coreferent based on the statistical information, therefore they are further processed in the next step of our algorithm.

5.2 Step 2: Overlapping

The attributes a^S and a^T, whose statistical properties are not similar enough, are considered in this step. More specifically, a lexical comparison of the subsets $dom'(a^S)$ and $dom'(a^T)$ of the attribute domains is conducted using soft strings comparison. For that purpose, the method compareDom is used which works as follows (lines 19–25 in Algorithm 2).

First, special characters that appear in the values of $dom'(a^S)$ and $dom'(a^T)$, i.e., dash, semicolon, dot, etc., are replaced by a space character, which results in strings of terms separated by space. This way, each attribute is described by a multiset of obtained terms (W_{a^S} and W_{a^T}, respectively). Next, the intersection I of these multisets is calculated according to formula (3). Thus, multiset I contains the common terms of W_{a^S} and W_{a^T}, which are assigned a PTV $(1, 0)$ and are the basis for further checking whether the particular attributes a^S and a^T are coreferent or not. Namely, these associated PTVs multiplied by the term multiplicity form a multiset $\tilde{P}$ which is used to construct a possibility distribution $\pi_\mathbb{N}$ (a fuzzy integer) introduced in Definition 5.

The fuzzy integer $\pi_\mathbb{N}$ of intersection I reflects the possibility that two attribute domains are coreferent. Hence, if $\pi_\mathbb{N}$ is greater than the threshold $\pi^{thr}_{(dom^S, dom^T)}$ with respect to the order relation of Definition 6, then the attributes $a_S \in A_S$ and $a_T \in A_T$ are considered to be potentially coreferent (candidate attributes, line 21 in Algorithm 2), and the established matching m between them is added to the set $M_V^{1:1}$ of matchings (lines 22–24 in Algorithm 2). The threshold $\pi^{thr}_{(dom^S, dom^T)}$ is a fuzzy integer and is dynamically calculated by the method getThr (lines 20 in Algorithm 2). This threshold depends on the particular attribute domains and the predefined parameter $P_V.thrOverlap$, which specifies the percentage of domain terms that overlap. More specifically, $\pi^{thr}_{(dom^S, dom^T)}$ is constructed from n PTVs $(1, 0)$, where n is calculated by the following equation:

$$n = \lfloor \min(|W_{a^S}|, |W_{a^T}|) \times P_V.thrOverlap \rfloor \tag{30}$$

However, if a fuzzy integer $m.\pi_\mathbb{N}$ of matching is not larger than the threshold $\pi^{thr}_{(dom^S, dom^T)}$, then the domains of the considered attribute a^S and a^T are analogously compared again, but the equalness relation, which decides on the coreference of the terms, is replaced by the low-level string comparison method proposed in [2]. This low-level comparison method estimates the possibility that two given terms (strings) are coreferent or not and is based on an approximation of *weak string intersections* which is the set of longest common subsequences. It uses the concept of a moving window to construct the intersection of the two input strings. More specifically, the algorithm starts at the beginnings of both strings of a pair and moves a window over each of them. Each time common characters are detected under the moving windows they are added to the intersection, which is the largest set (in terms of set cardinality) that is a subset of both strings.

For example, consider a pair of strings $s_1 = tracks$ and $s_2 = tracklist$. The construction of the intersection goes then as follows. We start with two one-character wide windows. Initially each window is at the beginning of a respective string and contains a character '*t*'. This character is common so it is added to the intersection and both windows move to their next position. Similarly for '*r*', '*a*', '*c*' and '*k*'. In the next step the windows contain different characters, '*s*' and '*l*', respectively. Thus, the window size is increased by one. This is repeated until the windows contain a common character, here '*s*', or there are no more characters in both of the strings. Next, the common character is added to the intersection, windows are shrunk to one character and moved to the position where the common character was found increased by 1. This construction of the intersection is repeated until the windows reach the ends of strings. Finally, the resulting intersection is '*tracks*'. The non common characters are counted (considered as errors) and decrease possibility that steps are coreferent.

This method marks out four different types of errors during comparison. These are *prefix*, *suffix*, *gap* and *mismatch*. The *prefix* is an error where one of the input strings contains a prefix before the matched substring, for instance a letter '*d*' is a prefix for *dtitle* and *title*. Analogously for *suffix*. The *gap* consists of missing characters in the middle of a string, for instance '*li*' is a gap and '*t*' is a suffix for strings *tracklist* and *tracks*. Finally, a *mismatch* is an unmatched character in both strings. These errors have different importance and influence on the final matching result. Because of that, the importance of each error type is expressed by predefined weights between 0 and 1 and are problem dependent. The higher the weight of an error the lower degree of matching of two strings for which such an error occurs. In our case, where abbreviations are very popular, the crucial error types are *prefix* and *mismatch* so their weights are set to 1, *gap* has a weight 0.3 and *suffix* 0.1 is the minor error.

Our algorithm then compares pairs of strings from the domains of the considered attribute a^S and a^T. It generates PTVs which express the uncertainty about the coreference of the compared strings as described above. The possibility that a proposition p, stating that two strings are coreferent, is true ($\mu_{\tilde{p}}(T)$) and the possibility that p is false ($\mu_{\tilde{p}}(F)$) are calculated by the following equations:

$$\mu_{\tilde{p}}(T) = \frac{possT}{factor} \tag{31}$$

$$\mu_{\tilde{p}}(F) = \frac{possF}{factor} \tag{32}$$

where *possT*, *possF* and *factor* equal:

$$possT = \frac{|intersection|}{\max(s_1.length,\ s_2.length)} \tag{33}$$

$$possF = \sum_{i=0}^{|errors|} (errors_i.size \times w_i) \tag{34}$$

$$factor = \max(possT, possF) \qquad (35)$$

where $|intersection|$ denotes the number of common characters, an $|errors|$ is the number of types of the errors, $errors_i.size$ is the number of the errors of a given type.

On the one hand, $possT$ is the ratio between the number of characters that are found to be common for a pair of strings (cardinality of the intersection) and the length of the longer string. On the other hand, $possF$ is computed as the sum of the product of the number of the errors of a given type (from errors that are found during comparison) and predefined weight w_i of specific error type. Finally, $factor$ is the maximum of $possT$ and $possF$ and is used to normalize both possibilities.

Two terms are considered as coreferent if $\mu_{\tilde{p}}(F)$ of the resulting PTV is lower than $\mu_{\tilde{p}}(F)$ of the predefined threshold $P_V.thr$ (see Sect. 3.3).

This terms comparison method thus takes into account misspellings and abbreviations and, moreover, has a low computational complexity. This technique was chosen due to its efficiency [2]. In the literature a multitude of algorithms for string comparison has been proposed and these may also be employed here. An example of an interesting survey concerning strings in general is [11]. An example of an approach employing fuzzy logic which might also be of interest to the reader is [31].

Example Let us consider the attribute a^S = "Address" from the source dataset in Table 1 and the attribute a^T = "City" from the target dataset in Table 2. Let us assume that these attributes have not been indicated as coreferent based on the analysis carried out in Step 1. After preprocessing the subset $dom'(a^S)$ of the attribute domain, the resulting multiset W_{a^S} contains the terms: "Sint-Baafsplein" (multiplicity 2), "9000" (3), "Ghent" (4), "Hoegaarden" (1), "St-Denijs-Westrem" (1), "Gentstraat" (1), etc. Whereas, the multiset W_{a^T} of the subset $dom'(a^T)$ of the attribute domain consists of the terms: "Gent" (4), "Hoegaarden" (1) and "St-Denijs-Westrem" (1). Next, the intersection I of the multisets W_{a^S} and W_{a^T} is calculated which contains two terms, "Hoegaarden" and "St-Denijs-Westrem", both with a multiplicity equal to 1 (the multiplicity of an element in the intersection of two multisets is the minimum of the multiplicities of that element in both multisets, see Sect. 3.2). Both returns are given an associated PTV $(1, 0)$. Thus, the fuzzy integer of this intersection is constructed from the multiset of PTVs $\{(1,0); (1,0)\}$ by Eq. 24.

Figure 2 (the top left-most graph) shows the multiset of possibilistic truth values, where a *circle* denotes the possibility of T and a *triangle* denotes the possibility of F. The derived possibility distribution $\pi_{\mathbb{N}}$ (the fuzzy integer) is shown below the possibilistic truth values. The middle graph of Fig. 2, in turn, shows the multiset of PTVs which are used to construct the threshold $\pi^{thr}_{(dom^S, dom^T)}$ which depends on the particular attribute domains and the predefined parameter $P_V.thrOverlap$ and is shown in the graph below the multiset of PTVs. Following the specification, the multiset of PTVs which is used to construct the threshold $\pi^{thr}_{(dom^S, dom^T)}$ consists of n PTVs equal to $(1, 0)$, where n is calculated by Eq. 30 with $P_V.thrOverlap = 0.35$ and equals to $n = \lfloor \min(33, 6) \times 0.35 \rfloor = 2$. The fuzzy integer $\pi_{\mathbb{N}}$ is not larger than $\pi^{thr}_{(dom^S, dom^T)}$ (w.r.t. Definition 6), because the 1-cuts of both fuzzy integers have the same supremum equal 2. So, the domains of the considered attributes a^S and a^T are

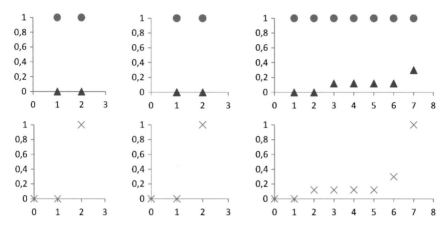

Fig. 2 Fuzzy integers derived from the possibilistic truth values of the attribute matching ("Address"; "City") based on the equality relation, the threshold $\pi^{thr}_{(dom^S,dom^T)}$ and the attribute matching ("Address"; "City") based on the low-level string comparison method

next compared using the low-level string comparison method with $\mu_{\tilde{p}}(F) = 0.5$ as the predefined threshold $P_V.t\tilde{h}r$. That comparison returns an intersection I, which consists of the following elements: ("Gentstraat", "Gent") with an associated PTV (1, 0.3) and multiplicity 1; ("Ghent", "Gent"), (1, 0.12), 4; ("St-Denijs-Westrem", "St-Denijs-Westrem"), (1, 0), 1; and ("Hoegaarden", "Hoegaarden"), (1, 0), 1. Figure 2 (the right-most top graph) shows the multiset of PTVs {(1,0); (1,0); (1,0.12); (1,0.12); (1,0.12); (1,0.3)}, which are used by Eq. 24 to construct a fuzzy integer $\pi_{\mathbb{N}}$ and is shown below the PTVs in Fig. 2. Now, it turns out that, the fuzzy integer $\pi_{\mathbb{N}}$ is larger than the threshold $\pi^{thr}_{(dom^S,dom^T)}$ (w.r.t. Definition 6), because the 1-cut of the right-most fuzzy integer has a higher supremum, equal 7, than the middle fuzzy integer threshold, which has the supremum 2. This is the same fuzzy integer threshold as above because it depends on the same particular attribute domains—we consider the same attributes. Thus, the attributes "Address" and "City" are considered as being potentially coreferent and the established matching m between them is added to the set $M_V^{1:1}$ of matchings.

5.3   Step 3: Generalization

The last step of the vertical matching phase derives a one-to-many schema matching $M_V^{1:n}$ by using the method called getGeneralization based on one-to-one schema matching $M_V^{1:1}$ (line 28 in Algorithm 2). Afterwards, the vertical schema matching M_V is composed of the schema matching $M_V^{1:n}$ and $M_V^{1:1}$ (line 29 in Algorithm 2). The method getGeneralization is implemented by the Algorithm 3 which works as follows.

The input schema matching $M_V^{1:1}$, which is generated in steps 1 and 2 (Sects. 5.1 and 5.2, respectively), is the basis to generate a set $M'_{1:n}$ of one-to-many matchings by combining a number of one-to-one matchings which have the same attribute from A_S or A_T, i.e., a one-to-many matching has either the form: (line 1 in Algorithm 3):

$$\left(A, a_j^T \right), \text{ where } A \subseteq A_S \wedge |A| \geq 2 \wedge \forall a_i^S \in A \ \exists \left(a_i^S, a_j^T \right) \in M_V^{1:1} \qquad (36)$$

or,

$$\left(a_i^S, A \right), \text{ where } A \subseteq A_T \wedge |A| \geq 2 \wedge \forall a_j^T \in A \ \exists \left(a_i^S, a_j^T \right) \in M_V^{1:1} \qquad (37)$$

Afterwards, for each matching $m_{1:n} \in M'_{1:n}$, the *extended* domains are compared by the compareDom method (line 3 in Algorithm 3). This is done analogously as in the "Overlapping" step of Sect. 5.2. An extended domain is constructed by the method getDom and contains concatenated values of all attributes that are specified in the parameters, i.e., of all attributes forming the set A in $m_{1:n}$ (cf. (36) and (37)). The values are concatenated one by one and separated with a white space into a new value which belongs to the extended domain.

Next, *alternative* matchings $M_{alt}^{1:1} \subseteq M_V^{1:1}$ are selected by the method getAlternatives (line 4 in Algorithm 3). An alternative matching $m_{1:1} \in M_{alt}^{1:1}$ for $m_{1:n}$ should have at least one attribute in common with the matching $m_{1:n} \in M'_{1:n}$, i.e., (a_k^S, a_l^T) is an alternative matching with respect to (A, a_j^T) if $a_k^S \in A$ or $a_l^T = a_j^T$, and similarly for (a_i^S, A). Finally, if the fuzzy integer $\pi_{\mathbb{N}}$ of the one-to-many matching $m_{1:n}$ is larger (w.r.t. Definition 6) than the fuzzy integer of any alternative matching $m_{1:1} \in M_{alt}^{1:1}$ (line 6 in Algorithm 3), then the matching $m_{1:n}$ is added to the schema matching $M_V^{1:n}$ (line 7 in Algorithm 3).

Algorithm 3 GENERALIZATIONALGORITHM

Require: Schema Matching $M_V^{1:1}$
Ensure: Schema Matching $M_V^{1:n}$
 1: Schema Matching $M'_{1:n} \leftarrow$ getCombination($M_V^{1:1}$)
 2: **for all** Matching $m_{1:n} \in M'_{1:n}$ **do**
 3:     $m_{1:n}.\pi_{\mathbb{N}} \leftarrow$ compareDom(getDom($m_{1:n}.A'_S$),getDom($m_{1:n}.A'_T$))
 4:     Schema Matching $M_{alt}^{1:1} \leftarrow$ getAlternatives($M_V^{1:1}$,$m_{1:n}$)
 5:     **for all** Matching $m_{1:1} \in M_{alt}^{1:1}$ **do**
 6:         **if** $m_{1:1}.\pi_{\mathbb{N}} \prec_{sup} m_{1:n}.\pi_{\mathbb{N}}$ **then**
 7:             $M_V^{1:n} \leftarrow M_V^{1:n} \cup m_{1:n}$
 8:             **break**
 9:         **end if**
10:     **end for**
11: **end for**

Remark. This generalization is specified for alphanumerical data, where numerical data are considered as character data. For numerical data more sophisticated concatenation method (such as aggregation or transformation function, e.g., a function

Table 5 Example of the vertical schema matching $M_V^{1:1}$

Matching m	Source	Target
1	Name	Type
2	Name	POI
3	Lon.	Geo2
4	Lat.	Geo1
5	Category	Type
6	Key	Id
7	Address	ZipCode
8	Address	Street
9	Address	City

which calculates average value) of values of all attributes that are specified in the matching parameters, i.e., of all attributes forming the set A in $m_{1:n}$ (cf. (36) and (37)), is required and it is out of the scope of this paper.

Example Let us consider the one-to-many matching $m_{1:n} \in M'_{1:n}$ as a combination of the one-to-one matching $M_V^{1:1}$ in Table 5. Namely, $m_{1:n}$ establishes a matching of the attribute $a^S = $ "Address" ($a^S \in A_S$) from the source dataset in Table 1 and the attributes $A'_T = \{$"Street", "City", "ZipCode"$\}$ from the target dataset in Table 2 ($A'_T \subseteq A_T$). This matching is derived as a combination of the one-to-one (candidate, alternative) matchings 7, 8 and 9 of Table 5.

First, the extended domains are constructed by using the method getDom. These domains contain the values of all attributes forming the set A in a one-to-many matching; cf. (36) and (37). The extended domain $dom_{ext}(A'_S) = dom(a^S) = dom($"Address"$)$ contains the values: "Sint-Baafsplein, 9000 Ghent", "Grotestraat 91, 7471 BL Goor", "Jan Breydelstraat 35, 9000, Ghent", etc. The extended domain $dom_{ext}(A'_T) = dom_{ext}(\{$"Street", "City", "ZipCode"$\})$ contains the concatenated values: "Emile Braunplein Gent 9000 BE, "Sint-Baafsplein Gent 9000 BE", "Jan Breydelstraat 35 Gent 9000 BE", etc. Both extended domains $dom_{ext}(A'_S)$ and $dom_{ext}(A'_T)$ are compared by the compareDom method just as in the "Overlapping" step in Sect. 5.2. More specifically, after preprocessing the attribute domains, the multiset $W_{A'_S} = W_{a^S}$ contains the terms: "Sint-Baafsplein" (multiplicity 2), "9000" (3), "Ghent" (4), "Hoegaarden" (1), "St-Denijs-Westrem" (1), "Gentstraat" (1), etc. The multiset $W_{A'_T}$ contains the terms: "BE" (6), "Gent" (4), "9000" (4) "Hoegaarden" (1), "St-Denijs-Westrem" (1), "Sint-Baafsplein" (1), etc. Next, the intersection I of these multisets is determined which contains the terms: "Kleine" with associated PTV $(1, 0)$ and multiplicity 1; "Gentstraat", $(1, 0)$, (1); "24", $(1, 0)$, (1); "5", $(1, 0)$, (1); "69", $(1, 0)$, (1); "3320", $(1, 0)$, (1); "9051", $(1, 0)$, (1); "Schouwburgstraat", $(1, 0)$, (1); "9000", $(1, 0)$, (4); "Stoopkensstraat", $(1, 0)$, (1); "Sint-Baafsplein", $(1, 0)$, (1); "Jan", $(1, 0)$, (1), "St-Denijs-Westrem", $(1, 0)$, (1); "Hoegaarden", $(1, 0)$, (1). All associated PTVs are $(1, 0)$ because all the compared terms are equal. Next, the fuzzy

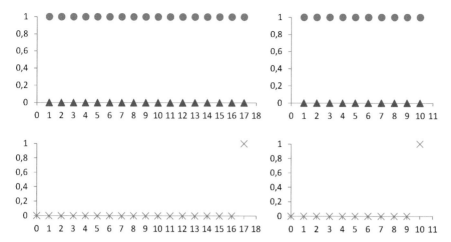

Fig. 3 Fuzzy integers derived from possibilistic truth values of the attribute matching ("Address"; "Street", "City", "ZipCode")

integer expressing the cardinality of the multiset of matching terms is constructed from the resulting multiset of PTVs multiplied by the term multiplicity by Eq. 24.

Figure 3 shows the set of possibilistic truth values, where a *circle* denotes the possibility of T and *triangle* denotes the possibility of F. The derived possibility distribution $\pi_{\mathbb{N}}$ (the fuzzy integer) is shown below the possibilistic truth values.

Next, *alternative* matchings $M_{alt}^{1:1} \subseteq M_V^{1:1}$ are selected by the method getAlternatives, i.e., one-to-one matchings that have at least one attribute in common with the constructed one-to-many matching $m_{1:n}$. Namely, ("Address"; "ZipCode"), ("Address"; "City"), and ("Address"; "Street"). Finally, it turns out that the fuzzy

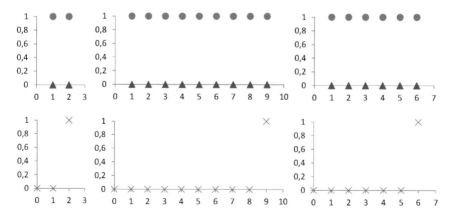

Fig. 4 Fuzzy integers derived from possibilistic truth values of the alternative attribute matchings for the attribute "Address": ("Address"; "City"), ("Address"; "Street") and ("Address"; "ZipCode")

integer $\pi_\mathbb{N}$ related to the one-to-many matching $m_{1:n}$ is larger than the fuzzy integer (w.r.t. Definition 6) related to each alternative matching in Fig. 4, because the 1-cut of the fuzzy integer resulting from concatenation $m_{1:n}$ has a higher supremum (17 in Fig. 3) than the fuzzy integers of the alternative matchings (2, 9, 6, respectively in Fig. 4). So, the matching $m_{1:n}$ is added to the schema matching $M_V^{1:n}$.

Finally, the union of sets of matchings $M_V^{1:1}$ and $M_V^{1:n}$ forms the final set of candidate matchings M_V, which is the basis to detect coreferent tuples and to establish the final matching between corresponding attributes in the next phase.

6   Phase II: Horizontal Matching

In this phase, the candidate vertical schema matching M_V from the previous phase is used to establish the horizontal schema matching M by the method getHorizontalMatchings (line 2 in Algorithm 1). More specifically, in step 1 of this phase the vertical schema matching M_V is used to efficiently detect coreferent tuples across heterogeneous data sources (Sect. 6.1). This significantly reduces the number of comparisons and the complexity of the approach and in turn is the basis for generating the final schema matching in the second step of this phase (Sect. 6.2). The following subsections describe both steps of the horizontal matching phase.

6.1   Step 1: Coreferent Tuples Detection

The coreferent tuples detection Algorithm 4 for schema matching searches for the n-most coreferent tuple pairs D across tuples in the source dataset S and the target dataset T using the candidate vertical schema matching M_V as follows. First, each tuple from the source is compared with each tuple from the target. More precisely, the values of the corresponding attributes, which are matched by M_V, are compared by the method compareTuples (line 3 in Algorithm 4) which inter alia calculates the possibility that two given tuples are coreferent (expressed by a PTV denoted as $\tilde{d}$) and returns a pair of coreferent tuples d. The details of this comparison are presented in Algorithm 5 and described in the next Paragraph "Tuples comparison". Next, coreferent tuple pair d is added to the set D of coreferent tuple pairs (line 4 in Algorithm 4). Finally, the detected coreferent tuple pairs $d \in D$ are sorted by $\tilde{d}$ using Eq. 11 and the $P_H.n$ most coreferent tuple pairs are the result of this algorithm (line 7 and 8 in Algorithm 4, respectively). Using a fixed threshold on the matching degree would be unreasonable because the (un)certainty of tuples coreference varies along with the number of corresponding attributes [1], i.e., if only a few attributes are truly coreferent then the certainty will be low. Thus, instead, our method ranks coreferent tuple pairs by their PTVs and gets the n-most coreferent tuple pairs. It has to be clear that the goal is not to detect all coreferent tuples. These coreferent tuple pairs serve as the basis to establish horizontal schema matching, what is discussed in the next Sect. 6.2.

Algorithm 4 COREFERENTTUPLEDETECTIONALGORITHM

Require: Dataset S, Dataset T, Schema Matching M_V, Parameters P_H
Ensure: n-most Coreferent Tuple Pairs D
 1: **for all** Tuple $t^S \in S$ **do**
 2:     **for all** Tuple $t^T \in T$ **do**
 3:         Coreferent tuple pair $d \leftarrow$ compareTuples(t^S, t^T, M_V, P_H)
 4:             $D \leftarrow D \cup d$
 5:     **end for**
 6: **end for**
 7: $D \leftarrow$ sort(D)
 8: $D \leftarrow$ getMostCoreferent($P_H.n,D$)

Tuples Comparison A comparison of two tuples is conducted by Algorithm 5 and works as follows. First, the input tuples t^S and t^T from the source dataset and the target dataset, respectively, form a pair of candidate coreferent tuples $(d.t^S, d.t^T)$ (line 1 and 2 in Algorithm 5, respectively). Second, a comparison of attribute values for each matching $m \in M_V$ computed in the previous step works as follows. An initial matching $d.m$ is initialized as a copy of a matching m (line 4 in Algorithm 5). Next, for each attribute(s) $m.A'_S$ ($m.A'_T$) of a candidate matching $m \in M_V$ the value(s) v^S or v^T from tuples $d.t^S$ and $d.t^T$ are respectively extracted (lines 5 and 6 in Algorithm 5). In case of 1:n matching of attributes (see Sect. 5.3), the extracted values can be vectors of values $v^S[]$ or $v^T[]$ whose coordinates are concatenated into v^S or v^T before they are compared. Afterwards, the extracted values v^S and v^T are compared using a data type-specific method which estimates the possibility $d.\tilde{m}$ that two given values are coreferent (line 7 in Algorithm 5). More precisely, a numerical and an alphanumerical matchers are considered as follows.

Numerical matcher. Numerical values are compared by a method which is based on the difference ($diff$) of the considered values and a difference threshold ($P_H.thrDiff$). The difference threshold is a real number which defines the maximum allowed difference of values, and depends on the range of values of a particular attribute and the predefined parameter $P_H.thrNum$, which specifies the percentage of difference for average range of the considered attributes $m.A'_S$ and $m.A'_T$ (1:n attribute matching does not apply for numerical data, thus, $m.A'_S = a^S$ and $m.A'_T = a^T$). More specifically, $P_H.thrDiff$ is calculated by the following equation:

$$P_H.thrDiff = \left\lfloor \frac{range(dom'(a^S)) + range(dom'(a^T))}{2} \times P_H.thrNum \right\rfloor \qquad (38)$$

where range is a difference between the maximum and minimum value of $dom'(a^S)$ or $dom'(a^T)$ from the source or the target. If the difference $diff$ between values is smaller than the difference threshold $P_H.thrDiff$, then the possibility that a proposition p stating that the two values are coreferent is true ($\mu_{\tilde{p}}(T)$) equals 1, and the possibility that p is false ($\mu_{\tilde{p}}(F)$) is a fraction of the difference *diff* and the difference threshold $P_H.thrDiff$; otherwise, a lack of coreference is declared, i.e., $\mu_{\tilde{p}}(T) = 0$ and $\mu_{\tilde{p}}(F) = 1$.

Alphanumerical matcher. Alphanumerical values are transformed into sets of substrings which are in most cases separate words. The string is split at the position of a white space, comma, dot or other special character. The usefulness of this approach follows from the fact that character-based methods are typically not well suited for longer strings. Next, the substrings are compared with one another by the low-level string comparison method [2] (see Sect. 5.2). This gives PTVs which are aggregated by the Sugeno integral [3, 4, 23]. Aggregation results in a single PTV which reflects the possibility that two given values are coreferent (see Sect. 3.3). More specifically, the aggregation operator for the comparison of two values, v^S and v^T, is defined by the Sugeno integral for PTVs, where:

- $P = \{p_i\}$ is a set of propositions stating coreference of pairs of substrings,
- $\tilde{P}$ is the set of selected PTVs corresponding to the above-mentioned propositions representing the uncertainty about their truth values computed by the low-level string comparison method,
- the fuzzy measure γ^T is defined by:

$$\gamma^T(Q) = \sum_{j=1}^{k} w_j, \qquad Q \subseteq P, Q = \{p_1, \ldots, p_k\} \tag{39}$$

where w_j is the weight of the jth pair (s_S^j, s_T^j) of substrings, computed by:

$$w_j = \frac{1}{|P|} \tag{40}$$

- the fuzzy measure γ^F is defined by

$$\gamma^F(Q) = \begin{cases} 1 & \text{if } Q = P \\ 0 & \text{otherwise} \end{cases} \tag{41}$$

what is implied by condition (20) is that for each Q which is a subset of P.

Moreover, pairs of substrings with the selected PTVs are grouped into two sets. The first set contains substrings which have an associated PTV that is larger than $P_H.\tilde{thr}$ w.r.t. Eq. 11, and are called coreferent tokens. The second set contains substrings which have an associated PTV that is not larger than $P_H.\tilde{thr}$ and are called non-coreferent tokens. These two sets are the basis for deriving the type of matching. The type *type* of matching $d.m$ (see Sect. 4.1) is specified based on the number of coreferent and non-coreferent tokens (substrings) of the values v^S and v^T by a method which is presented in the Sect. "Matching type of tuples" (line 8 in Algorithm 5).

Next, the matching $d.m$ is added to the set $d.M_V$ of matchings for the coreferent tuples pair d.

Finally, the set $d.M_V$ contains matchings $d.m$ of coreferent attributes A'_S and A'_T for the particular tuples $d.t^S$ and $d.t^T$. Each matching $d.m \in d.M_V$ has an associated PTV $(d.\tilde{m})$, which expresses the possibility that attributes A'_S and A'_T are coreferent for the particular tuples $d.t^S$ and $d.t^T$ based on their values, and is the basis for further checking whether the particular tuples $d.t^S$ and $d.t^T$ are coreferent or not. Namely, these associated PTVs form a multiset $d.\tilde{M}_V$ and are aggregated by the Sugeno integral [3, 4, 23]. The aggregation returns a single PTV $\tilde{d}$ which reflects the possibility that two given tuples are coreferent (line 11 in Algorithm 5). More specifically, the aggregation operator for the comparison of two tuples, $d.t^S$ and $d.t^T$, is defined by the Sugeno integral for PTVs, where:

- $P = \{p_i\}$ is a set of propositions stating coreference of attributes A'_S and A'_T for the particular tuples $d.t^S$ and $d.t^T$, represented by $d.M_V = \{d.m_i\}$,
- $\tilde{P}$ is the set of PTVs corresponding to the above-mentioned propositions, represented by $d.\tilde{M}_V$,
- the fuzzy measure γ^T is defined by:

$$\gamma^T(Q) = \sum_{j=1}^{k} w_j, \qquad Q \subseteq P, Q = \{p_1, \ldots, p_k\} \tag{42}$$

where w_j is the weight of the jth pair (A'_S, A'_T) of attributes for the particular tuples, computed by:

$$w_j = \forall d.\tilde{m} \in d.\tilde{M}_V : w_{d.\tilde{m}} = \frac{1}{|d.M_V|}. \tag{43}$$

These weights are equal for each PTV and depend on the number of matchings.
- the fuzzy measure γ^F is defined by

$$\gamma^F(Q) = \begin{cases} 1 \text{ if } Q = P \\ 0 \text{ otherwise} \end{cases} \tag{44}$$

what is implied by condition (20) is that for each Q which is a subset of P.

Matching Type of Tuples The matching types, which are defined in Sect. 4.1, depend on the factors *ratioS* and *ratioT*. These factors represent the completeness of each matching $d.m \in d.M_V$ for the particular tuples t^S and t^T and are based on the number of coreferent tokens (substrings) of compared values $v^S \in t^S$ and $v^T \in t^T$. The factor *ratioS* (*ratioT*) is the fraction of the number of coreferent tokens over the number of tokens of the value v^S (v^T, respectively). Thus, the following conditions have to be considered:

- if *ratioS* $= 0$ and *ratioT* $= 0$ then $m \in M$ is an "unknown" matching
- if *ratioS* $= 1$ and *ratioT* $= 1$ then $m \in M$ is a "full" matching

---

Algorithm 5 COMPARETUPLESALGORITHM

Require: Tuple t^S, Tuple t^T, Schema Matching M_V, Parameters P_H
Ensure: Coreferent tuple pair d
1: $d.t^S \leftarrow t^S$
2: $d.t^T \leftarrow t^T$
3: **for all** Matching $m \in M_V$ **do**
4:     Matching $d.m \leftarrow m$
5:     var $v^S \leftarrow d.t^S[m.A'_S]$
6:     var $v^T \leftarrow d.t^T[m.A'_T]$
7:     $d.\tilde{m} \leftarrow$ compare$(v^S, v^T, P_H.thr, P_H.thrNum)$
8:     $d.m.type \leftarrow$ mType(v^S, v^T)
9:     $d.M_V \leftarrow d.M_V \cup d.m$
10: **end for**
11: $\tilde{d} \leftarrow$ aggregate$(d.\tilde{M}_V)$

- if $ratioS = 1$ and $ratioT \neq 1$ then $m \in M$ is a "source is part of target" matching
- if $ratioS \neq 1$ and $ratioT = 1$ then $m \in M$ is a "target is part of source" matching
- if $ratioS \neq 1$ and $ratioT \neq 1$ then $m \in M$ is a "a common part" matching

Example Let us consider coreferent tuple pairs detection for the following case. The input is the vertical matching M_V of Table 6, which is the basis for detecting coreferent tuple pairs across the source and target datasets, of Tables 1 and 2, respectively. Each tuple from the source dataset is compared to each tuple in the target dataset which results in the set D of coreferent tuple pairs. For example, let us consider that tuple t^S

Table 6 Example of the vertical schema matching M_V

Matching m	Source	Target
m_1	Name	Type
m_2	Name	POI
m_3	Name	Type, POI
m_4	Name, Category	Type
m_5	Lon.	Geo2
m_6	Lat.	Geo1
m_7	Category	Type
m_8	Key	Id
m_9	Address	ZipCode
m_{10}	Address	Street
m_{11}	Address	City
m_{12}	Address	ZipCode, Street
m_{13}	Address	ZipCode, City
m_{14}	Address	Street, City
m_{15}	Address	ZipCode, Street, City

in row 2 in Table 1 is compared to tuple t^T in row 2 in Table 2. First, the values of the matched attributes (by M_V) are compared in sequence, e.g., the value "Saint Bavo" of the attribute "Name" in t^S is compared by the alphanumerical matcher to the following values in t^T w.r.t. the matchings m_1, m_2 and m_3 in Table 6: "Church" (of the attribute "Type"), "Sint-Bavokerk" (of the attribute "POI") and "Church Sint-Bavokerk" (of the concatenation of the attributes "Type" and "POI"). This comparison results in $d.M_V$ which contains all the attributes matchings $m \in M_V$, namely $d.m \in d.M_V$ (with associated PTVs $d.\tilde{m}$ which form a multiset $d.\tilde{M}_V$) for the particular tuples t^S and t^T, e.g., $d.\tilde{m}_1 = (0, 1)$, $d.\tilde{m}_2 = (1, 0.12)$, $d.\tilde{m}_3 = (1, 0.16)$, etc. Next, the matching type for each $d.m \in d.M_V$ for particular tuples is derived: $d.m_1.type$ is an "unknown" matching (no common tokens), $d.m_2.type$ is a "full" matching (all tokens are common), $d.m_3.type$ is a "source is part of target" matching (all tokens from the source are common, but not all from the target), etc. Next, the PTVs in $d.\tilde{M}_V$ are aggregated by the Sugeno integral with equal weights $w = 1/15$ (calculated by Eq. 43). This results in a single PTV $\tilde{d}$ that equals $(1, 0.5)$ which reflects the possibility that the two given tuples t^S and t^T are coreferent.

Finally, the detected coreferent tuple pairs D are sorted by $\tilde{d}$, and the $P_H.n$ (in our case 3) most coreferent tuple pairs are returned. Namely, the pairs of tuples from rows 2, 4 and 5 of Tables 1 and 2 are returned as the 3 most coreferent tuple pairs and are used to establish the final schema matching, what is discussed in the next section.

6.2 Step 2: Schema Matching

The n most coreferent tuples pairs of D detected in the previous step and the vertical schema matching M_V established in the first phase of our approach are used to infer the final horizontal schema matching M. Our novel approach is implemented by Algorithm 6, which works as follows. First of all, for each matching $m \in M_V$, the cardinality $card_m$ of this matching is calculated (lines 2–6 in Algorithm 6). This cardinality is the number of coreferent tuple pairs whose values of attributes $m.A_S'$ and $m.A_T'$ are coreferent. Hereby coreference is considered if $\mu_{\tilde{p}}(F)$ of $d.\tilde{m}$ is lower than $\mu_{\tilde{p}}(F)$ of the predefined threshold $P_H.thr\tilde{D}up = (0.5, 0.5)$ w.r.t. Eq. 11.

Next, if a particular matching m returns most of the coreferent tuple pairs (line 7 in Algorithm 6), i.e., $m.card_m/|D|$ is greater than the predefined threshold $P_H.thrMajority$, then m is added to the final schema matching M (line 8 in Algorithm 6). Next, the propositions evaluated using PTVs of the matching m across all coreferent tuple pairs of D (i.e., $\forall d \in D : d.\tilde{m}$) are aggregated by the Sugeno integral. This results in a possibility degree $\tilde{m}$ (PTV), which expresses the uncertainty of that matching $m \in M$ (line 9 in Algorithm 6) [3, 4, 23] (see Sect. 3.3).

More specifically, the aggregation operator for matching $m \in M$ is defined by the Sugeno integral for PTVs, where:

- $P = \{p_i\}$ is a set of propositions stating coreference of attributes $m.A'_S$ and $m.A'_T$ across all coreferent tuple pairs of D, represented by $D[m] = \{d_i.m\}$,
- $\tilde{P}$ is the set of PTVs corresponding to the above-mentioned propositions, represented by $\tilde{D}[m] = \{d_i.\tilde{m}\}$,
- the fuzzy measure γ^T is defined by:

$$\gamma^T(Q) = \sum_{j=1}^{k} w_j, \qquad Q \subseteq P, Q = \{p_1, \ldots, p_k\} \tag{45}$$

where w_j is the weight of the jth duplicate d_j, computed by:

$$w_j = \frac{1}{|D|}. \tag{46}$$

These weights are equal for each PTV $d_j.\tilde{m}$ and depend on the number of coreferent tuple pairs.
- the fuzzy measure γ^F is defined by

$$\gamma^F(Q) = \begin{cases} 1 \text{ if } Q = P \\ 0 \text{ otherwise} \end{cases} \tag{47}$$

what is implied by condition (20) is that for each Q which is a subset of P.

Moreover, a possibility degree $\tilde{m}$ of that matching $m \in M$ can be used to resolve schema matching *conflicts*. The schema matching M contains conflicts if there exists more than one matching $m \in M$ (called alternative matching) for any attribute $a^S \in A_S$ or $a^T \in A_T$. This means that a matching which is donated with the larger PTV than alternative matchings is preferable and another alternative matchings should be removed. However, the conflict resolution is subject to further research and outside the scope of this paper.

Finally, the matching types of the matching $m \in M$ across all coreferent tuple pairs D are unified by the method unifyType in line 10 in Algorithm 6. This method returns for each $m \in M$ the most popular matching type across all coreferent tuple pairs D. In the case of indistinguishable matching types, i.e., if maximum frequency of matching type for matching $m \in M_V$ over all coreferent tuple pairs of D is not unique, the matching type with the predefined lowest coverage level is selected (see Sect. 4.1). For example, if full matching (with coverage level 1) and inclusion matching (with coverage level 0.5) types are specified for the same number of coreferent tuple pairs then inclusion matching type is selected. The unified matching types inform about matching completeness and can be used to resolve schema matching conflicts but it is out of the scope of this paper.

Example Let us consider the coreferent tuple pairs in D which were detected in the previous step. The set D of coreferent tuples pairs consists of 3 pairs, each composed of rows 2, 4, 5 from Tables 1 and 2. The matching cardinality $card_m$ is calculated for each matching $m \in M_V$ in Table 6. For example, the $card_m$ of the matching

---

Algorithm 6 HORIZONTALMATCHINGALGORITHM

Require: Coreferent tuple pairs D, Schema Matching M_V, Parameters P_H
Ensure: Schema Matching M
 1: **for all** Matching $m \in M_V$ **do**
 2:     **for all** Coreferent tuple pair $d \in D$ **do**
 3:        **if** $d.\tilde{m} > P_H.thr\tilde{D}up$ **then**
 4:           $card_m \leftarrow card_m + 1$
 5:        **end if**
 6:     **end for**
 7:     **if** $card_m/|D| > P_H.thrMajority$ **then**
 8:        $M \leftarrow M \cup m$
 9:        $\tilde{m} \leftarrow$ aggregate($\tilde{D}[m]$)
10:        $m.type \leftarrow$ unifyType($D[m].type$)
11:     **end if**
12: **end for**

$m_{10} = \{$"Address"; "Street"$\}$ equals 2 because only the attribute values of two tuple pairs are coreferent for the threshold $P_H.thr\tilde{D}up$ equal to $(0.5, 0.5)$. More specifically, the value "Stoopkensstraat 24, 3320 Hoegaarden" is similar to "Stoopkensstraat 24", and "Kleine Gentstraat 69, 9051 St-Denijs-Westrem" is similar to "Kleine Gentstraat 69". The certainty as to their similarity is expressed for both of them by a PTV $(1, 0.33)$. The similarity of values "Sint-Baafsplein, 9000 Ghent" and "Sint-Baafsplein" is expressed by a PTV $(1, 0.5)$, but this does not exceed the threshold. In contrast, $card_m$ of the matching $m_{15} = \{$"Address"; "Street", "City", "ZipCode"$\}$ is equal to 3, because the attribute values of all coreferent tuple pairs are coreferent. This confirms that the concatenation of attributes makes sense.

Next, if $card_m$ of m satisfies the majority condition in line 7 of Algorithm 6, then the matching m is added to M. The predefined threshold $P_H.thrMajority = 0.3$ and $|D| = 3$, thus if $card_m$ of m is greater than 0.9, then m is considered as a matching in M. This means that M contains only matchings which are confirmed by at least one coreferent tuple pair. Matchings m_4, m_9 and m_{11} in Table 6 are not included in M because they do not satisfy this condition. Next, the PTVs for matching m across all coreferent tuple pairs D are aggregated by the Sugeno integral. For the matching m_{10}, the PTVs $(1, 0.33)$, $(1, 0.5)$, $(1, 0.33)$ are aggregated with equal weights $w = 1/|D| = 1/3$ to a single PTV equal to $(1, 0.33)$. This PTV reflects the possibility that the attributes "Address" and "Street" are coreferent. Finally, the matching types of the matching $m \in M$ across all coreferent tuple pairs D are unified by the method unifyType. For the matching m_{10}, the unified matching type is "t is part of s" (target is part of source) because the matching type of all considered coreferent tuple pairs is "t is part of s".

This gives us the schema matching M in Table 7 with alternative matchings (conflicts) for some of the attributes, e.g., m_{10}–m_{15} in Table 7. These conflicts can be resolved based on cardinality ($card_m$), certainty ($\tilde{m}$) and/or type of matching but it is out of the scope of this paper.

Table 7 Example of the schema matching M with alternative matchings

Matching m	Source	Target	$card_m$	$type_m$	$\tilde{m}$
m_1	Name	Type	1	t is part of s	(1, 0.67)
m_2	Name	POI	1	Has a common part	(1, 0.5)
m_3	Name	Type, POI	1	Has a common part	(1, 0.6)
m_5	Lon	Geo2	2	Full matching	(1, 0.33)
m_6	Lat	Geo1	2	Full matching	(1, 0.33)
m_7	Category	Type	1	Full matching	(1, 0.67)
m_8	Key	id	3	Full matching	(1, 0)
m_{10}	Address	Street	2	t is part of s	(1, 0.33)
m_{12}	Address	ZipCode, Street	3	Has a common part	(1, 0.03)
m_{13}	Address	ZipCode, City	1	Has a common part	(1, 0.5)
m_{14}	Address	Street, City	3	t is part of s	(1, 0.12)
m_{15}	Address	ZipCode, Street, City	3	s is part of t	(1, 0.12)

7 Evaluation and Discussion

In this section we describe an experimental evaluation of our method which shows the influence of the parameters and the benefits of using content data (compared to schema information-only-based methods).

7.1 Datasets

To illustrate the proposed approach we consider different real-world datasets, respectively, containing information about 'compact discs', 'restaurants' and 'points of interest'.

Compact disc (CD) data are contained in two datasets which are defined in two schemas. The first schema is extracted from FreeDB[2] and consists of 8 attributes. The second schema is extracted from Discogs[3] and consists of 24 attributes, of which 6 have been identified manually as being coreferent with the FreeDB schema attributes and act as the ground truth for our experiments. The FreeDB dataset contains 124 tuples which are extracted from the CD dataset,[4] while the Discogs dataset contains

---

[2]FreeDB, http://www.freedb.org/.

[3]Discogs, http://www.discogs.com/data/.

[4]http://hpi.de/naumann/projects/repeatability/datasets/cd-datasets.html.

132 tuples which are extracted from Discogs.[5] The number of coreferent tuples in these datasets is equal to 33, which are detected manually.

Restaurant data are represented by two famous datasets [28]. One dataset stems from the on-line guide 'Zagat', while the other dataset stems from the on-line guide 'Fodor'. Zagat contains 331 tuples and Fodor contains 533 tuples, where 112 coreferent tuples are counted (i.e., 112 restaurants occur in both lists). These datasets are defined in two pairs of schemas, called R1 and R2, respectively. Both schemas of the first pair R1 consist of 6 attributes, of which all 6 have been identified manually as being coreferent, i.e., each attribute in the Fodor schema corresponds to exactly one attribute in the Zagat schema, and vice versa. More specifically, 6 one-to-one matchings are established. The Fodor schema in the second schemas pair R2 is identical to the Fodor schema in the first schema pair R1. But the Zagat schema in the second schemas pair R2 consists of 5 attributes, of which the attribute "street-city" is a concatenation of the attributes "street" and "city". Thus, each of the 4 attributes in the Fodor schema corresponds to exactly one attribute in the Zagat schema, and the concatenation of the attributes "street" and "city" in the Fodor schema corresponds to the attribute "street-city" in the Zagat schema. So, four one-to-one matchings and one-to-many matching are present. The truly coreferent schema attributes act as ground truth for our experiments.

Points of interest data are represented by two datasets which are defined in two schemas. The first dataset is made available by the Belgian company RouteYou,[6] which is an on-line provider of cycling routes. In order to support their routing algorithms, RouteYou manages a database with POIs. This database is defined by a schema which consists of the attributes latitude, longitude, POI name, POI category, POI internal name (which is a copy of the POI name extended by location information) and the language in which the name and category are given. An important characteristic of the given POI database is that data is mostly contributed by independent users of the website. Hereby, a user can pinpoint a location on the map, type in the name of the POI he/she wants to add and associate one of the predefined POI categories to it. From the complete POI database, we inferred one dataset by selecting tuples in English and in a specific area: the center of Ghent. This resulted in the RouteYou dataset which consists of 136 tuples. The second dataset contains 945 tuples which were extracted from the Google Maps database and are related to the same specific area. The tuple extraction has been done using the Google Places API.[7] The resulting dataset is defined by a schema which consists of the attributes id, name, vicinity, lat, lng, googleId and type, of which 6 have been identified manually as being coreferent with the attributes of the RouteYou dataset.

Table 8 contains a summary of all datasets considered in the experiments. The number of attributes in the data varies between 5 and 24 (column 2 in Table 8), while the number of truly coreferent attributes varies between 5 and 6 (column 5 in Table 8). The number of detected coreferent tuple pairs varies between 33 and 112 (column

---

[5]Discogs, http://www.discogs.com/data/.

[6]http://www.routeyou.com.

[7]Google Places, http://developers.google.com/places/.

Table 8 Real-world datasets

Datasets	# attributes	# tuples	# dup	# coreferent attr.
S: CD.FreeDB	8	124	33	6
T: CD.Discogs	24	132		
S: R1.Fodor	6	533	112	6
T: R1.Zagat	6	331		
S: R2.Fodor	6	533	112	5
T: R2.Zagat	5	331		
S: POI.RouteYou	9	136	51	6
T: POI.Google	7	945		

4 in Table 8), while the number of tuples varies between 124 and 945 (column 3 in Table 8).

7.2 Evaluation Setting

To determine the quality of our approach, we compared its result against the manually derived results. Based on the standard confusion matrix, we will consider the following three sets. The first set, denoted as B, contains the truly coreferent objects which are discovered by our approach, i.e., so-called true positives. The second set, denoted as A, contains truly coreferent objects which are not identified, i.e., so-called false negatives. The last set, denoted as C, contains objects which are falsely identified as coreferent, i.e., so-called false positives.

Precision is defined as the fraction of truly coreferent objects among all objects classified by a given algorithm as being coreferent:

$$\text{Precision} = \frac{|B|}{|B| + |C|}. \tag{48}$$

Recall is another important quality measure which in our case can be defined as the fraction of true positive objects among all coreferent objects present in a test dataset:

$$\text{Recall} = \frac{|B|}{|A| + |B|}. \tag{49}$$

7.3 Experiment: Configuration of Parameters of the Vertical Matcher

Goal. Our vertical matching Algorithm 2 in Sect. 5 employs the parameters $P_V.thrStats$ and $P_V.thrOverlap$. $P_V.thrStats$ specifies the threshold above which statistical properties of attribute domains are considered as similar. $P_V.thrOverlap$ specifies the percentage of the domains overlap. This experiment evaluates the impact of these parameters on the precision and recall of the established vertical schema matching.

Procedure. For the parameter $P_V.thrStats$, a range from 0 to 1 with a step equal to 0.01 is considered. For the parameter $P_V.thrOverlap$, a range from 0 to 1 with a step equal to 0.1 is considered. Mean recall and precision for each value of $P_V.thrStats$ and $P_V.thrOverlap$ over all datasets are calculated. Statistical matcher is executed before the lexical matcher, thus, the overlap threshold does not have to be considered.

Result. Figure 5 shows the mean precision and recall over all datasets for the different values of the parameter $P_M.thrStats$ (uninterested results are omitted). For $P_M.thrStats$ values between 0.08 and 0.96 the statistical comparison of content data gives the highest precision of matching. In this case, precision is more important than recall because non matched truly coreferent attributes can be matched by the lexical matcher. Besides that, the criteria of the statistical matcher are strict (all properties have to be similar) because statistical information may mislead the matcher, i.e., there can exist domains which have similar properties but may describe non coreferent attributes. We choose $P_M.thrStats$ equal to 0.9 for the further evaluations.

Figure 6 shows the mean precision and recall obtained over all datasets for different values of the parameter $P_M.thrOverlap$ in the lexical matcher. For $P_M.thrOverlap$ values equal to 0.3 and 0.4 the lexical comparison of content data gives matchings with the highest precision and recall. The established matchings are the basis to detect coreferent tuple pairs, which in turn are used to derive the final schema matching,

Fig. 5 Mean precision and recall over all datasets in function of $P_V.thrStats$ and $P_M.thrOverlap$ equal to 0.3

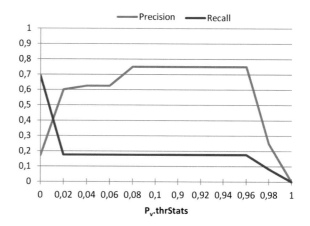

Fig. 6 Mean precision and
recall over all datasets in
function of $P_V.thrOverlap$
and $P_M.thrStats$ equal to 0.9

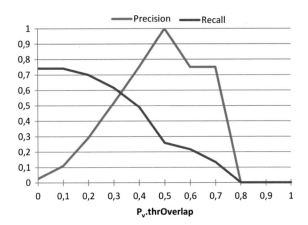

thus, the method has to derive as many as possible matchings at the expense of precision—the non coreferent matchings are eliminated by the horizontal matcher. Thus, $P_M.thrOvelap$ equal to 0.3 is selected for the further evaluations, because the smaller percentage of domain terms that overlap is easier to satisfy.

The combination of the matchings which are established by statistical and lexical matchers gives an average precision equal to 0.58 and recall equal to 0.79 over all datasets for $P_M.thrStats$ equal to 0.9 and $P_M.thrOvelap$ equal to 0.3.

7.4 Experiment: Configuration of Parameters of the Horizontal Matcher

Goal. The goal of this experiment is to show the impact of the number $P_H.n$ of coreferent tuple pairs, which is the basis to establish the schema matching, and the parameter $P_H.thrMajority$ of our horizontal matching algorithm in Sect. 6 on the precision and recall of the established schema matching. $P_H.thrMajority$ is a threshold which specifies that a matching is considered as a correct matching by the horizontal matcher.

Procedure. For the parameter $P_H.n$, a range from 1 to 10 is considered. For the parameter $P_H.thrMajority$, values 0.25, 0.5, 0.75 and 1 are considered. $P_H.thrMajority$ equal to 0.25 (0.5, 0.75 or 1) means that if any vertical matching $m \in M_V$ between the particular attributes is repeated by a quarter (two quarters, three quarters or all, respectively) of detected coreferent tuple pairs then m is added to the set M of horizontal matchings. The mean precision and recall for each value of $P_H.thrMajority$ and $P_H.n$ over all datasets are calculated.

Result. Figure 7 shows the mean precision and recall over all datasets for different values of the parameters $P_H.thrMajority$ and $P_H.n$.

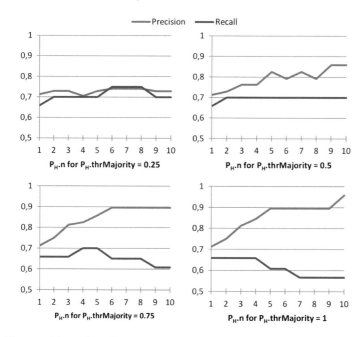

Fig. 7 Mean precision and recall over all datasets for different $P_H.thrMajority$ in function of $P_H.n$

Setting $P_M.thrMajority$ to 0.5 or 0.75, and $P_H.n$ to 5 or 6 is sufficient to establish the schema matching with high precision and recall. More coreferent tuple pairs ($P_H.n$ greater than 6) do not increase the precision significantly or can even decrease the recall for a large value of $P_H.thrMajority$, because $P_H.thrMajority$ equal to 1 forces that a particular matching has to be confirmed by all the selected coreferent tuple pairs. However, this may be unrealistic and difficult to satisfy. Besides that, our horizontal matcher is based on the n most coreferent tuples pairs so using many coreferent tuple pairs may result in coreferent tuples pairs having assigned low certainty (because the values of some attributes may not be coreferent). Thus, it is recommended to use only a few coreferent tuple pairs but then those that are assigned the highest certainty and $P_M.thrMajority$ between 0.5 and 0.75.

7.5 Benefits of Using Content Data

Content data-based schema matching approaches are able to establish semantical matchings of corresponding schema elements in those situations where the schema information-only-based methods can be ineffective. For example, a schema matching method which is based only on element names is not able to create a matching if the names of coreferent attributes are synonyms. This means that content data add additional information about the particular attribute which can be used to better infer the semantics of the attribute and, as a consequence, create a matching between coreferent attributes. Moreover, attribute names may even mislead the schema matching methods which are only based on schema information. Content data are a powerful and valuable source of information which can be used to considerably improve schema matching. In contrast, approaches which are based on schema information only can establish matchings of attributes which are not coreferent based on the content data, e.g., "id" attributes.

8 Conclusion

In this paper, a content data based schema matching algorithm has been proposed as a way to construct proper matching between corresponding schema elements of heterogeneous datasets. The algorithm is especially useful in cases where finding the correspondences between the schema elements based on schema information only is difficult or impossible. Our novel technique employs possibilistic truth values, to express certainty of matchings, similarities etc., and fuzzy integers, to express cardinalities of sets of true propositions based on the certainty of their truth expressed using PTVs. This allows us to explicitly cope with the (un)certainty of semantical one-to-one and one-to-many schema matchings which are set up by our novel techniques in an automated fashion. As a consequence, solutions to the attribute granularity problem and the data coverage problem are proposed. The behaviour of the novel schema matching algorithm has been evaluated on several real life datasets, thus providing us with a valuable insight into the influence of the different parameters. Moreover, it has been shown what are the advantages of the proposed approach compared with a schema matching approach based on schema information only.

Acknowledgments This contribution is supported by the Foundation for Polish Science under International PhD Projects in Intelligent Computing. Project financed from The European Union within the Innovative Economy Operational Programme 2007–2013 and European Regional Development Fund. This work was also partially supported by the National Science Centre (contract no. UMO-2011/01/B/ST6/06908).

References

1. Bilke, A., Naumann, F.: Schema matching using duplicates. In: Proceedings of the 28th International Conference on Data Engineering (ICDE) (2005)
2. Bronselaer, A., De Tré, G.: A possibilistic approach on string comparison. IEEE Trans. Fuzzy Syst. **17**(1), 208–223 (2009)
3. Bronselaer, A., De Tré, G.: Properties of possibilistic string comparison. IEEE Trans. Fuzzy Syst. **18**(2), 312–325 (2010)
4. Bronselaer, A., Hallez, A., De Tré, G.: Extensions of fuzzy measures and the sugeno integral for possibilistic truth values. Int. J. Intel. Syst. **24**(2), 97–117 (2009)

5. Calvo, T., Mayor, G., Mesiar, R. (eds.): Aggregation Operators: New Trends and Applications. Physica-Verlag GmbH, Heidelberg (2002)
6. Chua, C.E.H., Chiang, R.H.L., Lim, E.P.: Instance-based attribute identification in database integration. VLDB J. **12**(3), 228–243 (2003). Oct
7. de Cooman, G.: Towards a possibilistic logic. In: Ruan, D. (ed.) Fuzzy Set Theory and Advanced Mathematical Applications, International Series in Intelligent Technologies, vol. 4, pp. 89–133. Springer, US (1995)
8. Dhamankar, R., Lee, Y., Doan, A., Halevy, A., Domingos, P.: imap: discovering complex semantic matches between database schemas. In: Proceedings of the 2004 ACM SIGMOD International Conference on Management of Data, ACM Press (2004)
9. Do, H.h., Rahm, E.: Coma—a system for flexible combination of schema matching approaches. In: Proceedings of the VLDB 2002, pp. 610–621 (2002)
10. Doan, A., Domingos, P., Levy, A.Y.: Learning source description for data integration. In: WebDB (Informal Proceedings), pp. 81–86 (2000)
11. Elmagarmid, A., Ipeirotis, P., Verykios, V.: Duplicate record detection: a survey. IEEE Trans. Knowl. Data Eng. **19**(1), 1–16 (2007)
12. Hallez, A., De Tré, G., Verstraete, J., Matthé, T.: Application of fuzzy quantifiers on possibilistic truth values. In: Proceedings of EUROFUSE EURO WG on Fuzzy Sets, pp. 252–254. EXIT (2004)
13. Hastie, T., Tibshirani, R., Friedman, J.: The Elements of Statistical Learning. Springer Series in Statistics. Springer New York Inc, New York (2001)
14. Jain, A.K., Duin, R.P.W., Mao, J.: Statistical pattern recognition: a review. IEEE Trans. Pattern Anal. Mach. Intell. **22**(1), 4–37 (2000). Jan
15. Little, R.J.A., Rubin, D.B.: Statistical Analysis with Missing Data. Wiley, New York (1986)
16. Lu, H., Fan, W., Goh, C.H., Madnick, S., Cheung, D.: Discovering and reconciling semantic conflicts: a data mining prospective. In: Proceedings of IFIP Working Conference on Data Semantics (DS-7) (1997)
17. Madhavan, J., Bernstein, P.A., Rahm, E.: Generic schema matching with cupid. In: Proceedings of the 27th International Conference on Very Large Data Bases. pp. 49–58. VLDB '01, Morgan Kaufmann Publishers Inc., San Francisco, CA, USA (2001)
18. Mehdi, O.A., Ibrahim, H., Affendey, L.S.: Instance based matching using regular expression. Procedia CS **10**, 688–695 (2012)
19. Perkowitz, M., Doorenbos, R.B., Etzioni, O., Weld, D.S.: Learning to understand information on the internet: an example-based approach. J. Intel. Inf. Syst. **8**(2), 133–153 (1997). Mar
20. Prade, H.: Possibility sets, fuzzy sets and their relation to Lukasiewicz logic. In: Proceeding of 12th Int Symp on Multiple-Valued Logic. pp. 223–227 (1982)
21. Rahm, E., Bernstein, P.A.: A survey of approaches to automatic schema matching. VLDB J. **10**(4), 334–350 (2001). Dec
22. Reiss, R.D., Thomas, M.: Statistical analysis of extreme values: with applications to insurance, finance, hydrology and other fields. Birkhuser Basel, 3rd edn. (2007)
23. Sugeno, M.: Theory of Fuzzy Integrals and its Applications. Ph.D. thesis, Tokyo, Japan (1974)
24. Szymczak, M., Koepke, J.: Matching methods for semantic annotation-based XML document transformations. In: K. Atanassov, et al. (Eds.), New Developments in Fuzzy Sets, Intuitionistic Fuzzy Sets, Generalized Nets and Related Topics. Applications. Volume II. pp. 297–308. SRI PAS (2012)
25. Szymczak, M., Zadrożny, S., De Tré, G.: Coreference detection in XML metadata. In: Pedrycz, W., Reformat, M. (eds.) Proceedings of 2013 Joint IFSA World Congress NAFIPS Annual Meeting. pp. 1354–1359 (2013)
26. Szymczak, M., Bronselaer, A., Zadrożny, S., De Tré, G.: Semantical mappings of attribute values for data integration. In: Proceedings of NAFIPS 2014. pp. 1–8. IEEE (2014)
27. Szymczak, M., Zadrożny, S., Bronselaer, A., De Tré, G.: Coreference detection in an XML schema. Inf. Sci. **296**, 237–262 (2015)
28. Tejada, S., Knoblock, C., Minton, S.: Learning object identification rules for information integration. Inf. Syst. **26**(8), 607–633 (2001)

29. Yager, R.: On the theory of bags. Int. J. Gen. Syst. **13**(1), 23–27 (1986)
30. Zadeh, L.: Fuzzy sets as a basis for a theory of possibility. Fuzzy Sets Syst. **100**, 9–34 (1999). Apr
31. Zadrożny, S., Kacprzyk, J., Sobota, G.: Avoiding duplicate records in a database using a linguistic quantifier based aggregation—a practical approach. In: Proceedings of FUZZ-IEEE. pp. 2194–2201 (2008)

Prior-Art Relevance Ranking Based on the Examiner's Query Log Content

Jakub Wajda and Wlodek Zadrozny

Abstract This work belongs to the domain of technical information retrieval (IR) and, more specifically, patent retrieval. We show that the recorded history of patent examiner's search queries can be used to create a more effective method of finding prior art patents than search methods based on titles and claims. We verify the performance of the proposed method experimentally. Our experiments show that we can almost double the recall measure, compared to classical techniques based on titles and claims. The other contribution of our work is the creation of a database of over half a million patent examiners queries (recorded search activity over the patents prosecution process). The paper also discusses the limitations of the current work and the ongoing research to further improve the proposed approach.

Keywords Patent search · Prior-art search · Query log analysis

1 Introduction

Prior-art patent search, also called *patentability search*, aims to determine whether or not an invention described in a patent application is novel and, thus, if the patent in question should be granted. Hence, the goal is to identify and collect the most relevant resources helping to assess the novelty. These resources comprise related patents as well as all general information on the related topics, and are jointly named as the *prior-art*, with respect to a given patent application.

The prior-art is predominantly the textual information and thus the techniques of the information retrieval (IR) find here a natural application. They are used to help *examiners*, persons responsible for evaluating a submitted grant, to decide if

J. Wajda (✉)
Systems Research Institute, Polish Academy of Sciences, Warsaw, Poland
e-mail: wajdaj@ibspan.waw.pl

W. Zadrozny
University of North Carolina, Charlotte, NC, USA
e-mail: wzadrozn@uncc.edu

© Springer International Publishing Switzerland 2016
G. De Tré et al. (eds.), *Challenging Problems and Solutions in Intelligent Systems*, Studies in Computational Intelligence 634,
DOI 10.1007/978-3-319-30165-5_15

323

its novelty is clear enough to justify granting the patent. A patent application has usually a strictly defined structure and is composed of several parts. Among them, most often the title of the patent and the claims part (defining the precise scope of the protection which is being sought by the patent applicant) are used to form queries aimed at retrieving the prior art.

In this paper we propose a technique which may support an examiner in the process of collecting the prior-art. Namely, an examiner is usually executing a number of queries in search for relevant documents. We assume that these queries are recorded and available in the form of a *query log*. Then, we show that the recorded history of patent examiners' search queries can be used to create a more effective method of finding prior art patents than search methods based on titles and claims. In particular, it provides for a more effective ranking of the retrieved documents.

We verify the performance of compared methods using a collection of patents in a number of computational experiments. In comparison with selected other techniques [32], the results of our experiments show that we can approximately double the recall measure using a smaller amount of input data.

The other contribution of our work is the creation of a database of over half a million patent examiners queries (sessions histories of the patents prosecution process). We are planning to make it available to other researchers.

While this paper verifies that the historic data about examiner's queries are very useful in the considered context, we do not address the issue of how to automatically create such queries for a new patent application. This is the practical limitation of the current work, and we are planning to address it in the ongoing research using machine learning paradigm.

1.1 Some Characteristic Features of Prior-Art Examiner's Search Activity

Patentability search is an essential task for the patent examiners and also one of the most challenging problems of IR. The main features of the prior-art search may be described as follows.

The patent search process always starts with a document describing the invention being the essence of the patent to be examined. Patent search professionals are experts in creating complex queries with numerous keywords (terms) combined using advanced operators. Usually, they reformulate the queries multiple times until they obtain satisfactory results. Surveys of patent examiners behavior [4, 14] show that in a patent application validity checking task, an examiner on average spends about 12 h to become acquainted with about 100 relevant documents retrieved using about 15 different queries. Formulating a single query takes about 5 min-more or less as much time as it takes to judge a single document.

Most of the surveyed examiners agree that query features like: Translation, Expansion, Wildcard, Truncation as well as Distance, Field, and Boolean operators are important to formulate effective queries.

1.2 Boolean Character of Patent Searches

The most popular IR model among patent searchers is the Boolean one. A log of Boolean queries provides also a kind of a self-documented insight into the patent application processing.

According to [28], operators AND and OR occur in, respectively, 48–57 % and 22–30 % queries. In query logs we have analyzed, all Boolean operators appear explicitly in 55.85 % of queries. Proximity operators can be regarded as a more restrictive version of the AND operator, with special conditions which are described in Table 1. Operators of this type occur in 36.2 % of queries. In total, 71.54 % of queries contain one of these operators explicitly after expansion of references to other queries.

1.3 Improving Patent Queries

Despite the importance of Boolean queries in professional search, there has not been much research on helping information professionals formulate those queries. On the other hand, past few years have seen an explosive growth in scientific and regulatory documents related to the patent system.

In this paper, we propose a method to a prior-art relevance ranking based on the analysis of query logs.

Table 1 Boolean and Proximity operators after the expanding of references where Proximity operators are: *ADJn* Terms should be within n words of each other and in the order specified

AND	OR[a]	NOT	XOR
43.6 %	21.7 %	36.2 %	0.0018 %
ADJn[b]	NEARn[b]	WITH	SAME
13 %	5.2 %	15 %	16 %

[a]Used explicitly, by default represented as space character
[b]n is an optional natural number
NEARn Terms should be within n words of each other and in any order. *SAME* Terms must be in the same paragraph, in any order. *WITH* behaves identically to SAME which is recommended and supported for use in Thomson Innovation

2   Related Work

Information retrieval has been intensively researched to bridge the gap between users queries, or information needs, and relevant documents. In particular, there have been many diverse studies on information retrieval and query processing in general web search [13], including studies on query expansion [9] or query intent detection [12, 23]. However, although patent related information retrieval also touches all aspects of IR research [17], relatively little work has been done concerning patent retrieval, as well as, in fact, other domain-specific search environments such as, e.g., legal search or medical information search.

Although many successful general query expansion techniques have been proposed [9, 11, 18, 28], most of them are not easily applicable to our tasks because they are not able to deal with Boolean queries required for professional search environments. Due to the diverse objectives, the proposed methods differ, and so attempts to apply general query log analysis techniques in the context of patent IR usually refers to the automatic query generated from a piece of text using traditional TF-IDF based term weighting methods [6, 16, 32, 33]. The huge disadvantage of this solution is that this kind of input data "may contain many ambiguous and vague terms. It is particularly difficult to formulate a successful query because patent writers tend not to use standard terms to make them look novel" [20]. This difficulty with choosing proper keywords is ameliorated due to the fact that some keywords come from pattern-based semantic tagging [20], automatic patent annotation [1], subtopic extraction [26], automatic refinement of queries [19] or reduction [11].

Another improvement of text-based methods are citation-based methods [10, 21, 29–31] or non-patent references [8]. Their disadvantage is that, firstly, new articles often refer to classics in their field of study and secondly, there are most often different motivations and actors in the citing process in two consecutive publications [10].

In [27, 28] authors operate on small number (less than 500) of real United States Patent and Trademark Office (USPTO) query logs, where they group them by the class and the objective is to collect keywords statistics.

Of interest for our purposes may be also some studies of similarity of graph queries with edit distance constraints [34] which find application in different areas such as bioinformatics, chemistry, social networks, pattern recognition, etc.

3   Methods and Experiments

This section describes three methods. The first two are based on the analysis of logs of examiners' queries. They attempt to extract information from query logs and use it to retrieve and rank relevant documents. The second method is an enhanced version of the first one. Thus, we will focus in our description on the second method, showing at the same time how it extends the first one. The third method has been proposed in [33]. It consists in building queries for prior-art search based on particular sections of patent application.

3.1 Basic Keywords Extraction Based Method and Its TF-IDF Boosting Enhancement

The input is a query log, such as shown in Fig. 1, which contains a history of queries executed by examiners while searching prior-art of several patents, and the identifier of a patent in question, *PatID*. We process the query log, meant as a collection of queries Q, analyzing in a loop all queries related to patent PatID (Algorithm 1, line 2). Each related query is parsed and keywords occurring therein are extracted (line 3). In case of the simple keyword extraction based method (referred to earlier as the first method) that is all: the set of thus extracted keywords forms a query which is executed using a selected search engine.

Ref #	Hits	Search Query	DBs	Default Operator	Plurals	Time Stamp
L1	48824	(short adj messag$ or sms)	US-PGPUB; USPAT	OR	ON	2005/07/19 14:53
L2	4037	(short adj messag$ or sms) same (voice or speech or talk)	US-PGPUB; USPAT	OR	ON	2005/07/19 14:54
L3	1758	mobile same (short adj messag$ or sms) same (voice or speech or talk)	US-PGPUB; USPAT	OR	ON	2005/07/19 14:54
L4	93	mobile same (select$ or choos$3 or request$3) near4 (short adj messag$ or sms) same (voice or speech or talk)	US-PGPUB; USPAT	OR	ON	2005/07/19 14:59
L5	113	(select$ or choos$3 or request$3) near4 (short adj messag$ or sms) with (voice or speech or talk)	US-PGPUB; USPAT	OR	ON	2005/07/19 15:03
L6	3462	(select$ or choos$3 or request$3) near4 (short adj messag$ or sms)	US-PGPUB; USPAT	OR	ON	2005/07/19 15:38
L7	20570	(select$ or choos$3 or request$3) near4 (voice or speech or talk)	US-PGPUB; USPAT	OR	ON	2005/07/19 15:38
L8	321	l6 and l7	US-PGPUB; USPAT	OR	ON	2005/07/19 15:03
L9	635466	(mobile or wireless or cellular or radio)	US-PGPUB; USPAT	OR	ON	2005/07/19 15:04
L10	280	l8 and l9	US-PGPUB; USPAT	OR	ON	2005/07/19 15:05
L11	1314	(select$ or choos$3 or request$3) near4 (short adj messag$ or sms)	EPO; JPO; DERWENT; IBM_TDB	OR	ON	2005/07/19 15:38
L12	9936	(select$ or choos$3 or request$3) near4 (voice or speech or talk)	EPO; JPO; DERWENT; IBM_TDB	OR	ON	2005/07/19 15:38
L13	21	l11 and l12	EPO; JPO; DERWENT; IBM_TDB	OR	ON	2005/07/19 15:39

Fig. 1 An example of a part of a query log. Queries are built of keywords (terms) connected using: standard Boolean operators ("and" and "or"), some proximity operators (discussed in Table 1), as well as references to other queries (cf., e.g., query l13 which is a conjunction of queries l11 and l12)

The basic approach sketched above suffers from a generic weakness of the standard Boolean model, namely the importance of all extracted keywords in the resulting query is the same. On the other hand, it may be argued that, e.g., more frequent terms should be more emphasized in the resulting query. Thus, in the second method we add what we call the *TF-IDF boosting*, following the terminology used in the Apache Lucene query language [2]. It boils down to applying a popular vector space model term weighting technique of TF-IDF [25]. In the vector space model the TF-IDF weight of a keyword in a document is proportional to the number of its occurrences in the document (term frequency, TF) and inversely proportional to the number of documents in the whole indexed collection in which given keyword occurs at least once (inverse document frequency, IDF). In our case, a subset of queries related to a given patent i, denoted as $P_i \subseteq Q$, corresponds to a document. The set of all P_i's, for all patents for which queries are present in the whole query log Q, corresponds to the collection of documents. Thus, the term frequency (TF) of a keyword with respect to a patent $PatID$ is computed as the number of its occurrences in all queries belonging to P_{PatID}, normalized by the total number of occurrences of all keywords in P_{PatID}. On the other hand, the inverse document frequency (IDF) of a keyword k_j with respect to a patent $PatID$ is computed as $log \frac{N}{n_j}$, where N is the total number of patents represented in the query log Q and n_j is the number of patents for which the keyword k_j occurs in at least one query related to this patent. Finally, the TF-IDF weight (boost) is computed as the multiplication of the TF and IDF components (line 4).

Thus, resulting query, denoted $queryOut$, is composed of the keywords associated with their TF-IDF weights computed as discussed above (line 5). We assume that such a query is admitted by the search engine employed, as it is the case of the Apache Lucene used in our experiments (cf. Sect. 4).

Algorithm 1 TF-IDF boosting process

```
1: function KEYWORDS(PatId, Q)
2:     while query ← GETNEXT(PatId, Q) do
3:         while term ← TOKENIZE(query) do
4:             boost ← TF- IDF(term, P_PatID, Q)
5:             queryOut ← PAIR(term, boost)
6:         end while
7:     end while
8:     return queryOut
9: end function
```

3.2 Patent Field as Prior-Art Queries

According to the study [33] we transform claims and titles of patents into queries. We focus on the claim part of the patent because using it gave good results in terms of the recall in the experiments reported in [33]; in this study the background summary of a patent was giving the best results but it is not indexed in our system). Another motivation for choosing the claims part is an assumption that examiner's mostly considered this field to determine whether or not invention is novel and to identify the relevant resources. In our study we used Okapi BM25 [22] ranking function which gave in our experiments better results than and TF or TF-IDF in terms of the recall.

4 Computational Experiments

4.1 Data Set

The USPTO publishes the examiner's search history (strategy) and results for each patent application. The data is collected since 2003 and is publicly available. However, there is no mechanism available for downloading it on a large scale.

The data is available in the form of a PDF image, and requires OCR and other types of processing before it can be used, or imported into a database. This means substantial preparation effort is required and this technical issue seems to be the main barrier in the widespread use of these data in patent IR [15, 27].

Thus the data consists of scanned copies of documents of varying quality. They are diverse: within a single document one can find handwritten notes, screen-shoots, and other material related to a patent application. The OCR software is not always able to cope with this diversity of form and quality. Thus the validation and selection of textual data is an essential element of preprocessing. The data used in our experiments was limited to class 455—Telecommunications, subclasses 403–466 Radiotelephone system extended with some randomly selected non-design patents. After preprocessing and refinement we gathered 5,49,705 unique search queries, relating to the set of 19,360 patents in Telecommunications class.

4.2 The Settings

The general scheme for our experiments is the following. We select a patent, PatID, and construct a query according to the three methods described in Sect. 3. Then, we execute the query against a collection of the remaining patents using the Apache Lucene search engine [3]. This way we obtain a ranked list of patents deemed relevant for the patent PatID by the search engine based on the constructed query. We evaluate each method of query constructing computing various measures of retrieval effectiveness, notably the Recall@100 [5]. We assume that for a given patent PatID actually relevant are those retrieved patents which are cited by patent PatID. Thus, we have to further limit the data set of patents used as only a fraction of them contain citations of other patents. In total, we use 357 patents which contain at least one citation. Overall number of citations in this set amounts to 1064, i.e., there are on average 2.98 citations per patent. Thus, in our experiments we construct 357 queries and report the results of the retrieval averaged over all executed queries.

4.3   The Results

We have compared the three methods, first analyzing the Recall@N measures obtained for each method. The results are illustrated in Fig. 2. In this figure "N" is denoted as "Rank size". The values shown are averaged over all executed queries constructed for particular patents. Our method with boosting gives clearly the best results. It is worth noticing that this method produces much shorter query than the second best which is basically using all keywords present in the titles and claims parts of the patent description. Thanks to that, the execution time of the queries is shorter.

In Table 2 we show the values of three effectiveness measures computed for top 100 patents returned by each query, again averaged over all queries executed. Here, again, the method 2 is the best.

Figure 3 shows comparison of only two best methods in the form of classical Precision-Recall curve. Again, the superiority of our method is clearly visible.

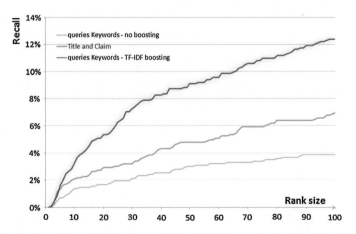

Fig. 2   Comparison of Recall@N for three methods; "N" is denoted as the "Rank size"; the results are averaged over all 357 queries

Table 2 Comparison of three methods in terms of the recall, precision and F-measure computed taking into account the first 100 patents returned by particular queries

Method	Recall (%)	Precision (%)	F-measure (%)
Queries keywords (no boosting)	3.868	0.134	0.259
Queries keywords (TFIDF boosting)	12.406	0.430	0.831
Queries based on document (title and claim)	6.955	0.039	0.482

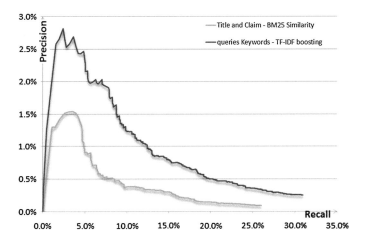

Fig. 3 The difference between compared algorithms (ranking size 1–1000)

5 Conclusions

We have proposed a method to retrieve relevant patent which may support a patent examiner. It constructs a query based on the query log of the examiner boosting extracted keywords using the TF-IDF keywords weighting method. It has been tested on a set of real patents. There are many aspects of the proposed method which can be improved. Also its thorough testing calls for a more sophisticated settings. This is left as the subject of further research.

Acknowledgments The first author contribution is supported by the Foundation for Polish Science under International PhD Projects in Intelligent Computing. Project financed from The European Union within the Innovative Economy Operational Programme (2007–2013) and European Regional Development Fund.

References

1. Agatonovic, M., Aswani, N., Bontcheva, K., Cunningham, H., Heitz, T., Li, Y., Roberts, I., Tablan, V.: Large-scale, parallel automatic patent annotation. In: Proceedings of 1st International ACM Workshop on Patent Information Retrieval—PaIR'08, pp. 1–8. California (2008)
2. Apache Lucene—Query Parser Syntax : Boosting a Term (2013). Accessed from http://lucene. apache.org/core/2_9_4/queryparsersyntax.html
3. Apache Lucene—Scoring (2013). Accessed from http://lucene.apache.org/core/3_5_0/ scoring.html
4. Azzopardi, L., Vanderbauwhede, W., Joho, H.: Search system requirements of patent analysts. In: Proceedings of the 33rd International ACM SIGIR Conference on Research and Development in Information Retrieval, pp. 775–776. ACM, New York (2010)
5. Baeza-Yates, R.A., Ribeiro-Neto, B.A.: Modern Information Retrieval —The Concepts and Technology Behind Search, 2nd edn. Pearson Education Ltd., Harlow (2011)
6. Bashir, S., Rauber, A.: Improving retrievability of patents in prior-art search. In: Advances in Information Retrieval, pp. 457–470. Springer, Heidelberg (2010)
7. Bravo-Marquez, F., L'Huillier, G., Ros, S.A., Velsquez, J.D.: A text similarity meta-search engine based on document fingerprints and search results records. In: Proceedings of the 2011 IEEE/WIC/ACM International Conferences on Web Intelligence and Intelligent Agent Technology-, vol. 01, pp. 146–153. IEEE Computer Society, Washington (2011)
8. Callaert, J., Van Looy, B., Verbeek, A., Debackere, K., Thijs, B.: Traces of prior art: an analysis of non-patent references found in patent documents. Scientometrics 69(1), 3–20 (2006)
9. Carpineto, C., Romano, G.: A survey of automatic query expansion in information retrieval. ACM. Comput. Surv. 44(1), 1–56 (2012)
10. Fujii, A.: Enhancing patent retrieval by citation analysis. In: Proceedings of the 30th Annual International ACM SIGIR Conference on Research and Development in Information Retrieval, pp. 793–794. The Netherlands, Amsterdam (2007)
11. Ganguly, D., Leveling, J., Jones, G.J.F.: United we fall, divided we stand: a study of query segmentation and PRF for patent prior art search. In: Proceedings of the 4th Workshop on Patent Information Retrieval, pp. 13–18. ACM, New York (2011)
12. Jansen, B.J., Booth, D.L., Spink, A.: Determining the user intent of web search engine queries. In: Proceedings of the 16th International Conference on World Wide Web, pp. 1149–1150. Banff, Alberta (2007)
13. Jiang, D., Pei, J., Li, H.: (2013). Mining search and browse logs for web search: a survey. ACM Trans. Intell. Syst. Technol. 4(4), 57:1–57:37
14. Joho, H., Azzopardi, L.A., Vanderbauwhede, W.: A survey of patent users: an analysis of tasks, behavior, search functionality and system requirements. In: Proceedings of the Third Symposium on Information Interaction in Context, pp. 13–24 (2010)
15. Jürgens, J., Hansen, P., Womser-Hacker, Ch.: Going beyond CLEF-IP: The reality for patent searchers? In: Catarci T., Forner P., Hiemstra D., PeAas A., Santucci G. (eds.) Information Access valuation. Multilinguality, Multimodality, and Visual Analytics, pp. 30–35 Springer, Heidelberg (2012)
16. Kim, Y., Seo, J., Croft, W.B.: Automatic boolean query suggestion for professional search. In: Proceedings of the 34th International ACM SIGIR Conference on Research and Development in Information Retrieval, pp. 825–834. Beijing, China (2011)
17. Lupu, M.: Patent information retrieval: an instance of domain-specific search. In: Proceedings of the 35th International ACM SIGIR Conference on Research and Development in Information Retrieval, pp. 1189–1190 (2012)
18. Magdy, W., Jones, G.J.F.: A study on query expansion methods for patent retrieval. In: Proceedings of the 4th Workshop on Patent Information Retrieval, pp. 19–24 (2011)
19. Mahdabi, P., Andersson, L., Keikha, M., Crestani, F.: Automatic refinement of patent queries using concept importance predictors. In: Proceedings of the 35th International ACM SIGIR Conference on Research and Development in Information Retrieval, pp. 505–514 (2012)

20. Nguyen, K.L., Myaeng, S.H.: Query enhancement for patent prior-art-search based on keyterm dependency relations and semantic tags. In: Larsen, Salampasis B. (ed.) Multidisciplinary Information Retrieval, pp. 28–42. Springer, Heidelberg (2012)
21. Oh, S., Lei, Z., Lee, W.C., Mitra, P., Yen, J.: CV-PCR: A context-guided value-driven framework for patent citation recommendation. In: Proceedings of the 22nd ACM International Conference on Conference on Information; Knowledge Management, pp. 2291–2296 (2013)
22. Prez-Iglesias, J., Prez-Agüera, J.R. Fresno, V., Feinstein, Y.Z.: Integrating the Probabilistic Models BM25/BM25F into Lucene. CoRR (2009)
23. Potey, M.A., Patel, D.A., Sinha, P.K.: A survey of query log processing techniques and evaluation of web query intent identification. In: Advance Computing Conference (IACC), 2013 IEEE 3rd International, pp. 1330–1335 (2013)
24. Sormunen, E.: A novel method for the evaluation of boolean query effectiveness across a wide operational range. In: Proceedings of the 23rd Annual International ACM SIGIR Conference on Research and Development in Information Retrieval, Special Issue of SIGIR Forum 34, pp. 25–32 (2000)
25. Sparck Jones, K.: A statistical interpretation of term specificity and its application in retrieval. J. Doc. **28**, 11–21 (1972)
26. Takaki, T., Fujii, A., Ishikawa, T.: Associative document retrieval by query subtopic analysis and Its application to invalidity patent search. In: Proceedings of the 2004 ACM CIKM International Conference on Information and Knowledge Management, pp. 399–405 (2004)
27. Tannebaum, W., Rauber, A.: Mining query logs of USPTO patent examiners. In: Forner P., Mller H., Paredes R., Rosso P., Stein B. (eds.) Information Access Evaluation. Multilinguality, Multimodality, and Visualization. Springer, Heidelberg pp. 136–142 (2013)
28. Tannebaum, W., Rauber, A.: Analyzing query logs of USPTO examiners to identify useful query terms in patent documents for query expansion in patent searching: a preliminary study. In: Proceedings of the 5th Conference on Multidisciplinary Information Retrieval, 127–136 (2012)
29. Tiwana, S., Horowitz, E.: Findcite: automatically finding prior art patents. In: Proceedings of the 2nd International Workshop on Patent Information Retrieval (2009)
30. Von Wartburg, I., Teichert, T., Rost, K.: Inventive progress measured by multi-stage patent citation analysis. Res. Policy. **34**(10), 1591–1607 (2005)
31. Wu, H.C., Chen, H.Y., Lee, K.Y., Liu, Y.C.: A method for assessing patent similarity using direct and indirect citation links. In: Proceedings of the 2010 IEEE International Conference on Industrial Engineering and Engineering Management (IEEM), pp. 149–152 (2010)
32. Xue, X., Croft, W.B.: Automatic query generation for patent search. In: Proceedings of the 18th ACM Conference on Information and Knowledge Management, pp. 2037–2040 (2009)
33. Xue, X., Croft, W.B.: Transforming patents into prior-art queries. In: Proceedings of the 32nd International ACM SIGIR Conference on Research and Development in Information Retrieval, pp. 808–809 (2009)
34. Zhao, X., Xiao, C., Lin, X., Wang, W., Ishikawa, Y.: Efficient processing of graph similarity queries with edit distance constraints. VLDB J. **22**(6), 727–752 (2013)

What NEKST?—Semantic Search Engine for Polish Internet

Dariusz Czerski, Krzysztof Ciesielski, Michał Dramiński, Mieczysław Kłopotek, Paweł Łoziński and Sławomir Wierzchoń

Abstract We introduce a new semantic search engine, developed at our institute. Its unique feature is the automatic construction of semantic resources, like discovery of millions of facts, IS-A relations and automated generation of sentimental analysis dictionaries. We developed a new method of document categorization. The engine can be queried in natural language and possesses interfaces to be used not only by humans but also by machines.

Keywords Search engine technology · Semantic transformations · Knowledge graph

1 Introduction

For decades, informatics managed to ignore the meaning of information it was claiming it was processing. So did e.g. data base management systems that left the interpretation of data to the database user, or data mining (DM) that insisted that the researcher has to interpret the outcome of DM algorithms. Hence for decades

D. Czerski (✉) · K. Ciesielski · M. Dramiński · M. Kłopotek · P. Łoziński · S. Wierzchoń
Institute of Computer Science, Polish Academy of Sciences,
ul. Jana Kazimierza 5, 01-248 Warszawa, Poland
e-mail: dcz@ipipan.waw.pl

K. Ciesielski
e-mail: kciesiel@ipipan.waw.pl

M. Dramiński
e-mail: mdramins@ipipan.waw.pl

M. Kłopotek
e-mail: klopotek@ipipan.waw.pl

P. Łoziński
e-mail: pawel.lozinski@ipipan.waw.pl

S. Wierzchoń
e-mail: stw@ipipan.waw.pl

© Springer International Publishing Switzerland 2016
G. De Tré et al. (eds.), *Challenging Problems and Solutions in Intelligent Systems*, Studies in Computational Intelligence 634,
DOI 10.1007/978-3-319-30165-5_16

335

variants of entropy proved to be quite sufficient measures of information content. The advent of World Wide Web has caused, however, that such a vision of information in computer science ceased to be sufficient. WWW broke several previously existing canons of information theory. E.g. for a long time it was believed that the entropy is the upper limit of the file compression. HTML document format proved this wrong by shifting the "ornamentation" issues to the browser. Similarly sound stream compression by MIDI files is impressive. Last not least visually lossless image compression beat Shannon's entropy limits by an order of magnitude. And the emerging concept of surplus entropy proves also Shannon wrong, pointing at the same time to the importance of syntactic, semantic, pragmatic and apobetic analysis of text.

Especially this last result, beside various practical considerations, underlines that there is no escape from semantic information analysis on the Internet. The WWW is particularly challenging in this context because we do not have any (topical or vocabulary) close world assumptions.

Our research aiming at construction of a semantic search engine was an attempt to face this challenge. We followed and developed the path of so-called semantic transformations in order to enable semantically enriched search. We restricted ourselves to the domain of Polish Internet (we automatically recognize the language of the document), but the task is still challenging as we know Polish WWW comprises over a billion Web pages (we actually collected nearly 500 millions of them). While doing so, we aimed at adaptation of state of the art search engine methodologies to the Polish language document collections. While some were straight forward, some other proved to be a challenge, like e.g. Hearst rules for automated taxonomy extraction. They work well for English, while it has a rigid grammatical structure, while Polish sentences are more flexible, and as shown below, more advanced parsing was necessary.

In this paper we will outline the architecture and capabilities of the system and explain rationale behind them.

2 Semantic Analysis for Search Engines

Within the NEKST project we approach the semantics of information in operational manner. We exploit the fact that both the sender and the receiver of information is a human being and therefore it is man and not the search engine that must really understand the text. It is really sufficient when search engine supports the man. This support is done by applying the so-called, semantic transformation.

Semantic Search Transformation is such a transformation of the content of the document and/or the content of the query, which allows the use of classical search engines in such a way that the response is semantically related to the original query and/or is exhibiting additional information about the document (e.g. abstract) even if (and especially if) this response cannot be derived from the literal content of the document/query itself. So in fact this means that we use some additional resources to respond to the query. These additional resources are called semantic resources, though they do not need to exhibit features of "semantic web resources".

In the framework of NEKST we implemented the following types of semantic transformations [8]: (1) prompts for the user, (2) substituting synonyms/hypernyms/hyponyms and other related concepts/words, (3) disambiguation of terms (the so-called identification of Synsets), (4) assigning the document to (one or more) categories of the taxonomy/ontology (a tree or an acyclic directed graph), (5) personalized PageRank computation [9], (6) and cluster analysis to assign keywords to focus the document (7) extraction of objects and relations between them in the form of a knowledge graph and search in the graph, (8) automatic construction of the graph concepts, (9) sentiment analysis, (10) the diversification of response, (11) the identification and classification of harmful content (12) dynamic summarizing.

These transformations rearrange the online resources and enable their more user-friendly sharing. We allocate automatically Internet resources into thematic groups, highlighting thematic channels on websites, and assign labels and categories to documents and their groups. From the point of view of the user, this translates into not only a more precise identification of documents of interest to the user, but also gives him the opportunity to search the context of both the individual documents and their groups, such as channels or services. It also provides the opportunity to diversify answers search engine.

Diversification [12] represents the diversity of responses in such a way that you saw not only the best papers, but also a variety of thematic or ambiguity, such as in a classical question about apples (fruits? computers?).

The need for taking into account semantic aspects for Web search becomes apparent when the document sought is understandable only in the context of other documents of a thematic channel. For example, asking a search query about battery capacity of mobile phones a user would expect in response links to specific pages with telephones even if these websites do not mention that they are about telephones. That is the information that something is a phone name should be extracted from some other documents.

While the above-mentioned semantic transformations (1)–(12) are nowadays well-known and appreciated, bringing them together and making them interact within a single system creates in our opinion a new quality. For example automated taxonomy construction and fact extraction is subject of intense research for years. But search engines like Google use rather a manually constructed knowledge graph. The NEKST search engine (http://neks.pl) applies the automatically extracted knowledge graph directly to finding documents, to presenting facts, to constructing a dynamic summary as well as in answering natural language queries and clustering of documents, last not least in response diversification. Clustering drives novel methods of personalized PageRank computation and generally ranking the responses. Term disambiguation serves both diversification and document categorization. And so on.

3 Selected Examples of Semantic Transformations

3.1 Assigning Categories to Documents

The NEKST categorization module assigns categories to Web documents exploiting a modified version of Polish Wikipedia graph [2]. We chose Wikipedia as a source of categories for open domain documents because of its richness in the sense of the number of terms and phrases that are assigned categories. We can assign therefore one or more categories to a vast majority of words occurring in typical Web document. The idea of document categorization algorithm is based on finding common categories among those assigned to the words. Weighting, disambiguation and other processes are necessary because of conflicts that usually occur. The advantage over other approaches to categorization is the locality—one can process one document at a time so massive parallelization is possible.

We developed also methods for supervised classification for limited sets of categories (e.g. geographical ones), based on the labeling using heterogeneous classifiers committees over baseline methods of categorization.

So when issuing a query "rowery górskie" (in Polish: "mountain bikes"), we get e.g. a page http://rowery.synergia-serwis.pl/, which will be assigned categories "rowery", "wypożyczalnie_rowerów" ("bicycles", "rent-a-bicycle-service"). Neither Google nor Yahoo search engines label the individual web pages. They associate the query, instead, with the general query terms like "tanie górskie rowery" ("cheap mountain bikes"), "rowery szosowe" ("street bikes").

A query "zamki śląskie" ("Silesian castles") provides with a page http://zamki.res. pl/ that is labelled with categories "fortyfikacje", "zamki", "dwory" ("fortifications", "castles", "manors"). Google associates with this query terms like "castle in Książ", "Silesian manors", but never "fortifications". Yahoo has no clue.

3.2 Diverse Methods of Searching for Documents

NEKST offers classic search method for documents in the form of a string of words, phrases (included in quotation marks), the words combined with the operators AND, OR, with the ability to build complex parenthesized expressions. The returned documents are stamped with a dynamic summary and labels in the form of associated categories of documents (Sect. 3.1). Ranking of documents is based on PageRank [16] at the level of services, boosted by the lexical word forms, presence of key-words of the document and the document cluster etc. Dynamic summary is built based on the weighing of sentences of the document, the weighting is based in turn on count of the words with the harmonized frequencies, the intensity of occurrence of words in the query, and the length of sentences and the position of sentences in the text. Words semantically related to the query terms (synonyms, hypernyms, hyponyms) are taken into account.

Alternatively, instead of the documents, the user can ask about the facts: e.g. Which poet was born in Krzemieniec? What is the capacity to cell phone batteries? What musician was awarded by the Queen? Answers to such questions are not documents, but specific sentences of the documents describing the facts sought. The engine "will know" that people like Mickiewicz and Słowacki are poets, Lennon is a musician, and Nokia is the name of a phone, without the category terms (poet, musician, phone) occurring in the actual text document, from which the piece of knowledge is extracted. The queries can be formulated in a simple abstract grammar or in natural language (Polish of course).

So we may ask in the natural language: "Deuter jest izotopem którego pier- wiastka?" ("What chemical element is deuteron isotope of?")

And get the reply e.g.: "Chlorek deuteru to nieorganiczny związek chemiczny, połączenie chloru z izotopem **wodoru**—deuterem" ("Deuteron chloride is an inor- ganic compound consisting of chlorine with hydrogen isotope called deuteron") which is not quite exactly what was asked for but it contains the response.

Another way of formulating queries is to use an abstract syntax of the form: subject-predicate-object or in our case N1-V-N2, for example: N1<Adam Mickiewicz> V<urodził> N2<CLASS{miasto}>. Here "Adam Mickiewicz" is the name of a renown Polish poet (the subject of the query), "urodził się" means "was born" and is the predicate. The construct CLASS{miasto} (CLASS{city}) is an abstraction telling that we seek an object from the class of cities.

And we get the response: "**Adam Mickiewicz urodził** się 24 Grudnia 1798 roku w Zaosiu nieopodal **Nowogródka**." ("Adam Mickiewicz was born on 24th December 1798 in Zaosie near Nowogródek).

The query N1<CLASS{poeta}> V<urodził> N2<Krzemieniec> ("poet born in Krzemieniec") yields the response [Juliusz Słowacki] [urodził się] [4 września 1809 roku w Krzemieńcu] ("Juliusz Słowacki was born on September 4th, 1809 in Krzemieniec"). Google's first response is the Kamieniec entry of Wikipedia and no reference to Słowacki in the summary. Yahoo returns a Krzemeniec youtube address, without any reference to Słowacki in the summary, though both Wikipedia and Youtube articles speak about this famous poet.

The query N1<CLASS{papież}>V<beatyfikował> ("Pope beatified") gives responses [Klemens ix 8 marca 1669] [beatyfikował] [Mikołaja z Flue] ("Clement IX beatified Nicolas of Flue on March 8th, 1669"), [W roku 1741 Benedykt xiv] [beat- yfikuje] [Aleksandra Sauli] ("In 1741 Benedict XIV beatifies Alexander Sauli") in spite of the fact that none of the referred pages contains the word "Pope". Neither Google nor Yahoo return among the first 20 responses pages containing the name of the Pope but not the word "Pope".

Query V<odznaczył>N2<CLASS{ksiądz}> ("priest was awarded") returns answers like [Prezydent Bronisław Komorowski] [odznaczył] [Krzyżem Wielkim Orderu Odrodzenia Polski biskupa Tadeusza Pieronka] (text containing "bishop" and not "priest"), [Krzyżem Oficerskim Orderu Zasługi RP] [odznaczył] [prezydent Bronisław Komorowski Kardynała Timothy' ego Dolana] (text containing word "car- dinal", but not "priest"). No such abstraction capabilities can be seen in Google nor in Yahoo.

Query N1<CLASS{telefon}>V<ma>N2<baterię CLASS{liczba}> ("phone has battery *number>* can be used to extract information numerically describing phone batteries. A response of the system NEKST includes: [w810i] [ma] [baterię trzymającą 2 - 3 dni] ("w810i has a battery holding 2–3 days"). Such a query is next to impossible to formulate in Google or Yahoo. These systems hardly associate w810i with a query concerning the term *phone*. Here the automated extraction of taxonomies is of particular importance. The growth of the market offers is so vast that manual maintenance of a taxonomy encompassing all products from all industry branches is difficult to imagine with a moderately sized staff of a search engine.

3.3   Extraction of Objects and Relationships Between Them in the Form of a Knowledge Graph

The fact search functionality may have been based e.g. on Wikipedia. But it turns out that the knowledge contained there is quite limited in scope compared to the common knowledge of humanity. There is neither time nor processing capabilities to keep the knowledge graph up-to-date in an open-domain search engine. Therefore in NEKST the semantic resources for fact base querying are built automatically directly from the document collection. We identify the IS-A hierarchy exploiting specific grammatical constructs obtained from shallow parsers. E.g. phrase "president Nixon" provides us with relation "Nixon" IS-A "president". We created a module responsible for the detection of entities appearing on websites, based on the SEAL algorithm [11] and implemented a module for the identification of proper names in the text, based on the tool NERF.

Approaches to IS-A extraction described in literature rely on evidence from pattern extraction and statistical information (cf. [4, 18]). Pattern-based methods rely heavily on so called Hearst patterns, first described in [5]. These are ways of expressing instance enumerations of a class in natural language. Typical forms are "c *such as* i1, i2 or i3" or "c, *for example* i1, i2 or i3". Terms extracted with such patterns may serve as input for elaborate taxonomy and ontology construction methods as e.g. [10]. While lexico-syntactic patterns proposed by Hearst prove quite useful for rigid order languages like English, they are unable to cover many useful patterns occurring in languages with inflection like Polish. Therefore in the NEKST system we introduce a novel method of IS-A relation extraction from patterns, which uses morpho-syntactical annotations along with grammatical case of noun phrases that constitute entities participating in IS-A relation. It is known that languages that have inflection and free word order are much harder for automatic analysis than e.g. English. We argue that inflection in language isn't only a drawback but can also be a great advantage if only the problems of morpho-syntactic tagging and dependency parsing can be solved with reasonable precision. Typical constructs that express explicitly the hypernymy relation in Polish language are:

$$NP_1^{Nom} \text{ to } NP_2^{Nom}$$

$$NP_1^{Nom} \text{ jest } NP_2^{Nom}$$

Both of them are a way of saying NP_1 is NP_2 and in both cases noun phrase NP_1 is expressed in nominative. They differ in grammatical case of NP_2, where in the first construct we have nominative and in the second: ablative. The second pattern has its equivalent for past tense (NP_1 was NP_2):

$$NP_1^{Nom} \text{ był/była/było } NP_2^{Abl}$$

Conducted experiments have shown, that combination of word and grammatical case pattern allows for relation extraction with quite high precision.

Additionally we also developed a method for increasing the number of extracted relations that we call *pseudo-subclass boosting* which has potential application in any pattern-based relation extraction method.

Our knowledge graph consists of facts of the form of triplets <noun phrase><verb phrase><noun phrase>, obtained by exploiting several methods and sources:

1. Shallow parsing of the text extracted from web documents with additional checking against existing of connections between noun and verb phrases based on result obtained from dependency parsing.
2. Rule based parsing method based on dependency parsing result designated to processing Wikipedia short definitions (in most cases starting at the beginning of every document).
3. Rule based parsing method based on dependency parsing result designated to processing simple sentences of certain form (contains predicate, subject and object element).

Facts supported by a sufficient number of occurrences are kept only. Queries may be formulated by presenting text fragments that should occur at one, two or three of these positions. Instead of text one can use an abstraction (the parent in the IS-A hierarchy) of the searched term.

3.4 Cluster Analysis and Keyword Assignment to the Documents

An important element of the semantic analysis of documents is detection of groups of similar documents. Clustering in NEKST system is done using algorithms like k-means and fuzzy c-means that operate in a distributed computing model, guaranteeing scalability of the solution [3, 15].

Grouped documents are then subject to assigning to the list of the most common and also the characteristic keywords (tfidf measure for groups). When the user issues a query, the documents belonging to the groups characterized by query terms are more likely to be found. This follows the intuition that a document from a topical group is more worthy than one that contains the query terms by chance. In practice clustering information is used as an additional feature in the feature vector used by the learning-to-rank algorithm. Among the almost 100 features this is the one of the most important. It gives an ability to model higher levels of topical abstraction of documents. Finally relevant documents can be found even though they don't contain query terms or queried words are not emphasized adequately. Throughout experiments we have found, that state of the art LambdaMart [17] learning-to-rank method achieves best results.

Moreover, to assist in the correct grouping and eliminating unjustified overemphasis, a duplicate detection and domains mirror discovery module has been implemented, based on hashing [13] (due to the vast resources to search, the method must be fast). It uses the so-called locality sensitive hashing, parameterized by probability of similarity between documents. Documents related above a certain threshold are pulled together. Hash function is the basis for assigning documents to the so-called buckets, which allow for next to linear identification of duplicates. Only documents belonging to the same bucket require pair-wise comparisons of text. In addition to grouping the documents are subject to cluster analysis of domains. Domains are treated as documents which are built *tfidf* vectors of words, which are the mean values of all documents' tfidf in a domain. Grouping and labeling is done with the same methods as for individual documents.

Several other methods of semantic clustering of documents were implemented.

4 Architecture and Functionality of the NEKST Search Engine

NEKST has a general outline of a classical search engine. There is a battery of parallel spiders, a distributed indexing system and also a distributed front-line for responding to queries, capable of handling of several hundreds of millions of documents. But it differs in many details. Each of the subsystems consists of a lot of modules, which in collaboration or separately are able to provide utility functions useful for various applications. The idea of these modules is built around mechanisms for semantic indexing of large collections of electronic resources, visual, semantic access to content in mass collections of documents of inter-and intra-network sites and massive parallelism of information processing.

Compared with the leading products, this search engine introduces a number of innovations. With the tools of semantic content analysis, search is made in an innovative way. Special indexing methods were designed to allow for search for documents not only lexically but also semantically similar to the query. In particular using abstract concepts of object classes. Through hierarchical clustering systematization of documents is introduced. Not only whole documents but also facts from these documents are subject of indexing and hence querying. In this way the functionality of fact search is introduced where a fact query is responded by both the document and the concrete document fragment where it can be found. Internal mechanisms exploit sentimental analysis, clustering, personalized PageRank, categorization, dynamic summarization to provide a new feeling in the response to queries.

The newly developed, advanced search system is ideal for the creation of specialized search engines for various purposes, e.g. (by adding the appropriate semantic resources) specialization for economical documents, legal documents etc.

Currently, internal testing of the system are performed using the 500 millions of documents collected so far.

The system operates as a stand-alone tool with a user interface for constantly updated database of documents covering the Polish Internet. It may be coupled to other systems which can query it automatically. NEKST provides not only the standard search service but also specialized ones like detection of copies of the documents or their parts, detection of mirror domains, document summarization etc. Also individual modules of the search engine, powered by the resources created by the NEKST mechanisms, may find application completely independent of their use in a search engine, such as modules dynamically generating summaries, constructing ontologies from the texts belonging to the field of ontology, geographical and general labeling (categorization), analysis of the emotional overtones of text documents relating to the organization or business, etc.

The system was designed as a parallel one from the offset because of the size of Polish internet and its dynamics. The basis of the parallelization of the processes at NEKST is a public system of Hadoop (Apache project). Most of the search engine algorithms have been designed in our system in accordance with the MapReduce paradigm. We use also other publicly available software components like ones originating from nutch, lucene projects, lematizers for Polish, German and English and a number of other tools for the analysis of the Polish language, including ones developed by the Institute of Computer Science of Polish Academy of Sciences and the Technical University of Wroclaw, which we adapt to the architecture of Hadoop.

From the hardware point of view, NEKST uses linked servers in a logical cluster where two sub-clusters are distinguished. The first one is intended for processing off- line data, preparing documents for search. The second sub-cluster is intended to be distributed to respond to requests, interacting with human and machine users, retrieving documents.

The off-line part is populated with spider and indexing modules. The data from the Internet is fed by a team of spiders, which harvest documents by crawling through links, acquire documents from special sources, including for example, RSS feeds, and finally keep them up-to-date. Spiders save documents in our own proprietary Hadoop-based distributed database, adapted for efficient storage and dispersion of documents.

Indexer module converts documents to the vector space representation and enriches the information about the document (e.g. language, domain, server, IP) and the words (tfidf statistics, parts of speech/grammatical basic forms, phrases, etc.). It also extracts the links, identifies stopwords, builds a dictionary of terms and sets a unique hash value for the terms and documents. Analyzer module, being an extension to the basic indexer, performs clustering of documents, creating new documents (like pseudo-documents representing a domain, a server or a group of documents) and attaches additional attributes to the primary and the derived documents, like group membership, keywords and categories, creates a frame for dynamic summarization, labels with family filter information (content harmful/harmless) and checks for spam. For each outgoing link its topicality is measured etc. IS-A hierarchy extraction and fact extraction are also parts of this off-line analysis. Various brands of PageRanks are computed [14], including personalized PageRank, TrustRank, DomainRank etc. Finally a distributed inverted index is constructed, based on the original method proposed by the doctoral dissertation of Dr. Czerski of the IPI PAN, defended in January 2010. The index covers both documents and facts.

Off-line processing is also applied to information obtained from users (query logs) and from other resources, like dictionaries, encyclopedias etc. The outcome of the analysis are feeds to services proposing the user query expansion or serving query responses. engaged in developing a basis for expansion of queries.

Due to the dimensionality of the problem, all modules have the ability to operate in incremental mode.

The online part works on products of the offline part of the system. As already mentioned it offers an interface to interact with users as well as services designed for interaction with other systems. The user interface is created as a combination of separate Web services that feed various parts of user interface. While interacting with the user the system tries to capture his intensions. When it recognizes a question in natural language, the related fact is returned. In case user's intension is finding a definition, it looks up encyclopedia first. If the intention is to find a translation, dictionaries are searched. The system tries also to guess if some words constitute a phrase (even if not enclosed into quotation marks). It tries also to expand known abbreviations. The ranking mechanism of responses tries to diversify the replies and to promote documents representative of a particular thematic group. The emphasis is on effective search pages semantically similar and presentation of search results discounted copies of documents and pages.

5   Conclusions and outlook

The NEKST system proposes a new vision of a Web search engine: a vision of a system trying to act as a partner in the search process who tries to understand both the content of the document and the user query. With this vision, the portfolio of potential applications of a search engine is broadened considerably.

The basic attribute of the search engine is the coverage of major portion of Polish Internet. So first of all it may provide citizens with any Poland related information they need.

With the anticipated broad usage new opportunities occur. Search engines can be used for continuous monitoring, e.g. health of the population. A sudden wave of requests for flu drugs can serve as an indicator of potentially impending flu epidemic. This of course will work if the system can understand that Gripex and other drugs are applied against flu. This means semantic information processing is needed here.

Any other kinds of authorities may benefit. For example, the Ministry of Finance can check whether all tax offices made available relevant form for taxing or distributed important announcements. Sentimental Analysis Module can be used to assess the mood of commentators on blogs relating to the organization of sporting events, administrative decisions, etc. The analysis of the profile and the number of social organizations (associations, societies, fraternities, clubs associations, etc.) can be useful for understanding of the needs of the population and the creation of a suitable policy.

Also the industry may benefit from semantically underpinned analysis of population searches. Just mention tourism or transportation benefiting to holiday related queries.

Both for the business of the company and the supervisory review of the administrative bodies there is a need to keep records on functioning enterprises, their offers and business profile. This information is simple to collect from the internet, but with big human effort. In contrast, search modules, supported by adequate resources, semantic (called ontologies, patterns and appropriate software, supporting semantic structure automatically recognition compounds) can maintain respective directories of firms. Similarly, it is possible to create and maintain resources on educational or medical services etc.

Do not forget the importance of search engines for Polish Science. Firstly one should mention its importance for all kinds of linguistic research, linguistic analysis of current social linguistic processes, frequency analysis, and the like. In particular, automatically detected triplets (<noun> <verb> <noun> phrases) or relationships of the type IS-A can become the basis for the verification of hypotheses, expanding/narrowing published dictionaries, etc. In addition, the magnitude of the collected resources will be of great importance for the development of tools and algorithms for language analysis and graph analysis of hypertext documents [7]. There will opportunities for research on the propagation of information on the Internet, its reproduction, etc.

The vast majority of potential applications listed above does not belong to the standard functionality of search engines. What is needed is access to both the resources and the individual modules functioning search engine. Therefore, the optimal solution is to develop and maintain open for research and business search engine with a rich set of semantic tools. Without extensively developed semantic tools would be impossible to achieve this functionality, delineated in this article.

NEKST search engine in this context is an important milestone on the one hand, on the other hand only a starting point that can and should begin the march towards creating an interesting, important and useful application based on already developed core search engine.

Modules constituting the search engine can find the application completely independent of their use in a search engine, such as modules dynamically generating summaries, constructing ontologies (i.e., a set of concepts from the field, along with description of the relationships between these concepts) from the texts belonging to the area of interest, geographic and thematic labeling or emotional overtones analysis of text documents relating to the organization or business.

References

1. Ciesielski, K., Czerski, D., Dramiński, M., Kłopotek, M., Wierzchoń, S., Sydow, M., Borkowski, P., Wajda, J., Chojnicki, S., Trojanowski, K.: Semantic Resources for Enhancing Search Engines 2012, 3 Polish Academy of Sciences Annual Report 2011
2. Ciesielski, K., Borkowski, P., Kłopotek, M., Trojanowski, K., Wysoki, K.: Wikipedia-Based Document Categorization. LNCS, vol. 7053, pp. 265–278. Springer, New York (2012)
3. Dramiński, M., Owczarczyk, B., Trojanowski, K., Czerski, D., Ciesielsdki, K., Kłopotek, M.: Stabilization of Users Priofiling Proceeded by Metaclustring of Web Pages. LNCS, vol. 7912, pp. 179–186. Springer, New York (2013)
4. Fountain, T., Lapata, M.: Taxonomy induction using hierarchical random graphs. In: Proceedings of the 2012 Conference of the North American Chapter of the Association for Computational Linguistics: Human Language Technologies, pp. 466–476. Association for Computational Linguistics (2012)
5. Hearst, M.A.: Automatic acquisition of hyponyms from large text corpora. In: Proceedings of the 14th Conference on Computational Linguistics - Volume 2, COLING'92, Stroudsburg, pp. 539–545. Association for Computational Linguistics (1992)
6. Kłopotek, M.: What is the Value of Information - Search engine Point of View. LNCS, vol. 8104, pp. 1–12. Springer, New York (2013)
7. Kłopotek, M.A., Wierzchoń, S.T., Ciesielski, K., Dramiński, M., Czerski, D.: Ocena wartości informacji hipertekstowej. Wydawnictwo Instytut Informatyki Uniwersytetu Śląskiego Systemy Wspomagania Decyzji, pp. 227–242 (2011)
8. Kłopotek, M.A., Wierzchoń, S.T., Ciesielski, K., Czerski, D., Dramiński, M.: Analiza semantyczna dokumentów dla wyszukiwarek internetowych. w: Alicja Wakulicz-Deja: Systemy Wspomagania Decyzji. Instytut Informatyki Uniwersytetu Śląskiego, Katowice. pp. 135–148 (2013)
9. Kłopotek, M., WSierzchoń, S., Czerski, D., Ciesielski, K., Dramniński, M.: A Calculus for Personalized PageRank. LNCS, vol. 7912, pp. 212–219. Springer, New York (2013)
10. Kozareva, Z.: Simple, fast and accurate taxonomy learning. Text Mining, pp. 41–62. Springer International Publishing, New York (2014)
11. Mironczuk, M., Czerski, D., Sydow, M., Klopotek, M.: Language-independent information extraction based on formal concept analysis. In: Proceedings of Second International Conference on Informatics, Applications (ICIA), pp. 323–329 (2013). ISBN 978-1-4673-5255-0
12. Sydow, M., Ciesielski, K., Wajda, J.: Introducing diversity to log-based query suggestions to deal with underspecified user queries. In: Proceedings of Joint International SIIS 2011 Conference, Revised Selected Papers, vol. 7053, pp. 251–264. LNCS/Springer (2012). ISBN 978-3-642-25260-0
13. Szmit, R.: Locality Sensitive Hashing for Similarity Search Using MapReduce on Large Scale Data. LNCS, vol. 7912, pp. 171–178. Springer, New York (2013)
14. Wierzchoń, S.T., Kłopotek, M.A., Ciesielski, K., Dramiński, M., Czerski, D.: Metody obliczeniowe dla wyznaczania wektora PageRank, Wydawnictwo Instytut Informatyki Uniwersytetu Śląskiego Systemy Wspomagania Decyzji, pp. 243–252 (2011)
15. Wierzchoń, S., Kłopotek, M.A., Ciesielski, K.: Community detection with spectral optimization. Security, Intelligent Information Systems XVI. In: Proceedings of the International SIIS:2011
16. Wierzchoń, S.T., Kłopotek, M., Ciesielski, K., Czerski, D., Dramiński, M.: Accelerating PageRank computations. Control Cybern. **40**(2), 377–400 (2011)

17. Wu, Q., Burges, C.J.C., Svore, K.M., Gao, J.: Ranking, boosting, and model adaptation. Technical report MSR-TR-2008-109, Microsoft Research (2008)
18. Wu, W., Li, H., Wang, H., Zhu, K.: Probase: a probabilistic taxonomy for text understanding. In: ACM International Conference on Management of Data (SIGMOD), pp. 481–492 (2012)

Printed in the United States
By Bookmasters